D1119915

REALITY RULES: I

REALITY RULES: I

Picturing the World in Mathematics —The Fundamentals—

JOHN L. CASTI
Technical University of Vienna
Vienna, Austria
and
Santa Fe Institute
Santa Fe, New Mexico

A Wiley-Interscience Publication
JOHN WILEY & SONS, INC.
New York / Chichester / Brisbane / Toronto / Singapore

Library of Congress Cataloging in Publication Data:

Casti, J. L.

Reality rules: picturing the world in mathematics/John L. Casti.

p. cm.

"Wiley-Interscience publication."

Includes index.

Contents: v. 1. The fundamentals – v. 2. The frontier.

1. Mathematical models. I. Title.

QA401.C3583 1992

511'.8–dc20 92-12213

ISBN 0-471-57021-4 (vol. 1) CIP

ISBN 0-471-57797-9 (set)

Printed in the United States of America

10 9 8 7 6 5 4 3 2 1

To the memory of my teacher

RICHARD E. BELLMAN,

who would have been interested

PREFACE

Mathematical modeling is about rules—the rules of reality. What distinguishes a mathematical model from, say, a poem, a song, a portrait or any other kind of "model," is that the mathematical model is an image or picture of reality painted with logical symbols instead of with words, sounds or watercolors. These symbols are then strung together in accordance with a set of rules expressed in a special language, the language of mathematics. A large part of the story told in the 800 pages or so comprising the two volumes of this work is about the grammar of this language. But a piece of the real world encoded into a set of mathematical rules (i.e., a model) is itself an abstraction drawn from the deeper realm of "the real thing." Based as it is upon a choice of what to observe and what to ignore, the real-world starting point of any mathematical model must necessarily throw away aspects of this "real thing" deemed irrelevant for the purposes of the model. So when trying to fathom the meaning of the title of this volume, I invite the reader to regard the word "rule" as either a noun or a verb—or even to switch back and forth between the two—according to taste.

The book you now hold in your hands started its life as a simple revision of my 1989 text-reference *Alternate Realities.* But like Topsy it just sort of grew, until it reached the point where it would have been a misnomer, if not a miscarriage of justice, to call the resulting book a "second" or "revised" or "updated" or even a "new" edition of that earlier work. And, in fact, the project grew to such an extent that sensibility and practical publishing concerns dictated a splitting of the work into two independent, yet complementary, volumes. But before giving an account of the two halves of my message, let me first offer a few words of explanation as to why a three-year old book needed updating in the first place.

Alternate Realities was a mathematical-modeling text devoted to bringing the tools of modern dynamical system theory into the classroom. As such, the focus of the book was on things like chaos, linear system theory, cellular automata, evolutionary game theory, q-analysis and the like. While the book itself was published in early 1989, the actual writing took place during late-1986. As a result, most of the material was based on what was current in the research literature *circa* 1985. In the intervening years research interest and results in dynamical system theory has been nothing short of explosive. The chart below gives some indication of the magnitude of this exponential growth of published research in just the fields of chaos and fractals. And work in the other topical areas addressed in *Alternate Realities* has certainly been no less intense. Hence, the call from both readers and my editor for an update.

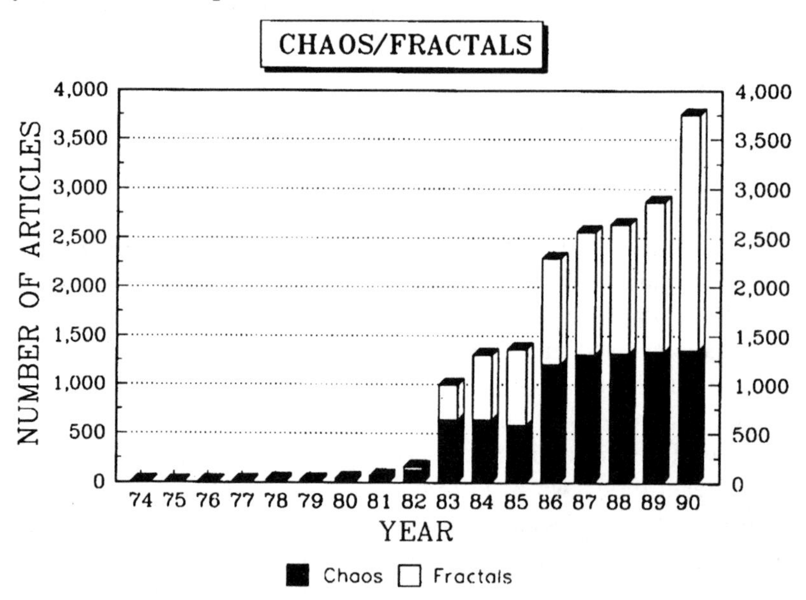

So what are you getting in the two halves of this book that's new besides some polishing of the reference lists to a little brighter shine and a tidying-up of typos and other literary infelicities in *Alternate Realities?* Briefly, here are the main attractions:

- *New chapter and sections*—Recognizing the fact that for a modern mathematician a computer program has become a legitmate answer to a mathematical question, *The Frontier* contains an entirely new chapter (Chapter 9) on computation and complexity. In particular, this chapter takes up the question of what can actually be done by way of formalizing real-world phenomena within the confines of a computer program, i.e., a set of rules. This chapter offers the student and researcher a point of contact with a host of matters of current intellectual concern running the gamut from Gödel's Incompleteness Theorem and its connection to work in artificial intelligence to the problem of NP-completness and the complexity of numerical algorithms.

In addition to the new chapter, a number of new sections have been added to the chapters originally constituting *Alternate Realities.* These sections introduce topical areas like artificial life that were only touched on briefly in that earlier work, or treat entirely new areas of concern like the relation between chaos and stock-price fluctuations that were yet to be born in 1985.

• *Exercises*—One of the main comments (and complaints) I heard from readers and users of *Alternate Realities* was that the problems were too hard. Too hard, anyway, to be used as drill exercises in a classroom setting. To remedy this defect, I have added Exercises of the drill-type at the end of almost every section in every chapter of the book (except for the final chapter, which is more a meditation on the philosophy of modeling than on the mathematics). These Exercises make the book much easier for instructors to use as a mathematical modeling text in the classroom, as well as provide drill problems for those using the book for self-study.

• *Examples*—As a further effort toward making the material accessible to students and researchers alike, many new examples have been sprinkled throughout the book at strategic locations. Not only do these examples introduce new applications of the theoretical results, they also serve to illustrate exactly **how** to use the theory in a diverse array of realistic situations.

• *Discussion Questions and Problems*—Several new Discussion Questions have been added to each chapter, both to extend the earlier material and to bring forth ideas nonexistent a few years ago. Moreover, the Problems sections have been augmented with new results from the research literature that for one reason or another didn't fit into the mainstream of the text.

• *Solutions Manual*—From an instructor's point of view, it's always a bit awkward, not to mention annoying, trying to teach from a book containing problems that you yourself can't solve! And I must confess that the problems in *Alternate Realities* were by and large pretty hard, most of them having been taken directly from the primary research literature. So to ease the pain of grappling with these research-type questions, I have prepared a Solutions Manual for the problems appearing in each chapter's *Problems* section. This manual is available at no charge from the publisher. To obtain a copy, ask your Wiley sales representative or write directly to Wiley-Interscience, 605 Third Avenue, New York, NY 10158–0012.

So there you have it. More than 300 pages of new points of contact between the worlds of nature and mathematics—in itself material constituting a fair-sized book. I think that after looking over this new material, readers of *Alternate Realities* and librarians will both agree that a new book was indeed in order. Now let me talk for a moment about the intended audience for this book and how it might be used.

One of the questions that always comes up in deciding on a text for a course is the puzzler: What are the prerequisites? Since this is a book on mathematical modeling, it goes without saying that in general terms the prerequisites for using the book are a *working knowledge* of basic undergraduate mathematics. This means that the reader should have been exposed to *and assimilated* the material typically found in one-semester courses in calculus, linear algebra and matrix theory, ordinary differential equations and, perhaps, elementary probability theory. Moreover, s/he should know about the basic vocabulary and techniques of mathematics. So, for example, things like "sets" and "equivalence relations" should be familiar territory, as should the idea of an inductive argument or a proof by contradiction. The emphasis here, of course, is on the phrase "working knowledge." To illustrate, it is definitely *not* sufficient to have had a course in matrix theory and have just *heard* of the Cayley-Hamilton Theorem. You must actually know what the theorem says and how to use it. Or, at the bare minimum, you should at least know where to go and look up the result. This is what I mean when I say the prerequisites for accessing this book are a working knowledge of basic undergraduate mathematics.

Now let's turn to how the book might be used as a text in a course in mathematical modeling. It's manifestly evident, I think, that there's far more material in these two volumes than can be comfortably addressed in even a one-year course, let alone in a single semester. Hence, one of the prime motivations for splitting the book into two pieces. Since what distinguishes *modern* mathematical modeling from its classical counterpart is its emphasis upon dynamics and nonlinearity, *The Fundamentals* contains the essentials of these matters. By way of illustration, let me briefly indicate how I have made use of this material in my own courses.

First of all, I always include the material from Chapter 1 as it sets the general framework for just about every type of mathematical modeling undertaking. I sometimes follow Chapter 1 with selected material from Chapter 8 on q-analysis. For some reason students seem to find it easier to grasp what's going on in the modeling game by starting with the context of sets and binary relations rather than jumping immediately into the thicket of dynamical systems and its daunting terminology and definitions. Moreover, the generality of the q-analysis idea lends itself to a number of interesting examples in art, literature and life that lie outside the domain of the "hard" sciences. Since just about everyone nowadays wants to know about chaos, my lectures usually continue with a selection of material from Chapters 2, 3 and 4. The second chapter contains the necessary background for dealing with dynamical systems, which is then used in Chapters 3 and 4 within the specific settings of cellular automata and chaotic processes. If there is still time remaining in the term, I conclude the course with a discussion of some of the more general philosophical issues linking science and other reality-generating mechanisms, as outlined in the book's final chapter. Since this program involves bits and pieces from *The Frontier,* it would be cruel and inhuman, not to mention imprudent, to close this Preface without saying a word or two about the second half of the book.

As I've already mentioned, the sheer size of this "magnum opus" dictated its division into two volumes. But unlike many multi-volume efforts, this split was not made on the basis of introductory versus advanced material and/or techniques. Quite to the contrary, in fact, as in putting this book together I tried valiantly to ensure that the difficulty level was as uniform as possible throughout. So the line of demarcation between the two volumes lies in a very different direction.

The Fundamentals contains the material that I feel is essential for anyone to consider himself or herself a player in the game of "modern" mathematical modeling. Just as it's difficult to conceive of writing a book without knowing the alphabet, there are certain concepts and results that anyone who wants to get involved with modeling must have at his or her disposal. *The Fundamentals* provides an account of this irreducible minimum of basic knowledge, material that can form the basis for a one-semester initial encounter with mathematical modeling.

For those already familiar with the essentials of system modeling, *The Frontier* introduces a number of application areas and/or associated techniques of modeling that complement the ideas presented in *The Fundamentals.* Chapter 5 shows how dynamical system theory and concepts from game theory can be brought together to shed new light on problems of population biology and ecology. This chapter also gives a mathematical account of the controversial problem of sociobiology, as well as taking a long look at the emergence of cooperative behavior in a world of egoists.

In Chapter 6 the notion of a control system is introduced within the cozy confines of linear processes. Of special importance in this chapter are the ideas of reachability and observability, concepts that form the basis for what we can hope to know about a system from measurements made upon it, as well as the degree to which we can alter a system's behavior by feeding inputs into it from the outside. By restricting the setting to linear structures, these notions are used to show how to construct "good" models directly from observed data. The chapter concludes with an indication of how these ideas can be extended to more general types of nonlinear processes.

If a system's behavior can be altered by applying inputs from the outside, how should those inputs be chosen so as to maximize or minimize some measure of system performance? This question is the central theme of Chapter 7. Following an introduction to the two principal techniques for calculating optimal control policies—the Minimum Principle and dynamic programming—the chapter presents an extended consideration of a novel class of adaptive control processes motivated by considerations arising in cellular biology. These so-called "(M, R)-systems" are then used to formalize ideas of self-reference and anticipation in application areas like manufacturing operations and input/output economics.

Shifting the emphasis from dynamics and continuity to statics and discrete combinatorial structures, Chapter 8 addresses the ways in which patterns in art, literature and other areas of human endeavor outside the natural sciences can be formulated in meaningful mathematical terms. Employing the ideas of q-analysis, which is a kind of extension of classical graph theory, this chapter develops a number of applications ranging from M. C. Escher engravings to humor and pathos in literature. These applications show how mathematical structures lurk at the heart of such reality-generation mechanisms heretofore thought to be beyond the bounds of mathematical analysis.

The thread running through every chapter of both volumes of *Reality Rules* is the idea that a mathematical model is a particular way of expressing a rule. In today's world, this rule is formulated in terms of a computer program more often than not. Chapter 9 is devoted to showing why there is no difference that matters between a computer program, a dynamical system and a deductive logical system. Exploiting the equivalence between these three seemingly distinct ways of formalizing a real-world situtation, the chapter shows how the results of Turing, Gödel and Chaitin lead us to consider the ultimate limitations of the scientific method as a procedure for telling us the way the world works.

Having spent nine chapters looking at the scientific way of getting at the scheme of things, the book's final chapter turns a bit philosophical and examines the degree to which the answers provided by science are in any way superior to those provided by other reality-generation procedures like religion, mysticism, music, art or literature. Following a discussion of science versus pseudoscience, the chapter concludes with the message that there are many realities, of which the scientific variety have no special claim to intrinsic superiority.

Taken together, it's my hope that the material of the two volumes making up *Reality Rules* will serve as both a text and a reference that students and researchers alike will be able to turn to as a source of inspiration and information as they make their way through the ever-shifting quicksand and minefields of the complex, weird and wonderful world we all inhabit.

JLC
Vienna, Austria
January 1992

CREDITS

Grateful acknowledgment is made to the following sources for permission to reproduce copyrighted material. Every effort has been made to locate the copyright holders for material reproduced here from other sources. Omissions brought to our attention will be corrected in future editions.

Cambridge University Press for Figure 1.2, which is reproduced from Thompson, D'Arcy, *On Growth and Form,* 1942, and for the figures on pages 132 and 149, which are reproduced from Saunders, P., *An Introduction to Catastrophe Theory,* 1980.

E. P. Dutton, Inc. for Figure 2.8, as well as the figures on pages 129 and 136, which are reproduced from Woodcock, T. and M. Davis, *Catastrophe Theory,* 1978.

MIT Press for Figures 2.12 and 2.13 found originally in Arnold, V. I., *Ordinary Differential Equations,* 1973.

American Physical Society for Figures 3.4, 3.5 and 3.8–3.10, which are reproduced from Wolfram, S., "Statistical Mechanics of Cellular Automata," *Reviews of Modern Physics,* 55 (1983), 601–644, and for Figure 4.20 from Eckmann, J. P., "Roads to Turbulence in Dissipative Dynamical Systems," *Reviews of Modern Physics,* 53 (1981), 643–654.

Corgi Books for Figure 3.11 which appeared in Gribbin, J., *In Search of the Double Helix,* 1985.

Basil Blackwell, Ltd. for Figures 3.12, which is taken from Scott, A., *The Creation of Life,* 1986.

Alfred Knopf, Inc. for Figures 3.14, 3.16 and 3.17 from Eigen, M. and R. Winkler, *The Laws of the Game,* 1981.

William Morrow and Co. for Figure 3.18 which appeared originally in Poundstone, W., *The Recursive Universe,* 1985.

Elsevier Publishing Co. for Figure 3.19 which appeared originally in Young, D., "A Local Activator-Inhibitor Model of Vertebrate Skin Patterns," *Mathematical Biosciences,* 72 (1984), 51–58, for Figures 4.17 and 4.18 taken from Swinney, H., "Observations of Order and Chaos in Nonlinear Systems," *Physica D,* 7D (1983), 3–15, and for the figure on page 419, which is reproduced from Barash, D., *Sociobiology and Behavior,* 1977.

Longman Group, Ltd. for Figures 3.20–3.23, which are reproduced from Dawkins, R., *The Blind Watchmaker,* 1986.

Springer Verlag, Inc. for Figure 4.3 taken from Arnold, V. I., *Mathematical Methods in Classical Mechanics,* 1978, and for Figure 4.5 found in Lichtenberg, A., and M. Lieberman, *Regular and Stochastic Motion,* 1983.

Mathematical Association of America for Figure 4.14 from Oster, G., "The Dynamics of Nonlinear Models with Age Structure," in *Studies in Mathematical Biology, Part II,* S. Levin, ed., 1978.

Academic Press, Inc. for Figure 5.1 from Vincent, T., and J. Brown, "Stability in an Evolutionary Game," *Theoretical Population Biology,* 26 (1984), 408–427.

Heinemann Publishing, Ltd. for Figure 8.10 from Atkin, R., *Mathematical Structure in Human Affairs,* 1974.

Houghton Mifflin Company for Figures 9.2 and 9.4, which are reproduced from Rucker, R., *Mind Tools,* 1987.

Professors E. C. Zeeman for Figures 2.17 and 2.18, as well as for the figures on pages 134 and *II–77*; Jeff Johnson for Figures 8.6 and 8.9, as well as the figures on pages *II–272* and *II–279;* Robert Rosen for Figures 1.7, 1.8 and 7.8; Thomas C. Schelling for Figures 3.2 and 3.3; Helen Couclelis for Figures 3.6 and 3.7; Christopher Langton for the figures on pages 239 and 240; Erik Mosekilde for Figures 4.10–4.13; Otto Rössler for Figures 4.24 and 4.25, as well as the figure on page 350; Douglas Hofstadter for Figure 9.5.

CONTENTS

CONTENTS OF THE FRONTIER

Chapter Five STRATEGIES FOR SURVIVAL: COMPETITION, GAMES AND THE THEORY OF EVOLUTION

Chapter Six THE ANALYTICAL ENGINE: A SYSTEM-THEORETIC VIEW OF BRAINS, MINDS AND MECHANISMS

REALITY RULES: I

CHAPTER ONE

THE WAYS OF MODELMAKING: NATURAL SYSTEMS AND FORMAL MATHEMATICAL REPRESENTATIONS

1. A Theory of Models

What do you think of when you hear the word "model"? An elegant mannequin from the pages of *Vogue* perhaps? Or maybe a miniature version of that super-exotic Ferrari or Lamborghini that you've been drooling over? To a system scientist or applied mathematician, a quite different picture comes to mind. For such practitioners of the "black arts," a *model* means an encapsulation of some slice of the real world within the confines of the relationships constituting a formal mathematical system. Thus, a model is a mathematical representation of the modeler's reality, a way of capturing some aspects of a particular reality within the framework of a mathematical apparatus that provides us with a means for exploring the properties of the reality mirrored in the model.

This book is about the ways and means of constructing "good" models of reality, the properties of such models, the means for encoding specific realities into definite formal systems and the procedures for interpreting the properties of the formal system in terms of the given real world situation. In short, we're interested in the ways of modelmaking. Before embarking upon a more detailed account of what we mean by the terms *model, encoding, formal system,* and so forth, let's first look at some basic epistemological and operational issues lying at the heart of what we shall term the *theory of models.*

According to the great nineteenth-century British physicist James Clerk Maxwell, "the success of any physical investigation depends upon the judicious selection of what is to be observed as of primary importance." This view suggests the notion that what constitutes one's reality depends upon one's capacity for observation. Here we adopt the position that since natural phenomena impinge upon our consciousness only through instruments of observation, then, to paraphrase Maxwell, "the success of any modeling venture depends upon a judicious selection of observables and means for encapsulating these observables within the framework of suitable formal mathematical systems."

As noted by Robert Rosen, in dealing with the idea of a natural system, we must necessarily touch on some basic philosophical questions of both an ontological and epistemological character. This is unavoidable in any case and must be addressed at the outset of a work such as this, because our tacit assumptions in these areas determine the character of our science. Many scientists find an explicit consideration of such matters irritating, just as many working mathematicians dislike discussions of the foundations of mathematics. Nevertheless, it's well to recall the remark of philosopher David Hawkins: "Philosophy may be ignored but not escaped; and those who most ignore escape least."

Our viewpoint is that *the study of natural systems begins and ends with the specification of observables describing such a system, and a characterization of the manner in which these observables are linked.* Purely theoretical issues may be pursued in the process of investigating a system, but ultimately contact with reality occurs through the observables. During the course of this book, we will argue that the concept of a model of a natural system N is a generalization of the concept of a subsystem of N, and that the essential feature of the modeling relation is the exploration of the idea that there is a set of circumstances under which the model describes the original system to a prescribed degree of accuracy. In other words, a particular facet of system behavior remains *invariant* under the replacement of the original system by a proper subsystem.

At this point it is well to consider why one constructs models of natural phenomena in the first place. Basically, the point of making models is to be able to bring a measure of order to our experiences and observations, as well as to make specific predictions about certain aspects of the world we experience. The central question surrounding the issue of model credibility is to ask to what extent "good" predictions can be made if the best the model can do is to capture a subsystem of N. The answer is wrapped up in the way in which the natural system N is characterized by observables, the procedure by which observables are selected to form the subsystem, and the manner in which the subsystem is encoded into a formal mathematical system F which *represents,* or "models," the phenomena of concern. These notions will be made more explicit later. But before doing so, let's consider a familiar example illustrating many of these points.

Consider an enclosed homogeneous gas for which we take the observables to be the volume V occupied by the gas, the gas pressure P and the temperature T. By this choice, the abstract states of the system, i.e., the actual *physical states* giving rise to the volume, pressure and temperature, are encoded into the three-dimensional space R^3, using the coordinates P, V and T. At equilibrium these three observables are not independent, but are linked by the relation (equation of state) $PV = T$, the *ideal gas law.* Of course, we know that such a selection of observables represents an abstraction in the sense that many other observables have been omitted that also influence the gas (external radiation, properties of the container, etc.). Experience has shown, however, that the subsystem consisting of the quantities P, V and T, together with its encoding into the region of R^3 defined by the ideal gas law, enables us to make very accurate predictions about the *macroscopic* behavior of the system. But should we desire to make predictions about the gas at the molecular level, it would be necessary to select a different set of observables. In the *microscopic* context, the positions and momenta of the 10^{24} or so molecules composing a mole of the gas would be a natural choice. And these observables would then be encoded into the space $R^{6\times10^{24}}$. Linkages between the observables in this case are specified by the laws of mechanics operating at this microlevel.

Our primary objective in this chapter is to provide a framework within which we can speak of fundamental issues underlying any theory of modeling. In this book we shall place primary emphasis on the following questions:

- What is a model?
- What features characterize "good" models?
- How can we represent a natural process N in a formal system F?
- What is the relationship between N and F?
- When does the similarity of two natural systems N_1 and N_2 imply that their models F_1 and F_2 are similar?
- How can we compare two models of the same natural process N?
- Under what circumstances can we consider a linkage between observables as constituting a "law of nature"?
- What procedures can we employ to identify key observables and thereby simplify a model?
- How does a given system relate to its subsystems?
- When can two systems that behave similarly be considered as models of each other?

Such a list (and its almost infinite extension) expresses basic issues in the philosophy of science and, in particular, the theory of models. No uniform and complete answers to these questions can ever be expected; the best we can hope for is to provide a basis for considering these matters under conditions appropriate to a given setting.

In what follows, we sketch a formalism suitable for studying the above questions and indicate by familar examples some of the advantages to be gained by looking at issues of system modeling in such generality. In essence, our argument is that the detailed study of a given natural system N cannot be suitably interpreted and understood by remaining at the level of the system N itself. A more general metalevel, together with a metalanguage provided by a theory of models, is required.

2. States, Observables and Natural Systems

Consider a particular subset S of the observable world, and assume that S can exist in a set of distinct states $\Omega = \{\omega_1, \omega_2, \ldots\}$. We call Ω the set of *abstract states* of S. Note that an observer probing the behavior of S may or may not be able to determine which of these states S is actually in at any particular moment. It all depends upon the resolution of the measuring instruments (observables) at the observer's disposal. Note also that the set Ω may be finite or infinite (uncountable, even). Here it's important to emphasize that what counts as an "observable state" is not an intrinsic property of the system itself, but depends crucially upon the observer and the ways he has of probing the system and distinguishing one state from another. Since this notion of state is crucial for all system modeling, let's look at some examples to firmly fix the idea.

Example 1: An On-Off Switch

1) If S is the on-off switch for the table lamp on my desk, then a reasonable set of abstract states for S might be

$$\Omega = \{\omega_1 = OFF,\ \omega_2 = ON\}.$$

Here Ω is a finite set, consisting of the two elements *ON* and *OFF*.

Example 2: An Equilateral Triangle

Let S be the equilateral triangle $\triangle$ with vertices denoted $[a, b, c]$, moving counterclockwise from the top. Assume that S can be distinguished in three forms corresponding to rotations by $0, 2\pi/3$ and $4\pi/3$ radians. Then one possibility for the set of abstract states Ω of the triangle is

$$\Omega_1 = \{\omega_1 = [a, b, c],\ \omega_2 = [c, a, b],\ \omega_3 = [b, c, a]\},$$

corresponding to an ordered labeling of the vertices of S. But an equally valid set of abstract states is

$$\Omega_2 = \left\{\omega_1 = 0,\ \omega_2 = \frac{2\pi}{3},\ \omega_3 = \frac{4\pi}{3}\right\},$$

or even

$$\Omega_3 = \{\omega_1 = 0,\ \omega_2 = 1,\ \omega_3 = 2\},$$

where each state $\omega_i \in \Omega$ is just a *label* for a *physical* state of S.

This example illustrates several vital points about abstract state spaces:

- A given system S usually has many different sets of abstract states Ω.
- The set of abstract states for S need not in general be a set of numbers.
- All abstract state sets are equivalent insofar as they characterize the distinct states of the system S. In nature there is no preferred space of states. But for modelers, some sets Ω may be more *convenient* than others.

Now let's turn to the notion of an observable. Assume that we are given the system S, together with a set Ω constituting the abstract states of S. Then a rule f associating a real number with each $\omega \in \Omega$ is called an *observable* of S. Thus, we see that an observable is just a way of generalizing the everyday idea of a measuring device. In what follows, it might be helpful for the reader to think of an observable as a kind of tape measure or thermometer by which we "measure" some aspect of the system. More formally, an observable is a map $f\colon \Omega \to R$.

Example: A Ball on an Inclined Plane

Let S be a ball constrained to roll along an inclined plane of unit length, and let $\Omega = \{\omega_1, \omega_2, \omega_3, \omega_4\}$, where ω_1=ball at the top of the plane, ω_2=ball one-third of the way down the plane, ω_3= ball two-thirds of the way down the plane, ω_4= ball at the bottom of the plane. Define the observable $f\colon \Omega \to R$ by the rule

$$f(\omega) = \text{distance of the ball from the top of the plane.}$$

Then

$$f(\omega_1) = 0, \quad f(\omega_2) = \tfrac{1}{3}, \quad f(\omega_3) = \tfrac{2}{3}, \quad f(\omega_4) = 1.$$

Using the same system S and state-space Ω as above, let the observable $g\colon \Omega \to R$ be defined by the rule

$$g(\omega) = \text{distance of the ball from the middle of the plane.}$$

Then

$$g(\omega_1) = \tfrac{1}{2}, \quad g(\omega_2) = \tfrac{1}{6}, \quad g(\omega_3) = \tfrac{1}{6}, \quad g(\omega_4) = \tfrac{1}{2}.$$

In this case it can be seen that the states ω_1 and ω_4, as well as ω_2 and ω_3, are indistinguishable using the observable g, while the observable f distinguishes (separates) all states. Consequently, we can "see" more of the system S using the observable f than by using g. This is a crucial point that we will return to often. Note also the *linkage* relationship between f and g given explicitly by the rule

$$f(\omega) = \begin{cases} \tfrac{1}{2} - g(\omega), & \omega = \omega_1, \omega_2, \\ \tfrac{1}{2} + g(\omega), & \omega = \omega_3, \omega_4. \end{cases} \tag{$*$}$$

Linkage generalizes the intuitive notion of one observable being a *function* of another. Here we see that although the observable g alone contains less information about the system than f, this lack of information can be compensated for if we know the linkage relation above. Speaking intuitively,

$$g + \text{linkage relation } (*) = f.$$

Generally, in order to "see" the complete system S we need an infinite number of observables $f_\alpha : \Omega \to R$, where α ranges over some possibly uncountable index set. Thus, the complete system S is described by Ω and the entire set of observables $\mathcal{F} = \{f_\alpha\}$. But for practical modeling purposes it's certainly inconvenient (and usually just plain impossible) to work with such a large set of observables. So we boldly just throw most of them away, focusing our attention on a proper subset A of $\mathcal{F}$. We call A an *abstraction* of S, since the view we have of S using the observables A is necessarily a partial view formed by abstracting, i.e., throwing away, all the information contained in the observables $\mathcal{F} - A$. It's an amusing aside to wonder why self-styled practically-oriented people reserve their greatest scorn for "useless abstractions," when the very essence of an abstraction is to reduce the description of a system to a simpler and, presumably, more tractable form. Thus, in many ways there is nothing more useful and practical than a good abstraction. This calls to mind Hilbert's dictum that "there is nothing more practical than a good theory." Much of our subsequent development is focused upon tricks, techniques and subterfuges aimed at finding good abstractions.

We are now in a position to put forth our notion of a *natural system N*. For us, N consists of an abstract state-space Ω, together with a finite set of observables $f_i : \Omega \to R$, $i = 1, 2, \ldots, n$. Symbolically,

$$N = \{\Omega, f_1, f_2, \ldots, f_n\}.$$

We employ the term "natural system" to distinguish N from the idea of a "formal mathematical system" to be discussed below. It should not be interpreted to mean the restriction of N to the class of systems studied in the "natural" sciences (chemistry, physics, astronomy, etc.); here "natural" includes social, behavioral and living systems as well. In all that follows, we shall use the term *natural system* in this extended sense.

Exercises

1. What would you consider to be good candidates for the abstract state space Ω of a national economy? A human brain? A natural language, like English or Spanish? An automobile? Construct at least two different state spaces for these natural systems and examine the relationship, if any, between the state spaces you propose.

2. Suppose you wanted to put a new carpet into your living room. In order to know what size carpet to buy, you use a meter stick (or a yardstick) to measure the size of the room. Define an abstract space of states and an appropriate observable on these states that characterizes this carpet-laying situation. Is there anything unusual about the state-set Ω you chose for this problem?

3. Consider traffic flow on a freeway. Suppose your job is to construct a graph showing the number of cars passing a particular point and their speeds over a 24-hour period. Suppose further that the data points for this graph are to be spaced at hourly intervals. Construct a state space and a set of observables that could be used for this job.

4. A particular kind of cell in a living organism can be expressed by many different observables: (i) its various biochemical inputs and outputs, (ii) the mechanical properties of its membranes, proteins and nucleotides, (iii) its functional activities metabolism and replication, or (iv) its physical position in relation to other cells. As an example, consider a human heart cell and create abstractions from these different sets of observables in order to characterize (a) the heart cell's biochemical activities, (b) its structural role in preserving the physical shape and position of the heart, and (c) its functional role in producing the pumping and self-repair activities of the heart.

3. Equations of State

The observables $\{f_1, f_2, \ldots, f_n\}$ provide the percepts by which we see a natural system N, the raw data, so to speak. But there is more to the "systemness" of N than just the separate observables by which we see it. Just as a book is more than the individual words that comprise it, the essential "systemness" of N is contained in the relationships linking the observables $\{f_1, f_2, \ldots, f_n\}$. We term such a set of relationships the *equation of state* for N. Formally, the equation of state can be written as

$$\Phi_i(f_1, f_2, \ldots, f_n) = 0, \qquad i = 1, 2, \ldots, m,$$

where the $\Phi_i(\cdot)$ are mathematical relationships expressing the dependency relations among the observables. We can write this more compactly as

$$\Phi(f) = 0.$$

Example: The Ideal Gas Law

Let N consist of an ideal gas contained in a closed vessel. Take Ω to be the positions and velocities of the molecules making up the gas, and define the three observables

$$\begin{aligned} P(\omega) &= \text{pressure of the gas when in state } \omega, \\ V(\omega) &= \text{volume of the gas when in state } \omega, \\ T(\omega) &= \text{temperature of the gas when in state } \omega. \end{aligned}$$

Then the ideal gas law asserts the single equation of state

$$\Phi(P, V, T) = 0,$$

where the specific functional form of Φ is

$$\Phi(x, y, z) = xy - z.$$

The foregoing example, simple as it is, serves to illustrate another deep epistemological issue in the theory of models, namely, the distinction between *causality* and *determinism*. An equation of state, $\Phi(f) = 0$, necessarily establishes a *deterministic* relationship between the observables. But it contains no information whatsoever about any possible *causal* implications among the elements of the set $\{f_i\}$. Thus, the ideal gas law asserts that once we know any two of the three observables P, V and T, the remaining observable is *determined* by the other two, but not *caused* by them.

The difficulty in assigning a causal ordering to the set of observables is one of the principal difficulties in economic modeling. For instance, we often have available theoretical and/or empirical equations of state, but little knowledge of how to separate observables into so-called exogenous and endogenous subsets. This same obstacle stands in the way of making effective use of system-theoretic techniques in many other areas of the social and biological sciences, and deserves far more formal attention than it has thus far received.

Now suppose that there are r observables whose values remain fixed for every state $\omega \in \Omega$. For simplicity of exposition, assume these are the first r observables. Such quantities are usually termed *parameters* and arise as a matter of course in describing almost every natural system. Typical examples are the stiffness constant of a spring, the prime interest rate in a national economy and the gravitational constant in orbital mechanics. If we let $f_i(\omega) = \alpha_i$, $\omega \in \Omega$, α_i a real number, $i = 1, 2, \ldots, r$, we can write the equation of state in this situation as

$$\Phi_{\alpha_1, \alpha_2, \ldots, \alpha_r}(f_{r+1}, f_{r+2}, \ldots, f_n) = 0, \tag{\dagger}$$

indicating explicitly the dependence of the description upon the values of the parameters $\alpha_1, \ldots, \alpha_r$. In other words, we have an *r-parameter family* of descriptions. For each set of values of the set $\{\alpha_i\}$, Eq. (†) describes a *different* system.

We now introduce the additional assumption that the last m observables $f_{n-m+1}, \ldots, f_n$ are functions of the remaining observables $f_{r+1}, \ldots, f_{n-m}$, i.e., we can find relations $y_i(\cdot)$, $i = 1, 2, \ldots, m$, such that

$$\begin{aligned} f_{n-m+1}(\omega) &= y_1(f_{r+1}(\omega), \ldots, f_{n-m}(\omega)), \\ &\vdots \\ f_n(\omega) &= y_m(f_{r+1}(\omega), \ldots, f_{n-m}(\omega)). \end{aligned}$$

If we introduce the notation

$$\begin{aligned}\alpha &\doteq (\alpha_1, \alpha_2, \ldots, \alpha_r),\\ u &\doteq (f_{r+1}, f_{r+2}, \ldots, f_{n-m}),\\ y &\doteq (f_{n-m+1}, f_{n-m+2}, \ldots, f_n),\end{aligned}$$

the equations of state (†) become

$$\Phi_\alpha(u) = y. \qquad (\ddagger)$$

In (‡) it's natural to think of the observables u as representing the *inputs* to the system, with the observables y being the resulting *outputs*. The vector α is, as before, the set of parameters. Note also that the assumption of being able to solve for the observables $f_{n-m+1}, \ldots, f_n$ in terms of the observables $f_{r+1}, \ldots, f_{n-m}$ introduces a *causal* direction into our previous acausal relationship (†). We can now think of the inputs as somehow "causing" the outputs. The problem, of course, is that there may be many ways of solving for some of the observables in terms of the others. So it's usually possible to have many distinct separations of the observables into inputs and outputs and, correspondingly, many different causal relationships.

As already noted, in some parts of the natural sciences (e.g., classical physics), there are natural and useful conventions that have been established for making this separation of observables; in other areas like the social and behavioral sciences, there is no clear-cut procedure or body of past evidence upon which to base such a classification of observables. Consequently, one ends up with many distinct, often contradictory, theories depending upon the choice that's made. For instance, is unemployment caused by inflation or is it the other way round, or perhaps neither? No one really seems to know. Yet far-reaching economic and social policy is made on the basis of one assumption or the other. We shall return to this crucial issue in various guises in almost every chapter of this book. But for now let's shift attention back to the parameterized equation of state (‡), giving an interpretation of it that will be of considerable intuitive use in our subsequent discussions.

If we adopt a biological view of (‡), it's natural to think of the parameter vector $\alpha = (\alpha_1, \alpha_2, \ldots, \alpha_r)$ as representing the "blueprint" or "program" for the system N. Biologically speaking, α represents the "genetic" make-up, or *genome*, of N. Similarly, the inputs u correspond to the *environment* in which N operates, and with this view of u we can only conclude that the output y corresponds to the "form" of N that emerges from the interaction of the genome with the environment. In more biological terms, y is the *phenotype* of N.

The foregoing development shows explicitly how we can start from the fundamental description of N given by the equations of state (†), and by making natural assumptions about the dependencies of observables upon the states of Ω and upon each other, arrive at a standard parameterized family of input/output descriptions of N that can be given several types of interpretations. We now explore the implications of this set-up for determining when two descriptions are the "same," along with the associated matters of system bifurcation, complexity and error.

Exercises

1. Consider the Moon as it moves around the Earth in accordance with Newton's laws of motion. (a) Construct the abstract state space for this situation. (b) What are the observables describing the position and velocity of the Moon? (c) Write down an equation of state for these observables (Hint: Use Newton's Second Law of Motion, $F = ma$).

2. In the ideal gas law, determine the causal relation that sees the pressure as being "caused" by the temperature and volume. Is this causal relation any more convincing than seeing the temperature as a consequence of the pressure and volume?

3. All that we can observe about another person is their physical form and behavioral patterns. Some investigators believe that these observable qualities arise as a direct consequence of an individual's genetic makeup. Develop the abstract state space, observables and equations of state describing this "hardline" sociobiological view. The other extreme is to assert that behavioral patterns, at least for humans, are culturally determined by the environment, with the genes playing almost no role whatsoever. Use the states and observables you have introduced in order to characterize this "humanistic" perspective on the nature/nurture question.

4. Equivalent Descriptions and the Fundamental Question of Modeling

The family of descriptions (‡) leads immediately to what we might term *The Fundamental Question of System Modeling:* "How can we tell if two descriptions of N contain the same information?" Or: "When are two descriptions of N equivalent?" In the pages that follow, we shall justify the lofty plane to which we've elevated this question by showing its central role in the analysis of *all* issues pertaining to matters of system complexity, bifurcation, error, self-organization, and adaptation. For now, let's focus upon the means for reformulating the Fundamental Question in more tractable mathematical terms.

To motivate the intuitive idea of two descriptions being equivalent, consider the descriptions of an ellipse in the (x, y)-plane. From elementary analytic geometry, we know that the expression

$$\frac{x^2}{a^2} + \frac{y^2}{b^2} = 1, \qquad a, b \in R,$$

describes a family of ellipses having semi-axes of lengths a and b [Fig. 1.1(a)]. In Fig. 1.1(b) we show the same ellipse rotated through an angle θ. On the one hand, it's clear that the two figures display different closed curves E and $\hat{E}$ in the plane; on the other hand, E and $\hat{E}$ are related to each other through the simple act of rotating the coordinate axes through an angle θ. That is, they can be made congruent to each other by rotating the axes we use to *describe* the curve. It is in this sense that the two descriptions

$$\frac{x^2}{a^2} + \frac{y^2}{b^2} = 1,$$

and

$$\frac{\hat{x}^2}{a^2} + \frac{\hat{y}^2}{b^2} = 1,$$

are *equivalent* through the coordinate transformations $\hat{x} = x\cos\theta - y\sin\theta$, $\hat{y} = x\sin\theta + y\cos\theta$. These two descriptions contain exactly the same information about the ellipse. This must necessarily be the case, since the curves E and $\hat{E}$ are the *same* curve. They appear different only because of the way we choose to look at them.

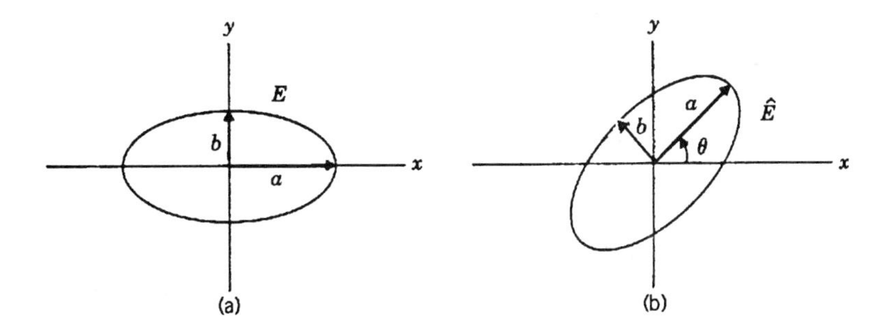

Figure 1.1. A Family of Ellipses

Note in the foregoing example that the numbers a and b are *intrinsic* properties of the curve; they remain the same in both descriptions. Such *invariants* represent properties of the system that are independent of the choice of coordinates used to describe the ellipse. It is the invariants that represent the true system-theoretic properties of the curve; all other coordinate-dependent properties are artifacts of the description, not properties of the system itself. We shall emphasize this point as we go along during the course of developing alternative descriptions later in the book.

In the preceding development, we have regarded the parameters a and b as being fixed. Assume now that we make the change $a \to \hat{a}$, $b \to \hat{b}$, i.e., we pass from a system (ellipse) with semi-axes a and b to one with semi-axes

$\hat{a}$ and $\hat{b}$. It's clear geometrically that if we start with the ellipse having semi-axes $\hat{a}$ and $\hat{b}$, we can make a coordinate transformation $x \to \hat{x}$, $y \to \hat{y}$ so that in the $(\hat{x}, \hat{y})$-plane the ellipse has semi-axes equal to a and b. In fact, from the formula

$$\frac{x^2}{\hat{a}^2} + \frac{y^2}{\hat{b}^2} = 1,$$

it follows that the required coordinate change is

$$\hat{x} = \left(\frac{a}{\hat{a}}\right) x, \qquad \hat{y} = \left(\frac{b}{\hat{b}}\right) y.$$

What has been accomplished by the change to the $(\hat{x}, \hat{y})$ coordinates is to neutralize the effect of the change of parameters $(a, b) \to (\hat{a}, \hat{b})$. When such a change of variables can be made, we call the two ellipses *equivalent.* Note that the above argument breaks down whenever any of the parameters $a, b, \hat{a}, \hat{b}$ are zero, i.e., when the family of ellipses degenerates into a different conic section. In these cases, such distinguished values of the parameters are called *bifurcation points.*

Generalizing the above set-up to our description of N given by (‡), we consider two descriptions $\Phi_\alpha, \Phi_{\hat{\alpha}}$ corresponding to the "genomes" α and $\hat{\alpha}$, i.e., we have the descriptions

$$\Phi_\alpha : U \to Y, \qquad \Phi_{\hat{\alpha}} : U \to Y,$$

where U and Y are the input and output spaces, respectively. We will consider these descriptions to be *equivalent* if it's possible to find coordinate changes (usually nonlinear) in U and/or Y such that the description Φ_α is transformed into $\Phi_{\hat{\alpha}}$. Diagrammatically, we seek bijections (one-to-one and onto) $g_{\alpha,\hat{\alpha}} : U \to U$ and $h_{\alpha,\hat{\alpha}} : Y \to Y$ such that the following diagram commutes

$$\begin{array}{ccc} U & \xrightarrow{\Phi_\alpha} & Y \\ {\scriptstyle g_{\alpha,\hat{\alpha}}}\downarrow & & \downarrow{\scriptstyle h_{\alpha,\hat{\alpha}}} \\ U & \xrightarrow[\Phi_{\hat{\alpha}}]{} & Y \end{array}$$

i.e., starting in the upper left-hand corner, it can be traversed along either of the paths from U to Y. What we are saying here is that the change from the system described by Φ_α to the system described by $\Phi_{\hat{\alpha}}$ can be "undone" or "neutralized" by a corresponding change of coordinates in U and/or Y. Biologically speaking, this says that two organisms are equivalent if a genetic change can be reversed by means of a suitable change in the environment and/or the phenotype. Since the idea of a commutative diagram will play a crucial role at various points in our narrative, let's pause for a moment to examine what's involved in a bit more detail.

Suppose we are given sets A, B, C and D, together with well-defined maps f, g, α and β acting in the following manner:

$$f : A \to B, \qquad g : C \to D,$$
$$\alpha : A \to C, \qquad \beta : B \to D.$$

Pictorially, it's convenient to represent this situation with the diagram

$$\begin{array}{ccc} A & \xrightarrow{f} & B \\ {\scriptstyle\alpha}\downarrow & & \downarrow{\scriptstyle\beta} \\ C & \xrightarrow[g]{} & D \end{array}$$

We say the above diagram is *commutative,* or *commutes,* if the various maps satisfy the following relation: $g \circ \alpha = \beta \circ f$. In other words, the diagram is commutative if we can start at the upper left-hand corner (i.e., with the set A) and traverse the diagram to D by following either the path ABD or the path ACD. The general idea can be extended to cover diagrams involving an arbitrary number of sets and maps, forming the key idea in the mathematical area called *category theory.* Readers interested in knowing more about this kind of abstract "diagram-chasing" are urged to consult the chapter Notes and References, as well as the treatment given in the Discussion Questions and Problems sections for this chapter.

Example 1: Noncommutativity of Diagrams

Sometimes it's easier to see what a concept is by looking at an example of what it isn't. So just to fix the idea of a commutative diagram, here is an illustration of a situation in which two maps cannot be made to look the same by means of a coordinate transformation.

Let the space X be the plane R^2, and consider the matrices

$$F = \begin{pmatrix} 2 & -1 \\ -1 & 0 \end{pmatrix}, \qquad \bar{F} = \begin{pmatrix} 0 & 1 \\ -1 & 2 \end{pmatrix}.$$

Now we ask if there exists a coordinate transformation T: $R^2 \to R^2$ which makes the diagram

$$\begin{array}{ccc} X & \xrightarrow{F} & X \\ {\scriptstyle T}\downarrow & & \downarrow{\scriptstyle T} \\ X & \xrightarrow[\bar{F}]{} & X \end{array}$$

commute. In other words, is there a nonsingular matrix T such that $TF = \bar{F}T$?

The simplest way to see that no such coordinate change T exists is to calculate the characteristic polynomials of the matrices F and $\bar{F}$. Carrying out this calculation, we find

$$\chi_F(z) = z^2 - 2z - 1,$$
$$\chi_{\bar{F}}(z) = z^2 - 2z + 1.$$

So, since the characteristic polynomials are different, the two matrices cannot be similar, i.e., they have different Jordan normal forms. This is the same thing as saying that there does not exist a nonsingular matrix T making the above diagram commute.

A slightly more tedious way of arriving at the same conclusion is by a direct calculation involving the elements of any such candidate matrix T. After a small amount of algebra, it's easy to see that the only possibility for a matrix T satisfying the equation $TF = \bar{F}T$ is if the matrix is singular, i.e., $\det T = 0$. But a singular matrix is definitely not a coordinate transformation (since it's not one-to-one). Thus, again we conclude that there is no such matrix T making the above diagram commute.

Example 2: d'Arcy Thompson's Theory of Biological Transformations

In the early part of this century, d'Arcy Thompson proposed a theory of biological structure whose main thrust was to regard a biological organism as a geometrical object. Each phenotype was associated with a point in a space of "forms" Y, and Thompson considered two organisms to be "close" in this space if there was a *homeomorphism* (continuous coordinate change) in Y that mapped them one to the other. Put more loosely, "phenotypes of closely related organisms can be deformed continuously into each other."

To make contact with our earlier ideas, let two organisms be "close" if their genomes are close. Assuming we have a measure on the space of genomes, we can rephrase Thompson's requirement in more specific terms using our diagram for equivalence. Consider two organisms described by the equations of state

$$\Phi_\alpha : U \to Y,$$
$$\Phi_{\hat{\alpha}} : U \to Y,$$

where U is the space of environments and Y is the topological space of phenotypes. The genomes α and $\hat{\alpha}$ are assumed to be close, i.e., $\|\alpha - \hat{\alpha}\| < \epsilon$, where $\epsilon > 0$ is some prescribed distance representing genomic "closeness." Then the phenotypes Φ_α and $\Phi_{\hat{\alpha}}$ are "close" *in a fixed environment U,* if there exists a homeomorphism $h_{\alpha,\hat{\alpha}}: Y \to Y$ such that the diagram

$$\begin{array}{ccc} U & \xrightarrow{\Phi_\alpha} & Y \\ {\scriptstyle id_U}\downarrow & & \downarrow{\scriptstyle h_{\alpha,\hat{\alpha}}} \\ U & \xrightarrow[\Phi_{\hat{\alpha}}]{} & Y \end{array}$$

commutes, i.e., a small genetic change $\alpha \to \hat{\alpha}$ can be offset by a continuous phenotypic deformation. Here the condition that the environment remains fixed requires that we take our earlier transformation $g_{\alpha,\hat{\alpha}}$ = identity on $U \doteq id_U$.

As a concrete illustration of this theory, consider the skull phenotypes displayed in Fig. 1.2. Here Fig. 1.2(a) represents a chimpanzee's skull, while Fig. 1.2(b) shows the skull of a baboon. From the superimposed curvilinear coordinate frames, it's clear that the baboon skull is obtained from the chimpanzee's by stretching and squeezing the coordinates used to describe the two shapes, i.e., a nonlinear transformation of the type $h_{\alpha,\hat{\alpha}}$ discussed above. Many other examples of this sort are found in d'Arcy Thompson's classic work cited in the Notes and References.

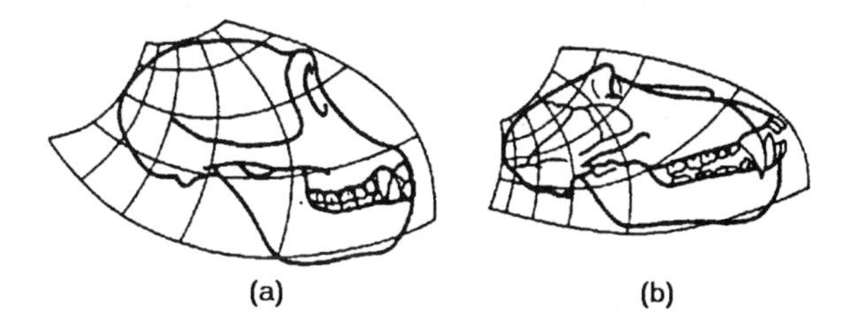

Figure 1.2. The Skulls of a Chimpanzee and a Baboon

Example 3: Forestry Yields

An empirical relation often used in forestry models for expressing timber yield as a function of tree diameter and height is

$$V = \epsilon + \beta D^2 H,$$

where

V = total tree volume exclusive of bark (in m^3),

D = tree diameter at breast height (in cm),

H = tree height from breast height (in m),

ϵ, β = parameters characterizing the specific tree type.

The yield relationship assumes our standard form if we set $U = (D, H)$, $Y = V$, $\alpha = (\epsilon, \beta)$ and use the description

$$\Phi_\alpha : U \to Y,$$
$$(D, H) \mapsto \epsilon + \beta D^2 H.$$

A change of parameter $\alpha \to \hat{\alpha}$ in this situation represents a change from one tree type to another, and it's of interest to ask whether such a change in tree type can be offset by a change in volume and/or diameter to keep the timber yield constant. Translating this question into a diagram, we ask if there is a map $g_{\alpha,\hat{\alpha}} : U \to U$, such that the diagram

$$\begin{array}{ccc} U & \xrightarrow{\Phi_\alpha} & Y \\ {\scriptstyle g_{\alpha,\hat{\alpha}}}\downarrow & & \downarrow{\scriptstyle id_Y} \\ U & \xrightarrow[\Phi_{\hat{\alpha}}]{} & Y \end{array}$$

commutes. If we demand that $g_{\alpha,\hat{\alpha}}$ be a smooth coordinate change, i.e., infinitely differentiable, then on the basis of arguments that will be developed in Chapter 2, it can be shown that there exists no such g. We conclude that for any tree species α, there is another species $\hat{\alpha}$ arbitrarily nearby, such that there is no way to "deform" the diameter and height observables to preserve the same yield.

Exercises

1. Define a relation R on the set of integers $\mathbb{Z}$ by the rule: "Two integers x and y are R-related if they leave the same remainder after division by 2." Thus, xRy if and only if rem $[x/2]$ = rem $[y/2]$. Show that the relation R is an *equivalence relation*, i.e., R is symmetric ($xRy \Rightarrow yRx$), reflexive (xRx) and transitive (xRy and $yRz \Rightarrow xRz$).

2. Show that any equivalence relation separates the set $\mathbb{Z}$ into subsets that form a *partition*. That is, the relation separates the integers into disjoint subsets such that every integer is in one and only one of the subsets and the union of all these subsets is the set $\mathbb{Z}$ itself. Now show that the partition of $\mathbb{Z}$ generated by the relation R of the preceding problem contains exactly two subsets, the even integers and the odd integers. (The set consisting of these two subsets is denoted $\mathbb{Z}/2$. It plays an important role in many areas of number theory, cryptography, and the theory of computation.)

3. Probably the most familiar change of coordinates is the transformation from rectangular coordinates in the plane to polar coordinates. In

this situation, a point whose cartesian coordinates are (x, y) is described by its distance r from the origin and by the angle θ of the radius vector from the origin to the point. The transformation representing this coordinate change is given by $r = (x^2 + y^2)^{\frac{1}{2}}$, $\theta = \arctan y/x$. (a) Find the inverse of this transformation, i.e., the rule expressing x and y in terms of the coordinates r and θ. (b) Show that these transformations are *smooth* coordinate changes, i.e., they are one-to-one, onto and infinitely differentiable. (Hint: Consider the matrices of the transformations).

4. Consider the curve in the plane given in rectangular coordinates by $(x^2 + y^2 + ay)^2 = a^2(x^2 + y^2)$, where $a > 0$. (a) Show that in polar coordinates this curve has the description $r = a(1 - \sin\theta)$ (technically, such a curve is called a cardioid). (b) Interpret the two descriptions of the curve as maps $f, g \colon R^2 \to R^1$, and show that the change of coordinates T from rectangular to polar coordinates makes the following diagram commute:

$$\begin{array}{ccc} R^2 & \xrightarrow{f} & R^1 \\ {\scriptstyle T}\downarrow & & \downarrow{\scriptstyle id_{R^1}} \\ R^2 & \xrightarrow[g]{} & R^1 \end{array}$$

5. *Bifurcations and Catastrophes*

In the preceding discussions we have seen that there can exist values of the system parameters α such that for all $\hat{\alpha}$ nearby to α, there do **not** exist coordinate changes

$$g_{\alpha,\hat{\alpha}} \colon U \to Y, \qquad h_{\alpha,\hat{\alpha}} \colon Y \to Y,$$

making the diagram

$$\begin{array}{ccc} U & \xrightarrow{\Phi_\alpha} & Y \\ {\scriptstyle g_{\alpha,\hat{\alpha}}}\downarrow & & \downarrow{\scriptstyle h_{\alpha,\hat{\alpha}}} \\ U & \xrightarrow[\Phi_{\hat{\alpha}}]{} & Y \end{array}$$

commute. We call such values of α *bifurcation points*; all other values are called *regular* or *stable points.*

It's important here to note that whether a given point α is a bifurcation point or not depends upon the space of maps in which we seek the coordinate changes g and h. For instance, α may be a bifurcation point if we seek g and h in the space of *smooth* coordinate changes or *diffeomorphisms,* but may well be a regular point if we look for g and h in the much larger space of continuous coordinate changes. The particular space we choose is usually dictated by the analytic nature of the description Φ_α, as well as by the structure imposed upon the spaces U and Y by the type of questions being considered. Illustrations of various cases will appear throughout the book, although we will usually ask that g and h be infinitely differentiable coordinate changes (diffeomorphisms) unless stated otherwise.

Example 1: Continuous versus Smooth Equivalence

To illustrate the foregoing point, consider the two functions $f(x) = x^2$ and $g(x) = x^2 - x^4$. If we change to a new variable y given by

$$y(x) = \frac{x}{|x|}\sqrt{\frac{1-\sqrt{1-4x^2}}{2}},$$

then it's easy to see that $g(y(x)) = f(x)$, i.e., the functions are continuously deformable one to the other via the change of variable $x \to y$. So the functions are topologically equivalent. However, this coordinate change is not smooth, since the derivative $y'(x)$ fails to exist at the origin. Thus, f and g are not smoothly equivalent (diffeomorphic). Further details on this question are given in Problem 17 of the next chapter.

If we assume that the possible parameter values α belong to some topological space A so that it makes sense to speak of points of A being "close," then we see that a point $\alpha \in A$ is a bifurcation point if Φ_α is *inequivalent* to $\Phi_{\hat{\alpha}}$ for all $\hat{\alpha}$ "near" to α, i.e., for all $\hat{\alpha}$ contained in a neighborhood of α. On the other hand, a point α is a regular point if Φ_α is equivalent to $\Phi_{\hat{\alpha}}$ for all $\hat{\alpha}$ in a neighborhood of α. Under reasonably weak conditions on the set A and the maps Φ_α, g and h, it can be shown that the bifurcation points form a "thin slice" of the set A (technically, a nowhere dense subset of A). Therefore, almost all points $\alpha \in A$ are regular points, and the qualitative changes of behavior that can emerge only at bifurcation points are rare events as we move about in the space of parameters A.

Example 2: The Cusp Catastrophe

Here we let

$$\begin{aligned} U &= \{\text{all real cubic polynomials in a single indeterminate } z\}, \\ &= \{z^3 + az + b\colon a, b \text{ real}\}, \\ Y &= \{1, 2, 3\}, \\ A &= \{(a, b)\colon a, b \text{ real}\}, \end{aligned}$$

together with the map

$$\begin{aligned} &\Phi_{(a,b)}\colon U \to Y, \\ &z^3 + az + b \mapsto \text{the number of real roots of the cubic polynomial.} \end{aligned}$$

Results from elementary algebra show that the root structure of the polynomial partitions the parameter space A as shown in Fig. 1.3.

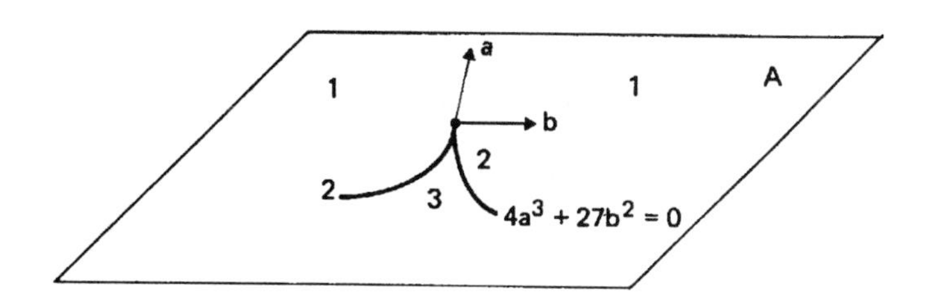

Figure 1.3. Cubic Polynomial Root Structure

For this example, the bifurcation set consists of the points on the curve $4a^3+27b^2 = 0$. The space A is partitioned into three disjoint regions marked "1", "2" and "3" in Fig. 1.3, corresponding to the number of real roots of $z^3 + az + b$ in the particular region of A. Here two cubics are equivalent if they have the same number of real roots. Since the set Y is discrete, two points $y_1, y_2 \in Y$ are "close" only if $y_1 = y_2$. Thus, we can have two cubics being close in the sense of their coefficients (a, b) being close in A, but they may be inequivalent in terms of their root structure. This can happen, however, only if one of the cubics corresponds to a bifurcation point and the other corresponds to a regular point, i.e., the coefficients (a^*, b^*) and (a, b) lie in differently numbered regions of the space A.

For reasons that will be made clear in the next chapter, the set-up just described serves as the prototype for the development of Thom's theory of *elementary catastrophes.* Our case of cubic polynomials corresponds to the most well-known elementary catastrophe, the *cusp.* In Thom's theory, the set of bifurcation points is usually termed the *catastrophe set* or *catastrophe manifold.* The other elementary catastrophes (fold, butterfly, etc.) correspond to polynomials of different degrees, with their catastrophe sets determined in a fashion similar to what's been done above. Many more details on this example, as well as this entire circle of ideas, are given in the next chapter.

Example 3: Stellar Structure

We consider a simplified version of star formation. Let's focus upon the ellipsoidal states of relative equilibrium of a rotating liquid mass. Assume the liquid has uniform density ρ, with semi-axes a, b and c in the x, y and z directions, and let the mass be rotating about the z-axis with angular velocity W (think of a rugby ball or an American football placed on the ground and given a spin). The mass

$$M = \frac{4}{3}\pi abc\rho,$$

and the angular momentum

$$L = \frac{1}{5}M(a^2 + b^2)W \doteq IW$$

are conserved, with the effective potential energy being given by

$$V = \frac{1}{2}\int \rho g\, dv + \frac{L^2}{2I},$$

where g is the gravitational potential.

The equilibrium configurations of this rotating mass are given by the stationary values of V. To compute these states, it's convenient to introduce the new variables

$$x_1 = \left(\frac{a}{c}\right)^2, \qquad x_2 = \left(\frac{b}{c}\right)^2, \qquad \lambda = 25\left(\frac{4\pi\rho}{3M}\right)^{\frac{1}{3}}\left(\frac{L^2}{3GM^3}\right),$$

where G is the gravitational constant. In these variables, it can be shown that the energy function takes the form

$$V = \frac{3}{10}\left(\frac{4\pi\rho}{3M}\right)^{\frac{1}{3}}\left(-(x_1x_2)^{1/6}\int_0^\infty [(1+s)(x_1+s)(x_2+s)]^{-\frac{1}{2}}\, ds + \frac{\lambda(x_1x_2)^{\frac{1}{3}}}{(x_1+x_2)}\right).$$

If we let $\alpha_1 = a/c$, $\alpha_2 = b/c$, $\alpha_3 = \lambda$, the above relation assumes the form

$$\Phi_\alpha\colon U \to Y,$$

where $U = R^3, Y = R$, $A = R^3$. There are two sets of solutions to the equilibrium equations

$$\frac{\partial V}{\partial x_1} = \frac{\partial V}{\partial x_2} = 0,$$

one given by Maclaurin spheroids with $a/c = b/c$, and the other by Jacobi ellipsoids with $a \neq b$. The Jacobi ellipsoids *bifurcate* from the Maclaurin spheroids at the point $\alpha_1^* = \alpha_2^* = 1.716$, $\lambda^* = 0.769$. It can be shown that the Jacobi ellipsoids are stable everywhere they exist, whereas the Maclaurin spheroids are stable only for $\lambda < 0.769$.

Exercise

1. Consider the quadratic $p(z) = z^2 + a$, where a is a real number. Following the same procedure as in the text for the cusp catastrophe, prove that this polynomial leads to the so-called "fold" catastrophe. Draw the the catastrophe set.

6. Complexity

The idea of one description of a system bifurcating from another also provides the key to begin unlocking one of the most important, and at the same time perplexing, problems of system theory: characterization of the complexity of a system. What does it mean when we say a collection of atoms or a national economy or a vertebrate eye constitutes a "complex" system? Intuitively, it seems to mean that the system contains many subsystems interacting in counter-intuitive ways and/or that we cannot offer an explanation for the system's behavior in terms of its observable subsystems. Or, equivalently, that there is no compact way of describing what the system is doing. All of these intuitive notions, as well as many others, have been offered up as the basis for a theory of system complexity, and all have failed in the sense that there exist important natural systems that by any intuitive interpretation would be termed "complex," but that fall outside the scope of the proposed measure of complexity. So why does it seem so difficult to capture formally what appears to be such an easily grasped intuitive concept?

The view that we champion here is that the basis of the complexity difficulty lies in the tacit assumption that complexity is somehow an inherent or intrinsic property of the system itself, i.e., that the complexity of a system is independent of any other system that the target system of interest is interacting with. In particular, we question the claim that the system complexity is independent of any observer/controller that may be influencing the system. If we reject this unwarranted independence hypothesis, we arrive at the relativistic view of system complexity depicted in Fig. 1.4.

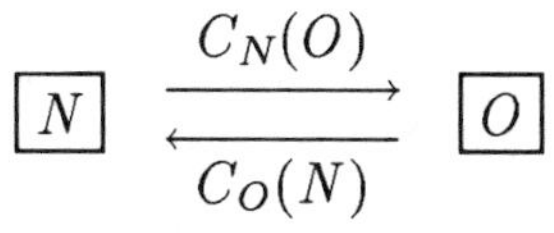

$$\boxed{N} \quad \underset{C_O(N)}{\overset{C_N(O)}{\rightleftarrows}} \quad \boxed{O}$$

Figure 1.4. The Relative Complexity of a System and its Observer

Here we have the natural system N being observed by (interacting with) the system O. The two arrows labeled $C_N(O)$ and $C_O(N)$ represent the complexity of O as seen by N and the complexity of N as seen by O, respectively, using whatever measure of complexity one chooses. The important point here is that the complexity of a given system is *always* determined relative to another system with which the given system interacts. Only in extremely special cases, where one of these reciprocal interactions is so much weaker than the other that it can be ignored, can we justify the traditional attitude regarding complexity as an intrinsic property of the system itself. But such situations are almost exclusively found in physics and engineering, which tends to explain why most approaches to the complexity problem emphasize rather classical concepts like entropy and information. As soon as we pass from physics to biology and on to the social and behavioral sciences we see the obvious flaws in the traditional view, finding that it's the reciprocal relation between the quantities $C_O(N)$ and $C_N(O)$ that is of real interest, not one or the other in isolation.

But how should we measure $C_O(N)$ and $C_N(O)$? If we want to explore the relationship between them, we must first find some common basis for their expression and comparison. At this point we can bring to bear our earlier ideas of system equivalence. Let's define the quantity $C_O(N)$ as

$$C_O(N) \doteq \# \text{ of inequivalent descriptions of } N \text{ formed by an observer } O.$$

The quantity $C_N(O)$ is defined analogously. Now let's make this definition a bit more explicit.

From O's perspective, we have the family of descriptions of N given as $\Phi_\alpha \colon U \to Y$, $\alpha \in A$. If O can "see" the entire set A, then $C_O(N)$ is just equal to the number of equivalence classes into which A is split by the induced relation $\sim$ given by

$$\alpha \sim \alpha' \text{ if and only if } \Phi_\alpha \sim \Phi_{\alpha'}.$$

Formally,

$$C_O(N) = \text{card } (A/_\sim).$$

However, if O can only see a proper subset $\hat{A}$ of A, then

$$C_O(N) = \text{card } (\hat{A}/_\sim) \leq \text{card } (A/_\sim),$$

since $\hat{A} \subset A$. Thus, the complexity of N as seen by O is, in general, less when O can form fewer descriptions of N. Of course, this conclusion is in complete accord with our feeling that the more we can "see" ("know") of a system, the finer discriminations we can make about its structure and behavior, and consequently, the more complex it will appear to us.

To illustrate the this point, think of a stone lying on the street. To an unskilled observer there are very few inequivalent ways to interact with such a stone. We can weigh it, throw it, break it apart—and that's about it. On the other hand, to a geologist such a stone will be much more complex, since such a trained observer has available many more inequivalent ways by which to interact with the stone. For example, the geologist can perform chemical analyses on the stone or x-ray it, both modes of interaction unavailable to the layperson. So we see that the complexity of the stone is not as much a property of the stone as it is a property of the interaction between the system and another system that is observing it.

At first glance, it may appear that the complexity of many systems will be infinite, since a great many situations don't lead to a finite classification of the points of A under the relation $\sim$. But such situations are, for the most part, exactly like that faced in matrix theory when we classify the real $n \times n$ matrices according to similarity. In that case, there is an uncountable infinity of classes, each class being determined by the characteristic values and their multiplicities (technically, the Jordan blocks) of the individual matrices in $R^{n \times n}$. So even though there are an infinite number of inequivalent matrices, we can label each class by a finite set of numbers (the characteristic values and their multiplicities) and operate to advantage in the set $R^{n \times n}/_{\sim}$ to determine system invariants, canonical forms and the like.

Example: The Cubic Polynomials

In the last section we considered the family of descriptions (relative to ourselves as observers) of cubic polynomials

$$\Phi_\alpha : U \to Y, \qquad \alpha \in A,$$
$$z^3 + az + b \mapsto \text{number of real roots},$$

where $A = R^2$. In Fig. 1.3 we saw the set A partitioned into equivalence classes by the cusp curve $4a^3 + 27b^2 = 0$. Each class is labeled by one of the integers "1," "2" or "3." Thus, if the natural system N is taken to be the set of cubic polynomials, our definition yields the complexity of N as $C_O(N) = 3$.

Exercises

1. What is the complexity of the integers $\mathbb{Z}$ using the equivalence relation R given in Exercise 2 of §4?

2. The example of the text suggests that a measure of the complexity of a polynomial $p(z)$ is just the degree of p. Using the same line of reasoning, show that the complexity of an nth order *linear* differential equation with constant coefficients equals n.

7. Error and Surprise

Another concept that is often associated with complexity is to say that a complex system "makes errors" or displays "surprising" or "unexpected" behavior. All of these notions ultimately derive from the idea of one description bifurcating from another. Let's illustrate this claim with an important situation in numerical analysis.

Example: Computational Roundoff Error

Consider the situation in which we have the system state-space $\Omega = R$, the real numbers, with the observables $f = (f_1, f_2, \ldots, f_n)$ defined by

$$f_i\colon R \to R, \qquad i = 1, 2, \ldots, n,$$
$$r \mapsto i\text{th coefficient in the decimal expansion of } r.$$

Then clearly $r_1, r_2 \in R$ are equivalent with respect to the set of observables f whenever r_1 and r_2 agree in the first n terms of their decimal expansions. Let's choose numbers r_1^*, r_2^* such that

$$r_1 \sim_f r_1^*, \qquad r_2 \sim_f r_2^*,$$

i.e., r_1 and r_1^* agree in their first n places, as do r_2 and r_2^*. Now let the 1-system interact with the 2-system through multiplication; that is, form the products $r_3 = (r_1 r_2)$ and $r_3^* = (r_1^* r_2^*)$. In general, we find that $r_3 \not\sim_f r_3^*$. In other words, the equivalence classes under f are split by the interaction, i.e., by the dynamics. The interaction generates a bifurcation of the f-classes, a bifurcation we usually term *roundoff error* in this context. It's instructive to examine the source of this so-called "error."

To see the way the error is introduced, let's look at a numerical example. Suppose

$$r_1 = 123, \quad r_1^* = 124, \qquad \text{and} \qquad r_2 = 234, \quad r_2^* = 235.$$

We use $f = (f_1, f_2)$; that is, the equivalence relation generated by f says that two numbers are equivalent if they agree in their first two places. Upon multiplying, we obtain

$$r_1 r_2 = 28{,}782 \not\sim_f r_1^* r_2^* = 29{,}140,$$

which disagrees with our expectation based upon f-equivalence. Our surprise at finding $r_1 r_2 \not\sim_f r_1^* r_2^*$ occurs because the set of observables $f = (f_1, f_2)$ is too small, thereby causing an unrealistic expectation concerning the interaction between the 1- and 2-systems. If we expand our set of observables to $\hat{f} = (f_1, f_2, f_3)$, then no such discrepancy occurs since with this set of observables r_1 and r_1^* are not equivalent. So we see that the real source of our observed error is due solely to the incompleteness in the way we chose to describe the system.

The preceding arguments are completely general: error or surprise always involves a discrepancy between the objects (systems) *open* to interaction and the abstractions (models, descriptions) *closed* to those same interactions. In principle, the remedy for closing this gap is equally clear: augment the description by including more observables to account for the unmodeled interactions. In this sense, error and surprise are indistinguishable from bifurcation. A particular description is inadequate to account for uncontrollable variability in equivalent states, and we need a new description to counteract the error.

It's interesting to note that since bifurcation and error/surprise are the same idea in disguise, while complexity arises as a result of the potential for bifurcation, we must conclude that complexity implies surprise and error. That is, to say a system displays counterintuitive behavior amounts to the same thing as saying that the system has the capacity for making errors—although the error is not *intrinsic* to an isolated system but occurs when the system interacts with another.

The concepts of bifurcation, complexity, similarity, error and surprise have all been introduced using the bare-bones minimum of mathematical machinery—essentially just the idea of a set and a mapping between sets. This is indeed a primitive level of mathematical structure, and we can certainly expect far more detailed and precise results if we make use of more elaborate and sophisticated mathematical representations. To this end, let's introduce the idea of a *formal* mathematical system.

Exercises

1. Characterize "surprise" sitations like the Chernobyl disaster, the October 1987 stock market crash, and the breakdown of the Iron Curtain in terms of bifurcations of descriptions.

2. Many FM radios now have digital tuners allowing a selection of frequencies like 102.1MHz, 102.3MHz, and so on. Each of these frequencies ends in an odd number. If you take such a radio to a country in which the FM band consists only of channels broadcasting on frequencies ending in even numbers, such a radio will be useless. In short, it will make "errors" in tuning-in to the local stations. Interpret this situation in terms of observables and interactions.

8. Formal Systems

In simplest possible terms, a formal mathematical system F is a collection of abstract symbols, together with a set of rules (a grammar) expressing how strings of these symbols can be combined in order to create new symbol strings. In addition, as part of the definition of F we also include a set of symbol strings that are taken to be theorems (i.e., they are assumed without proof). To complete the definition, we give a set of rules of logical inference enabling us to generate new grammatically correct strings (theorems) from earlier ones.

Example 1: Addition of Integers

Let the symbol strings of the system F be all finite strings of dashes, e.g., $- - --$ and $--$ would be the strings of this system, where the single symbol of F is just the dash $-$. Let the rule of combination be given by

$$\underbrace{- - \cdots -}_{p \text{ times}} \oplus \underbrace{- - \cdots -}_{r \text{ times}} = \underbrace{- - \cdots -}_{p+r \text{ times}},$$

$p, r \geq 0$. Further, take the single axiom to be $- \oplus 0 = -$.

This example serves as a formal *model* for the addition of nonnegative integers if we interpret the strings of dashes to be integers, i.e.,

$$\underbrace{- - \cdots -}_{r \text{ times}} \doteq (r) = \text{the integer } r,$$

and interpret the rule of combination as the usual rule for addition of whole numbers, viz. $(p) + (r) = (p + r)$. But this is just an *interpretation* of the symbols, and the formal system could also serve to model other situations. For example, it could be used as a model for the addition of apples and oranges or for counting the numbers of cars on a highway. It's important to keep in mind the fact that a formal system doesn't really represent *anything* until its symbols, rules and axioms are interpreted; until that time F is literally nothing but a collection of marks on a piece of paper and a prescription for creating new sequences of marks from old.

The big advantage of a formal system is that it can be interpreted in many ways. Thus, it can serve as a model for a variety of situations. In addition, a formal system contains an automatic mechanism for creating new theorems from old: the rules of logical inference. Furthermore, the need to interpret the symbols unambiguously and the need to specify precisely all rules of grammar and logical inference force us to be very specific and exact about our assumptions when using F to represent a natural system N.

Example 2: Euclidean Geometry

Certainly the most common example of a formal system is the set-up we're all familar with from high-school geometry. In Euclid's view, the world of geometry starts with a few simple postulates (axioms) about points, lines and circles, together with the usual rules of logical inference such as the replacement of equals by equals and so forth. The symbols of Euclid's system are just the usual symbols of the English language (ancient Greek, in

Euclid's case), along with the Greek letters, as well as the symbols of mathematics and logic like $\measuredangle, \doteq$ and Δ. The grammar by which these symbols are combined into meaningful strings is just the usual rules of English, along with the rules telling us how the mathematical symbols may be used. In elementary geometry we start with one of Euclid's axioms and proceed to apply the rules of logical inference to derive new sentences, each of which is then added to the set of theorems of the system.

Example 3: Computer Programming Languages

Consider a typical high-level computer programming language like Fortran. Here the symbol strings (words) of the formal system are the individual statements that can be written in the language, such as GO TO x, WRITE y, STOP, etc. The number of such words is usually restricted to a very small vocabulary of a couple dozen words or so. Similarly, the grammar of such a language is also very restrictive, admitting such combinations of words as IF(x.LT.y).AND.IF(z.GT.r)GO TO g, but not statements such as WRITE-READ. The individual words of the language constitute the axioms, while the theorems correspond to all correctly or well-formed programs.

Incidentally, the above example shows clearly the enormous difference in subtlety and quality between a computer programming language and a natural human language, underscoring the dubious nature of proposals often seen in universities nowadays to substitute knowledge of a computer language for a foreign language in Ph.D. language competency requirements. Even the most elaborate and sophisticated computer languages pale by comparison with natural languages in their expressive power, not to mention the enormous cultural content that's carried along almost for free with a natural language. For reasons such as these, many see such proposals to replace knowledge of a natural language by knowledge of a computer language as at best a shallow, anti-intellectual joke, and the theory of formal systems provides a basis for construction of rational arguments against their adoption. But to develop this line of argument would take us too far afield in a book of this type, so let's get back to system theory.

Exercises

1. Construct formal systems characterizing the common board games chess, checkers and Scrabble.

2. Consider a formal system whose strings consist of the three symbols ★ (star), ✠ (maltese cross), and ☁ (cloud). Let the two-element string ✠☁ be the sole axiom of the system. Letting x denote an arbitrary finite string of stars, crosses, and clouds, take the transformation rules of this system to be given by:

Rule I: x☁ $\longrightarrow$ x☁★
Rule II: ✠x $\longrightarrow$ ✠xx
Rule III: ☁☁☁ $\longrightarrow$ ★
Rule IV: ★★ $\longrightarrow$ —

In these rules, $\longrightarrow$ means "is replaced by." So, for instance, Rule I says that we can form a new string by appending a star to any string that ends in a cloud. The interpretation of Rule IV is that anytime two stars appear together in a string, they can be dropped to form a new string. Starting with the single axiom ✠☁, prove that the string ✠★☁ is a theorem of this system.

3. In the foregoing system, show that every string will necessarily begin with a ✠. Is every string beginning with a ✠ a theorem of the system? In other words, given an arbitary string beginning with a ✠, can we always find some sequence of Rules I–IV that will transform the single axiom of the system into the given string? (For the answer to the second part of this question, see Chapter 9.)

9. Modeling Relations

Let's now consider the notion of a formal system and its role in system modeling. Loosely speaking, we begin with a collection of symbols together with a finite set of rules for assembling such symbols into strings of finite length, as well as a set of axioms for the system. The axioms, together with the rules of grammar and logical inference, constitute our formal system F. Usually in modeling a system N, we take familiar mathematical objects such as finite groups, topological spaces, directed graphs and differential equations as the formal systems of interest. The important point to keep in mind regarding formal systems is that they are entirely constructions of the human mind. Unlike the situation with natural systems, which are defined by physically realizable observables, formal systems are defined solely in terms of symbols and the rules for their manipulation.

Formal systems provide the basis for making *predictions* about N. If we can establish a "faithful" correspondence between the observables and linkages of N and the elements of the formal system F, then we can use the rules of inference in F to derive, in turn, new theorems interpretable as relations between the observables of N. Under such conditions, we call F a *formal description* of N. Missing from this program is a specification of how to construct a faithful mapping, or *encoding,* of the system N into a particular formal system F.

The final step in the process of translating a specific natural system N into a particular formal system F is an *encoding map* $\mathcal{E}: N \rightarrow F$. In rough terms, $\mathcal{E}$ provides a "dictionary" associating the observables of N with the objects of F. Probably the most familiar example of such an encoding is when F is a system of ordinary differential equations, in which case $\mathcal{E}$ usually associates the observables of N with the dependent variables (coordinate functions) of the system of equations.

Since the encoding operation $\mathcal{E}$ provides the link between the real world of N and the mathematical world of F, it's essential that we be able to *decode* the theorems (predictions) of F if we hope to interpret those theorems in terms of the behavior of N. Basically, what we must try to ensure is that our encoding is consistent in the sense that the theorems of F become predictions about N that may be verified when appropriately decoded back into relations in N. This modeling relation is depicted schematically in Fig. 1.5.

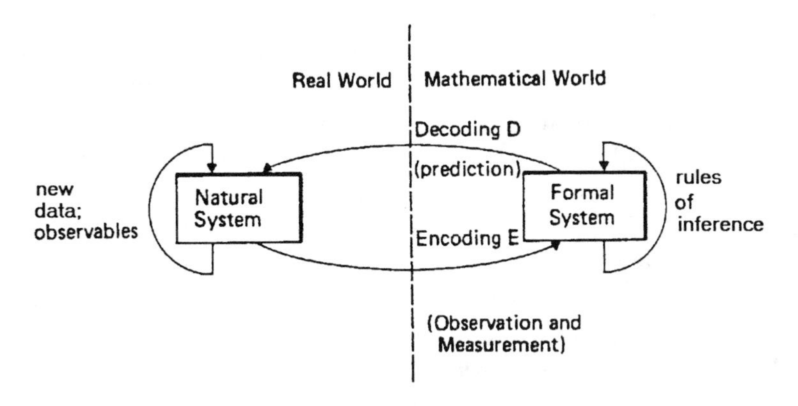

Figure 1.5. The Modeling Relation

In Fig. 1.5 we have explicitly indicated the separation between the real world of N and the mathematical world of F by the vertical dotted line. This diagram also makes explicit the role of the system scientist: he is neither a natural scientist living on the left side of the diagram nor the mathematician living on the right. Rather, the system scientist is the keeper of the abstract encoding/decoding operations $\mathcal{E}$ and $\mathcal{D}$. In short, the system scientist creates and maintains the dictionaries by which we translate back and forth between the real world of N and the mathematical world of F. Of course, in order to carry out this function, the system scientist has to be knowledgeable about both N and F—but not necessarily expert in either. His or her expertise is centered upon making adroit choices of $\mathcal{E}$ and $\mathcal{D}$, so that in the transition from N to F and back again as little information is lost as possible. To illustrate the steps in the modeling process, let's consider some familiar examples.

Example 1: Orbital Mechanics

A familiar setting in which to expose the general ideas outlined above is the special case of classical Newtonian mechanics, where we consider a system of material bodies moving in accordance with the force of gravity.

Our first order of business is to specify the fundamental observables that are to be encoded and to define the corresponding state space. Newton proposed that these fundamental observables could be taken to be: (1) the displacement of the bodies from some convenient reference point, and (2) the velocities of these bodies. Thus, given a system of N gravitating bodies, each abstract state of the system is represented by a set of $6N$ numbers. This gives an encoding of the abstract states into the points of R^{6N}, the *phase space*—or, more generally, the *state space*—of an unconstrained classical N-body mechanical system.

The next step in the Newtonian encoding is to postulate that every other observable of this system of material bodies can be represented by a real-valued function on this phase-space. Consequently, every other observable is totally linked to these state variables on a set Ω of abstract states. This postulate removes the necessity of ever referring back to the set Ω of abstract states. We can now confine ourselves entirely to the state-space into which Ω is encoded by state variables.

Since the Newtonian framework assumes that *every* observable pertaining to the system is represented by a function of the state variables, it must be the case that the *rates* at which state variables themselves change are also functions of the state alone. And if such rates are themselves observables, then there must be equations of state expressing the manner in which the rates of change of the state variables depend on the values of the state variables. It is this fact that leads directly to the entire dynamical apparatus of Newtonian physics.

Example 2: Keynesian Dynamics and Equivalent Economies

Motivated by the Newtonian framework, let's now consider its application to an economic situation. Suppose we have the economic system depicted in Fig. 1.6 which embodies the basic ideas underlying Keynes' theory of economic growth.

We begin by assuming that economic activity depends upon the rate at which goods are purchased. These goods are of two kinds: capital goods and consumer goods. The next step is to recognize that capital goods are another form of investment. Finally, we note that money is the driving force to buy both kinds of goods. This money can be obtained (as profits and wages) by manufacturing the goods. However, *all* profits and wages do not automatically go back as investment; some money is saved. The system is prevented from grinding to a halt by new investment. Thus, the levels of economic activity and employment depend upon the rate of investment.

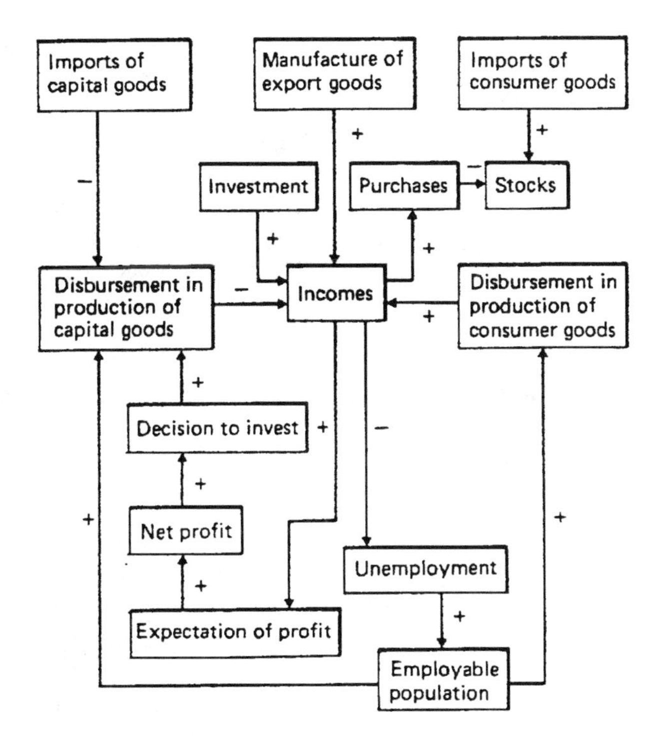

Figure 1.6. A Keynesian Economy

If we assume that the signed arcs in the diagram indicate a linear relationship between the origin and destination of the arc, and if we regard investment as the driving force, with measurable outputs being the imports of consumer and capital goods as well as manufacture of export goods, a dynamic Keynesian economy can be analytically represented by the linear system

$$\begin{aligned} \dot{x} &= Fx + Gu, \\ y &= Hx, \end{aligned} \qquad (F_1)$$

where $u(t)$ and $y(t)$ are the input and outputs, respectively, $x(t)$ is a vector representing the other components of the economy, and F, G and H are constant matrices whose entries are determined by the flows along the system arcs. Specification of the initial state of the economy $x(0)$, as well as the pattern of investment $u(t)$, then determines the future course of economic activity, including the observed outputs $y(t)$. Thus, the linear system F_1 represents an *encoding* of the economy into the formal mathematical system represented by the differential and algebraic relations F_1.

An alternate encoding is to associate the economy with a *transfer function* matrix $Z(\lambda)$, a rational matrix in the complex variable λ. Formally, if we let $U(\lambda)$ and $Y(\lambda)$ denote the Laplace transforms of the system input and output, respectively, then $Z(\lambda)$ expresses the linear relation between the economy's input and output, i.e.,

$$Y(\lambda) = Z(\lambda)U(\lambda). \qquad (F_2)$$

The encoding of the economy into the transfer matrix comprising the formal system F_2 is an alternate model of the economy.

In Chapter 6 we'll see that there is a relation between the formal systems F_1 and F_2 given by

$$Z(\lambda) = H(\lambda I - F)^{-1}G.$$

This relation forms one of the cornerstones of the theory of linear systems, enabling us to relate the input/output behavior of the economy to its internal behavior.

Example 3: Global Models

The global modeling movement, initiated by Jay Forrester and Dennis Meadows about two decades ago, provides another illustration of the principles of this chapter. All global modeling attempts begin by abstracting a certain finite collection of observables from the almost limitless number of observables defining the *world problematique.* For his original model, Forrester chose five observables: Population, Natural Resources, Capital, Pollution and Fraction of Capital Devoted to Agriculture. Various linkages were then postulated among these observables, leading to a formal system composed of a set of coupled finite-difference equations. These equations contained various parameters representing elements of the system such as birth and death rates, pollution generation rates, and land yields. It's important to recognize that these parameters were not part of the basic observables (state variables) of the model. As a result, global modelers could create *families* of models by selecting different values for these parameters. This simple observation provides the key to understanding one of the most severe criticisms aimed at such modeling efforts, namely, their almost pathological sensitivity to small changes in the defining parameters. We examine this issue from the vantage point of our general modeling framework.

Denote the abstract state space of the global model by Ω, and let A be its parameter space. Generally, A is some subset of a finite-dimensional space like R^n. Then the class of models is represented by the Cartesian product $\Omega \times A$. (Such a class of models is a special case of what in mathematics is termed a *fiber bundle,* with base A and fiber space Ω). Pictorially, we can envision the situation as in Fig. 1.7. For each fixed $\alpha^* \in A$, the linkages between the observables on Ω are used (via the encoding into difference equations) to obtain a time-history of the observables. In other words, we obtain a fibering of the state-space Ω into trajectories determined by solutions of the equations of motion associated with the corresponding choice

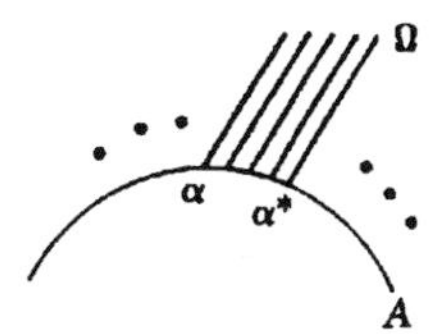

Figure 1.7. A Fiber Bundle of Global Models

of α. It then becomes an interesting question to relate the fiberings corresponding to different values $\alpha, \alpha^* \in A$. This leads to a special case of the overall question of *structural stability,* which we'll consider more fully in the next chapter.

To each $\alpha \in A$ we associate a one-parameter family of transformations

$$T_t^{\alpha} : \Omega \to \Omega,$$

representing the dynamics of the system in the state-space Ω. If $\alpha^* \in A$ is another point, then it will also be associated with a corresponding one-parameter family

$$T_t^{\alpha^*} : \Omega \to \Omega.$$

Roughly speaking, if the effect of replacing α by α^* can be annihilated by coordinate transformations in Ω, then, by definition, the dynamic imposed by the difference equations represented by T_t is *structurally stable* with respect to the perturbation $\alpha \to \alpha^*$. Another way of looking at this question is to ask if we can find mappings g_{α,α^*} and h_{α,α^*} such that the following diagram commutes

$$\begin{array}{ccc} \Omega & \xrightarrow{T_t^{\alpha}} & \Omega \\ {\scriptstyle g_{\alpha,\alpha^*}}\downarrow & & \downarrow{\scriptstyle h_{\alpha,\alpha^*}} \\ \Omega & \xrightarrow[T_t^{\alpha^*}]{} & \Omega \end{array}$$

Since the situation described here is so important, let's recast it into a picture analogous to that of Fig. 1.7. Figure 1.8 allows us to enlarge somewhat on our previous remark that $\Omega \times A$ itself possesses the structure of a fiber bundle. On each fiber Ω, there is imposed a group of transformations, namely, the corresponding dynamics T_t^{α}. The notion of structural stability allows us to consider mappings of the *entire* space $\Omega \times A$ that: (1) leave A fixed, and (2) preserve dynamics on the fibers. Actually, it's also possible to consider a broader class of mappings that do not leave A fixed. Such mappings are intimately connected with what physicists call *renormalization,* but we shall not go into this detail here. However, it should be noted that

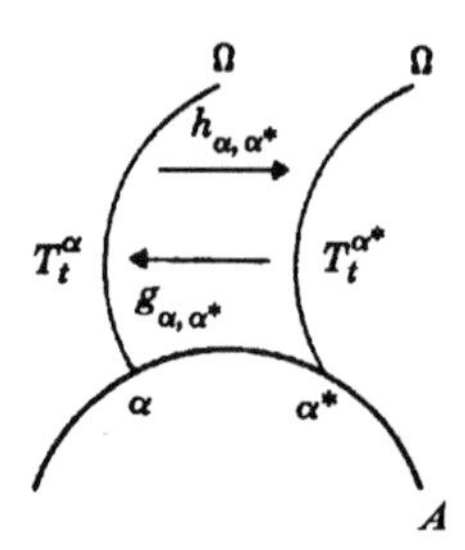

Figure 1.8. Fibering of the State Space

a great deal of the *technical* criticism received by the Forrester-Meadows-Mesarovic exercises could have been avoided by their paying more thorough attention to the structural stability of their models.

In closing, let us again emphasize that the structural stability of any model is contingent upon the choice of both Ω and A, as well as upon the dynamics T_t. Thus, the entire issue depends upon the abstractions that are made and the encodings that are chosen. No natural process is *absolutely* stable or unstable, and its stability depends initially upon nonmathematical considerations. To this extent, the matter of stability is a premethodological question.

To illustrate this point at a nontechnical level, some of the most damaging critiques of the Forrester-Meadows global models are the claims that the modelers omitted variables (observables) that were centrally important for understanding the behavior of the observables that actually were employed. Thus, in the Forrester model there is no accommodation made for energy supply and consumption, an omission that dramatically affects agriculture, population and pollution. Here we see how the *abstraction* operation can go wrong. The modelers threw away observables vitally important to the goals of the model, and as a result no amount of technical virtuosity using the observables that were retained allows any sort of meaningful conclusions to be drawn. In short, the models were *invalidated* as a consequence of faulty abstractions.

Exercises

1. The description of a swinging pendulum involves several parameters: the mass of the bob at the end, the length of the pendulum's arm, the initial position and velocity of the bob and so forth. Each choice of these parameters determines a *specific* pendulum. Characterize this sitation as a fiber bundle of pendulums.

2. Abstractly, a pendulum and a vibrating spring are isomorphic systems, since the behaviors of the pendulum bob and the weight at the end of the spring both satisfy the same differential equation $\ddot{x} + ax = 0$, where a is a parameter representing the stiffness of the spring or the reciprocal

of the length of the pendulum. Show how the trajectory in $(x, \dot{x})$-space determined by this equation corresponds to the one-parameter family of transformations T_t^α discussed in the text. What does α correspond to?

Discussion Questions

1. Consider the situation depicted below in which two natural systems, N_1 and N_2, code into the same formal system F. The specific features of this situation depend upon the degree of overlap in F of those propositions in $\mathcal{E}_1(N_1)$ and $\mathcal{E}_2(N_2)$ that represent properties of N_1 and N_2.

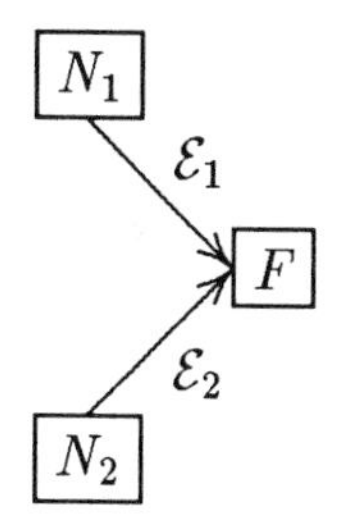

a) What does it mean if we have $\mathcal{E}_1(N_1) = \mathcal{E}_2(N_2)$? In this case, N_1 and N_2 share a common model. This situation is the basis for the idea of system *analogy*.

b) If we define the relation $N_1 \sim N_2$ if and only if N_1 and N_2 are analogous, show that $\sim$ determines an *equivalence* relation on the set of natural systems, i.e., $\sim$ is reflexive, symmetric and transitive.

c) What does the relation of system analogy have to do with schemes for analog or digital simulation?

d) Consider the cases $\mathcal{E}_1(N_1) \subset \mathcal{E}_2(N_2)$ and $\mathcal{E}_1(N_1) \cap \mathcal{E}_2(N_2) \neq \emptyset$. Give interpretations of these situations in terms of subsystem analogies.

e) Discuss these ideas in the specific setting when N_1 is a vibrating spring, N_2 is a swinging pendulum, and F is the differential equation $d^2/dx^2 + ax = 0$.

2. Suppose we have the modeling relation

$$\boxed{N_1} \xrightarrow{\mathcal{E}_1} \boxed{F_1}$$

$$\boxed{N_2} \xrightarrow{\mathcal{E}_2} \boxed{F_2}$$

in which two systems encode into different formal systems.

a) Consider the case in which $\mathcal{E}_1(N_1)$ and $\mathcal{E}_2(N_2)$ are *isomorphic,* i.e., there exists a one-to-one, onto, structure-preserving mapping between the sets of encoded propositions $\mathcal{E}_1(N_1)$ and $\mathcal{E}_2(N_2)$. Here a mapping is considered to be structure-preserving if it can be extended in a unique way to all inferences in F_1, F_2 that can be obtained from $\mathcal{E}_1(N_1)$ and $\mathcal{E}_2(N_2)$, respectively. What interpretation can you give to the relationship between N_1 and N_2 in this case? Does the relationship between N_1 and N_2 depend upon the isomorphism between $\mathcal{E}_1(N_1)$ and $\mathcal{E}_2(N_2)$? How?

b) Suppose $\mathcal{E}_1(N_1)$ and $\mathcal{E}_2(N_2)$ satisfy a weaker relation than isomorphism. For example, they may share a property like "stable" or "finite" or "finite-dimensional." In such cases, we have the basis for *metaphorically* relating N_1 and N_2 through their models sharing this common property.

In mathematics, the set of all objects possessing a given property forms what is termed a *category.* Thus we have the category "GROUPS," or the category "LINEAR SYSTEMS," and so forth. Typically, along with such objects come appropriate families of structure-preserving maps, providing us with the concept of morphism between objects of the category. For instance, the set of group homomorphisms are the morphisms in the category GROUPS. Show that the idea of a system metaphor defines a categorical relation upon the set of all natural systems.

3. Assume we have a single system N that encodes into two different formal systems, i.e.,

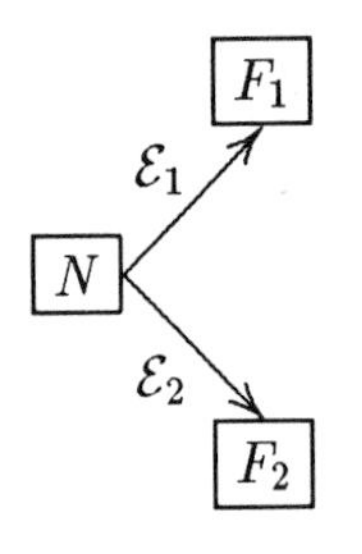

Here we want to consider the extent to which uncoded linkages of observables in N manifest themselves in the form of mathematical relations in F_1 and F_2.

a) *Case I:* $F_1 \rightarrow F_2$. In this situation, every proposition in F_1 can be associated with a unique proposition of F_2, but not necessarily conversely. Intuitively we would say that F_2 provides a more comprehensive description of N than does F_1; in fact, the encoding $\mathcal{E}_1(N)$ in F_1 can be *reduced* to the encoding $\mathcal{E}_2(N)$ in F_2.

What does the above situation have to do with the idea of *reductionism,* which asserts that there is a universal way of encoding every natural system into a formal system, and that every other encoding is somehow reducible to the universal encoding? What conditions would have to be demonstrated in order to establish reductionism as a valid scientific tool? If such a universal encoding were displayed, do you think it would serve any *practical* purpose? Why? (Remark: In this connection, think of the practical importance of the Turing machine, considered in Chapter 9, which serves as a universal model of the process of computation.)

b) *Case II:* $F_1 \cap F_2 = \{\emptyset\}$. In this event, there are no nontrivial relations between F_1 and F_2, and the encodings $\mathcal{E}_1(N)$ and $\mathcal{E}_2(N)$ cannot, in principle, represent any common linkages between observables in N. What does this case have to do with the problem of "complementarity" in quantum mechanics, which asserts that an object can display wave-like behavior or particle-like behavior, depending upon how we look at it, but not both simultaneously? Discuss also the relationship of this case to the concept of system complexity introduced in the text.

4. The situation described in the preceding question can also serve as a point of departure for discussing problems of bifurcation. Suppose that a relation can be established between a *subset* of F_1 and a *subset* of F_2. Here certain features or properties of N encoded into $\mathcal{E}_1(N)$ are related to corresponding properties encoded into $\mathcal{E}_2(N)$. Thus, this mapping establishes a logical relation between the encodings, but one that does not hold universally. A *partial* reduction is possible, and outside the domain in which this partial relation holds, F_1 and F_2 are logically independent and the linkage between the properties of N that are encoded is broken. Discuss this situation in relation to the process of bifurcation developed in the text.

5. Suppose we have a system N characterized by the variables (observables) $x = (x_1, x_2, \ldots, x_n)$ and a particular set of parameters $\alpha = (\alpha_1, \ldots, \alpha_p)$. Let the model describing the relationship between these quantities be given by the smooth (i.e., infinitely differentiable) function

$$f\colon R^n \times R^p \to R,$$

i.e., the system's equation of state is

$$f_\alpha(x) = \gamma, \qquad \alpha \in R^p,\ x \in R^n,\ \gamma \in R.$$

Catastrophe theory addresses the question: What does f look like in the neighborhood of a critical point, i.e., a point x for which $\text{grad}_x f_\alpha = 0$? We have already seen that f_α "looks like" f_{α^*} if we can find smooth coordinate changes g_{α,α^*} and h_{α,α^*} such that the following diagram commutes:

$$\begin{array}{ccc} X & \xrightarrow{f_\alpha} & R \\ {\scriptstyle g_{\alpha,\alpha^*}}\downarrow & & \downarrow{\scriptstyle h_{\alpha,\alpha^*}} \\ X & \xrightarrow[f_{\alpha^*}]{} & R \end{array}$$

Those α for which there is no such change of variables g, h for all α^* in a local neighborhood of α are the bifurcation points.

a) Relate the above catastrophe theory setup to the general situation described in Discussion Question 4.

b) Suppose you want to study how the quality of life changes in an urban area as a function of factors like climate, budgetary levels, property tax rates, zoning regulations and recreational facilities. How would you define the urban "genotype," "environment" and "phenotype" in order to formulate this question in catastrophe-theoretic terms?

6. We have seen that the equation of state $\Phi(f_1, f_2, \ldots, f_n) = 0$ establishes a mathematical relationship among the observables of a system N. Given such a relation, what additional criteria would you impose to distinguish the relation as a *law of nature,* as opposed to just an *empirical relationship?* (Hint: Consider the fact that, in general, when we change the encoding of N into the formal system F, we also change the linkages between the observables of N.)

7. Some of the observable characteristics used to single out "complex" systems are that they exhibit counterintuitive behavior, have many different types of interactions, possess many feedback-feedforward loops, involve a high degree of decentralized decisionmaking and are indecomposable. Give examples of natural systems displaying such features. Do these examples agree with your feeling as to what's involved in calling something "complex"? How can you relate these properties to the notion of complexity as being a relative concept depending upon interactions between systems?

8. The main goal in creating models in F to represent a natural process N is to be able to make predictions about the behavior of N via the processes of logical deduction in F. To what extent can this program be effectively carried out if the best that F can do is capture a *subsystem* of N? Discuss this problem in connection with determining the *credibility* of a model.

9. There are two main strategies that can be employed for theory development: *reductionism* and *simulation.* In reductionism, we try to explain the behavior of N in terms of *structural* subunits. But in simulation we attempt to understand a limited family of behaviors as manifested simultaneously by a family of *structurally diverse* systems, i.e., the units of analysis are behaviors, not material units. Discuss these dual modes of analysis, and give examples of situations in which one or the other would appear to be the strategy of choice.

10. Generally speaking, the abstract strings of a formal system are related to reality by giving them an *interpretation* within a given context.

In this way, we can speak of a string as being TRUE or FALSE depending on how that string's interpretation matches up to the real-world situation it represents. So, for example, consider a formal system representing the game of chess, in which every string is just a description of a position that the pieces could take on the board. Then we might label a particular string **s** as being TRUE if the position **s** is a win for White, while **s** would be FALSE if the position is a win for Black. Of course, **s** could be a stalemate position, in which case it would be neither TRUE nor FALSE. Our objective in constructing an interpretation of the strings is to make the the theorems of the system all correspond to TRUE statements in the interpretation, those strings which can be shown to **not** be theorems of the system, i.e., strings that can never be the endpoint of a proof sequence, are to correspond to FALSE statements.

We call a formal system (under a given interpretation) *consistent* if the statements X and not-X do not both correspond to theorems of the system. Loosely speaking, then, a formal system is consistent if it is not possible to prove contradictory theorems in the system. This means that if there is a theorem that when interpreted says "x is TRUE," then there cannot be another theorem of the system whose interpretation says "x is FALSE." We call the system *complete* if every TRUE statement corresponds to a theorem of the system. In other words, the system is complete if the set of theorems exhausts the set of TRUE statements.

Kurt Gödel considered all formal systems and their interpretation in the realm of whole numbers, i.e., arithmetic. He showed that for any formal system whose grammar was rich enough to allow us to express all the statements that can be made about numbers (for example, statements like "the sum of two odd numbers is even" or Goldbach's Conjecture that "every even number is the sum of two primes"), that it was impossible to find such a system that was both consistent and complete. In short, syntactical manipulation of symbols is not enough to allow us to capture all the true statements that can be made about numbers (Gödel's Incompleteness Theorem).

a) Consider the implications of Gödel's result to the broader philosophical claim that "there's more to life than following rules."

b) Gödel also showed that it was impossible to use the machinery of a given formal system F to prove its own consistency. Discuss the ramifications of this result and the Incompleteness Theorem for the ultimate limits to theorem-proving machines.

11. In the theory of algorithmic complexity, the complexity of a number is measured by the length of the shortest computer program that will print out that number (Chaitin-Kolmogorov complexity). This definition leads to the idea of a number as being random if the length of the shortest com-

puter program producing the number is no shorter than the number itself. By this definition, almost every real number is random (i.e., has maximal complexity).

a) How does this notion of algorithmic complexity, which seems to be observer-independent, link-up with the idea of complexity as a relative property as discussed in the text?

b) It's often argued that natural selection proceeds by eliminating the "simple" organisms, replacing them with more complex versions. If true, this interpretation would suggest that increased complexity is of some reproductive advantage. Discuss this claim within the context of living organisms and contrast your ideas with the same claim in the context of other kinds of objects that evolve like computer programs, fashions and languages.

12. In the process of abstraction we discard observables of the original system S, thereby obtaining a new system N that is *closed* to interactions with the outside world that S is open to, i.e., S can interact with other systems through the observables thrown away in forming N. In this sense, we can think of N as being a *closed* approximation to S.

a) Discuss the proposition that "surprise" is a result of the discrepancy between the behavior of a system open to interactions with the outside and the behavior of a similar system closed to those same interactions.

b) Consider the degree to which the system N formed from S by abstraction can be used to predict the occurrence of "surprising" behavior in S. Does the very idea of predicting surprises involve a contradiction in terms?

Problems

1. Suppose we have the two commutative diagrams

$$\begin{array}{ccc} A & \xrightarrow{f} & B \\ \varphi \downarrow & & \downarrow \psi \\ A & \xrightarrow[f']{} & B \end{array} \qquad \text{and} \qquad \begin{array}{ccc} B & \xrightarrow{g} & C \\ \varphi' \downarrow & & \downarrow \psi' \\ B & \xrightarrow[g']{} & C \end{array}$$

That is, f is equivalent to f' and g is equivalent to g'. Show that the composite map $g \circ f\colon A \to C$ is not, in general, equivalent to the map $g' \circ f'\colon A \to C$. Thus, a model of individual subsystems doesn't necessarily allow us to model the composite system formed by putting these subsystems together.

Show also that even if $g \circ f$ is equivalent to $g' \circ f'$, we will usually not have φ and ψ' being maps of $A \to A$ and $C \to C$, respectively. (This exercise bears upon the issue of reductionism vs. holism, since if $g \circ f$ were equivalent to $g' \circ f'$, then it would always be possible to reconstruct the behavior of the entire system by piecing together the behaviors of its component subsystems.)

2. Given a smooth equation of state $\Phi(f_1, f_2, \ldots, f_n)$: $R^n \to R$, show that there exists an integer p, $0 < p < n$, such that Φ can be rewritten as a *stable* p-parameter family, but not for any integer $k > p$, i.e., we can always stabilize any equation of state (Buckingham's Theorem). (Recall: A map Φ_α: $R^m \to R$ is called stable if for any $\hat{\alpha}$ in a neighborhood of α, we can find smooth maps $g_{\alpha,\hat{\alpha}}$, $h_{\alpha,\hat{\alpha}}$ such that the following diagram commutes:

$$\begin{array}{ccc} R^m & \xrightarrow{\Phi_\alpha} & R \\ {\scriptstyle g_{\alpha,\hat{\alpha}}}\downarrow & & \downarrow{\scriptstyle h_{\alpha,\hat{\alpha}}} \\ R^m & \xrightarrow[\Phi_{\hat{\alpha}}]{} & R \end{array}$$

3. Let

$$\begin{aligned} \Omega &= \{\text{all cubic curves in one variable } z\}, \\ &= \{z^3 + az + b\colon a, b \text{ real}\}. \end{aligned}$$

Consider two descriptions (parametrizations, coordinatizations) of Ω given by the observables

$$\begin{aligned} f_1 : \Omega &\to S = R^2, \\ \omega &\mapsto (a, b), \end{aligned}$$

and

$$\begin{aligned} f_2 : \Omega &\to \{1, 2, 3\} = \hat{S}, \\ \omega &\mapsto \text{the number of real roots of } \omega. \end{aligned}$$

Call a point $\alpha \in S$ (or $\hat{S}$) *generic* if there exists an open neighborhood U of α such that all $\hat{\alpha} \in U$ have the same root structure as α; otherwise, α is a *bifurcation point.*

Show that the generic points under the description f_1 are $\{(a, b)\colon 4a^3 + 27b^2 \neq 0\}$, while the generic points under f_2 are $\{\emptyset\}$, the empty set.

4. Consider two observables $f: \Omega \to S$ and $g: \Omega \to \hat{S}$. We say that f *bifurcates* from g on those points in S that are generic relative to f, but bifurcation points relative to g. Similarly, g bifurcates from f on those points in $\hat{S}$ that are generic relative to g, but not generic relative to f.

a) Prove that f is equivalent to g if and only if the bifurcation sets in S and $\hat{S}$ relative to the descriptions (g, f) and (f, g), respectively, are empty. That is, f and g differ only on bifurcation sets.

b) If f and g are not equivalent, then we say one description *improves* upon the other. Show that f improves upon g if and only if all $\alpha \in S$ are generic relative to g, while the bifurcation set in $\hat{S}$ relative to f is nonempty.

c) Consider the two descriptions f_1 and f_2 of cubic curves given in Problem 3. Define the product description

$$\begin{aligned} f_1 \times f_2 : \Omega &\to S \times \hat{S}, \\ \omega &\mapsto (f_1(\omega), f_2(\omega)). \end{aligned}$$

Show that $f_1 \times f_2$ improves upon either f_1 or f_2 taken separately.

5. Let Ω be a set of states, and let $f, g: \Omega \to R$ be two observables. Define equivalence relations R_f and R_g on Ω by the rule:

$$\omega \, R_f \, \omega' \text{ if and only if } f(\omega) = f(\omega'),$$

with R_g defined similarly. Call the equivalence classes $[\omega]_f$ and $[\omega]_g$, respectively. Consider the set Z of R_g classes that intersect $[\omega]_f$. We say that:

- g is *totally linked* to f at $[\omega]_f$ if Z is a single g-class.
- g is *partially linked* to f at $[\omega]_f$ if Z is more than one g-class but not all of Ω/R_g.
- g is *unlinked* to f at $[\omega]_f$ if $Z = \Omega/R_g$.

a) Show that g is totally linked to f if and only if R_f refines R_g. (Recall: A relation R' *refines* a relation R if $\omega \, R' \, \bar{\omega}$ implies $\omega \, R \bar{\omega}$.)

b) Prove that f and g are totally unlinked if and only if every R_f-class intersects every R_g-class, and conversely.

(Note: The idea of linkage is crucial when we want to speak of prediction. If g is linked to f at a state $[\omega]_f$, then we obtain information about $g(\omega)$ by knowing $f(\omega)$. In fact, if g is totally linked to f at $[\omega]_f$, then we can say that there exists a map $h: R \to R$ such that $g(\omega) = h(f(\omega))$, i.e., g is a *function* of f. So in this case we can obtain $g(\omega)$ solely by computation; we need not measure g, at all. In this sense, computation and measurement are identical processes, both of which depend critically on the existence of linkage relations among observables.)

6. A *dynamic* on a state-space Ω is a one-parameter family of transformations $T_t\colon \Omega \to \Omega$, $t \in R$. Given an observable f on Ω, we say that f is *compatible* with T_t if the dynamic preserves the equivalence classes under f. That is, for any two states $\omega, \bar{\omega}$ such that $f(\omega) = f(\bar{\omega})$, we have $T_t(\omega) = T_t(\bar{\omega})$ for all $t \in R$. Thus, f and T_t are compatible if T_t preserves the equivalence classes of R_f. In other words, we cannot use the dynamics on Ω to distinguish states that cannot be distinguished by f itself.

Now let g be another observable on Ω that is also compatible with the same dynamic T_t. In order for the two observers to conclude that they are actually observing the same system, there must be a mapping

$$\varphi\colon \Omega/R_f \to \Omega/R_g$$

such that the diagram

$$\begin{array}{ccc} \Omega/R_f & \xrightarrow{T_t^{(f)}} & \Omega/R_f \\ \varphi\downarrow & & \downarrow\varphi \\ \Omega/R_g & \xrightarrow[T_t^{(g)}]{} & \Omega/R_g \end{array}$$

commutes for all $t \in R$.

a) Show that the only candidate for such a map φ is the map that makes the following diagram commute:

$$\begin{array}{ccccc} & & \Omega & & \\ & \swarrow^{\pi_f} & & \searrow^{\pi_g} & \\ \Omega/R_f & & \xrightarrow[\varphi]{} & & \Omega/R_g \end{array}$$

and π_g is defined similarly, i.e., π_f and π_g are the natural projections of Ω onto its equivalence classes under f and g, respectively.

b) Show that such a map φ exists if and only if R_f refines R_g, or vice versa, i.e., each observer must acquire (or have) enough observables of the right type so that each of them sees the same relation R which simultaneously refines both R_f and R_g. Otherwise, f and g are *inequivalent* observers, and their descriptions of the system must necessarily bifurcate with respect to each other.

7. Consider the one-parameter family of *linear* maps $\Phi_\alpha\colon R^n \to R^m$, $\alpha \in R$. We can represent Φ_α by an $m \times n$ real matrix A_α, depending upon

the single parameter α. Show that the bifurcation points of the diagram

$$\begin{array}{ccc} R^n & \xrightarrow{A_\alpha} & R^m \\ g\downarrow & & \downarrow h \\ R^n & \xrightarrow[A_{\hat{\alpha}}]{} & R^m \end{array}$$

are exactly those α for which rank $A_\alpha \neq$ rank $A_{\hat{\alpha}}$. (Note: Here the transformations g and h are also assumed to be linear.)

Consider the same situation but for square matrices $A_\alpha \in R^{n \times n}$, in which case we must take $g = h$. In this special case, show that the bifurcation points consist of all $\alpha \in R$ such that $\mathcal{J}_i(A_\alpha) \neq \mathcal{J}_i(A_{\hat{\alpha}})$ for *some* $i = 1, 2, \ldots, n$, where $\mathcal{J}_i(A) \doteq i$th Jordan block of the matrix A.

8. In the situation of the last problem (with $g \neq h$), show that the *complexity* of the matrix A equals the rank of A. Thus, the complexity equals the number of independent "pieces" of information contained in the description A.

Notes and References

§1. The quote involving the need for a philosophical attitude toward system modeling is taken from

Rosen, R., *Anticipatory Systems*, Pergamon, Oxford, 1985.

Part of the thesis of this book is that there is a genuine need for developing an actual **theory** of models, employing a metalanguage that goes beyond the level of the phenomenon itself. It's just not possible to speak about the relationships among different models of a given situation by using terms appropriate to the level of the models themselves. We must go beyond this level to a metalevel in which the language is constructed specifically to address issues involving properties of models, and not the properties of systems being represented by the models. This idea has been a central theme in the philosophy of science for some time, and is well-summarized by the following statement of Wittgenstein's taken from the penultimate section of the *Tractatus*:

> My propositions serve as elucidations in the following way: anyone who understands me eventually recognizes them as nonsensical, when he has used them—as steps—to climb up beyond them. (He must, so to speak, throw away the ladder after he has climbed up it.) He must transcend these propositions, and then he will see the world aright.

For an account of these matters, see

Oldroyd, D., *The Arch of Knowledge,* Methuen, London, 1986,

Philosophy and Science, F. Mosedale, ed., Prentice-Hall, Englewood Cliffs, NJ, 1979,

Putnam, H., "Models and Reality," *J. Symbolic Logic,* 45 (1980), 464–482.

§2. The idea of linkage was introduced by Rosen to serve as a generalized version of the more traditional notion of the dependency of variables. The difference is that the linkage concept allows us to speak of various degrees of dependence, whereas the classical idea is an all-or-nothing notion. For a detailed account of the linkage concept and its use in describing the dynamics of interaction between systems, see

Rosen, R. *Fundamentals of Measurement and Representation of Natural Systems,* Elsevier, New York, 1978.

From the time of Newton, the concept of a system state has always been a somewhat murky one, and it's only in recent years that the real role of the system state-space has been clarified. The modern view of a "state" is that it is purely a mathematical construction introduced to mediate between the system inputs and outputs, and that it is a mistake to attach any intrinsic system-theoretic significance to the notion of state. In particular, it is only a coincidence, and of no special consequence, that on occasion we can give a physical interpretation to the state. What's important is the physically observed quantities, the inputs and outputs, not the states. This point is made clear by the triangle example of the text, which was used originally in

Krohn, K., J. Rhodes and R. Langer, "Transformations, Semigroups and Metabolism," in *System Theory and Biology,* M. Mesarovic, ed., Springer, New York, 1968.

§3. Problems of causation have bedeviled philosophers from the time of Aristotle, who elucidated the four causal categories (material, efficient, formal, final) in his *Metaphysics.* The basic problem of causation as currently seen by philosophers is to understand whether there is any necessary connection between a cause and its effect. As noted in the text, this question is sometimes confused with the issue of determinism which, philosophically speaking, involves issues pertaining to free will and really has nothing to do with the much more restricted notion of causal connections. Good introductory accounts of the problems of causation are found in

Taylor, R., "Causation," in *The Encyclopedia of Philosophy,* P. Edwards, ed., Macmillan, New York, 1967,

Popper, K., *The Logic of Scientific Discovery,* Hutchinson, London, 1959,

Troxell, E., and W. Snyder, *Making Sense of Things: An Invitation to Philosophy,* St. Martin's Press, New York, 1976.

For an interesting account of the causality/determinism issue in the light of modern physics and computing, see

Manthey, M., "Non-Determinism Can be Causal," *Int. J. Theoretical Physics,* 23 (1984), 929–940.

The existence of a deterministic relationship between system observables also introduces the question of whether or not every such relationship constitutes a "law" of nature, or whether there are additional conditions that must be imposed before we would dignify such a relationship by calling it a "law." A spectrum of discussions of this issue is given by

Ayer, A., "What is a Law of Nature?", in *The Concept of a Person,* A. Ayer, ed., London, 1963,

Lakatos, I., *Mathematics, Science and Epistemology,* Cambridge University Press, Cambridge, 1978,

Eddington, A., *The Nature of the Physical World,* University of Michigan Press, Ann Arbor, MI, 1958,

Casti, J., "System Similarities and the Existence of Natural Laws," in *Differential Geometry, Topology, Geometry and Related Fields,* G. Rassias, ed., Teubner, Leipzig, 1985, pp. 51–74.

§4. The classic work outlining how biological species can be transformed into one another by smooth coordinate changes is

Thompson, d'Arcy, *On Growth and Form,* Cambridge University Press, Cambridge, 1942, abridged ed., 1971.

See also

Rosen, R., "Dynamical Similarity and the Theory of Biological Transformations," *Bull. Math. Biology,* 40 (1978), 549–579.

It is impossible to overemphasize the importance of the concept of system equivalence in addressing basic questions of modeling. In this regard, the quantities that remain invariant under coordinate changes are the only aspects of the system that have any right to be termed *intrinsic* system-theoretic properties, and we should ideally make use only of the invariants in answering questions about the system. A formal account of some of these matters is found in

Kalman, R., "Identifiablity and Problems of Model Selection in Econometrics," *Fourth World Congress of the Econometric Society,* France, August 1980.

§5. The term "bifurcation" is used in several different ways in mathematics. In particular, in the theory of ordinary differential equations it's usually taken to mean the value of a parameter at which the solution to a set of ordinary differential equations becomes multiple-valued. Although this usage is consistent with that employed in the text, our interpretation of the term is of much broader currency, incorporating other types of "discontinuous" changes in system behavior as well.

The stellar stability example is taken from

Thompson, J. M. T., *Instabilities and Catastrophes in Science and Engineering,* Wiley, Chichester, 1982.

The treatment of the so-called "nonelementary" catastrophes quickly leads one into the deep waters of modern singularity and bifurcation theory. For an account of some of this work, see Chapter 17 of

Gilmore, R., *Catastrophe Theory for Scientists and Engineers,* Wiley, New York, 1981,

as well as the work that sparked off the interest in catastrophe theory,

Thom, R., *Structural Stability and Morphogenesis,* Benjamin, Reading, MA, 1975.

§6. An introductory account of the "relativistic" view of system complexity is given in the articles

Casti, J. L., "On System Complexity: Identification, Measurement and Management," in *Complexity, Language and Life: Mathematical Approaches,* J. Casti and A. Karlqvist, eds., Springer, Berlin, 1986,

Rosen, R., "Information and Complexity," *op. cit.*

A variety of approaches to measuring system complexity, under the assumption that complexity is an intrinsic property of the system itself, are described in

Casti, J., *Connectivity, Complexity and Catastrophe in Large-Scale Systems,* Wiley, New York, 1979.

The question of simple versus complex has been one that has troubled system thinkers and philosophers for a number of years. Here is a somewhat eclectic list giving a flavor of these musings:

Quine, W. v. O., "On Simple Theories of a Complex World," in *Form and Strategy,* J. Gregg and F. Harris, eds., Reidel, Dordrecht, 1964,

Simon, H. A., "The Architecture of Complexity," in *Sciences of the Artificial,* 2d ed., MIT Press, Cambridge, 1981,

Beer, S., "Managing Modern Complexity," *Futures,* 2 (1970), 245–257,

Ashby, W. R., "Some Peculiarities of Complex Systems," *Cybernetic Medicine,* 9 (1973), 1–8,

Gottinger, H., *Coping with Complexity,* Reidel, Dordrecht, 1983,

Warsh, D., *The Idea of Economic Complexity,* Viking, New York, 1984.

§7. It can safely be asserted that the problem of identifying and controlling roundoff error is one of the principal pillars upon which much of modern numerical analysis rests. A good treatment of many of the problems encountered (and some of the remedies) can be found in the classic work

Wilkinson, J., *The Algebraic Eigenvalue Problem,* Oxford University Press, Oxford, 1965.

The issue of "surprise" as a system property ultimately centers upon the discrepancy between a particular model of reality and reality itself. In turn, this gap can be represented by those observables with which the system Σ is in interaction, but which are not accounted for in our description of Σ. In short, surprise comes about from interactions to which the system is open, but to which the model is closed. The traditional way to account for this type of uncertainty is by classical probability theory, assuming the unmodeled interactions to be represented by random variables having appropriate distribution functions. Such an approach has come under increasing attack, primarily because it is based upon the dubious assumption that randomness and uncertainty are somehow the same thing—or at least that we can represent all forms of uncertainty by various types of randomness. To address this difficulty, a variety of alternatives to probability theory have been advanced, each of which purports to represent uncertainty in a nonprobabilistic fashion. Some of these alternatives are presented in

Zadeh, L., "The Concept of a Linguistic Variable and its Application to Approximate Reasoning—I, II, III," *Information Sciences,* 8, 9 (1975), 199–249, 301–357, 43–80,

Kickert, W. J. M., *Fuzzy Theories on Decision-Making,* Nijhoff, Leiden, 1978,

Klir, G. J., and T. A. Folger, *Fuzzy Sets, Uncertainty and Information,* Prentice-Hall, Englewood Cliffs, NJ, 1988,

Klir, G. J., and E. C. Way, "Reconstructability Analysis: Aims, Results, Open Problems," *Systems Research,* 2 (1985), 141–163.

§8. An outstanding introduction to the ideas underlying formal systems, Gödel's Theorem, thinking machines and much, much more is the Pulitzer prize-winning book

Hofstadter, D., *Gödel, Escher, Bach: An Eternal Golden Braid,* Basic Books, New York, 1979.

Another introductory account of algorithms, Turing machines, Gödel's Theorem and the meaning of mathematics is available in Chapter Six of

Casti, J., *Searching for Certainty: What Scientists Can Know About the Future,* Morrow, New York, 1991.

A somewhat more technical account of formal systems, but still at the university textbook level, is

Davis, M. and E. Weyuker, *Computability, Complexity and Languages,* Academic Press, Orlando, FL, 1983.

There is a very deep and perplexing connection between the idea of a formal system and the problem of the mechanization of human thought processes. This question has a very extensive literature, some of which is summarized in the Hofstadter book cited above. The original paper in this area, and still required reading for anyone even remotely interested in the topic, is

Turing, A., "Computing Machinery and Intelligence," *Mind,* 59 (1950), (reprinted in *The Mind's I,* D. Hofstadter and D. Dennett, eds., Basic Books, New York, 1981, pp. 53–67).

Following Turing's paper, a virtual torrent of work has appeared in the scientific as well as philosophical and popular press arguing the pros and cons of the "Can machines think?" question. A representative sampling of these arguments is provided in the following collection:

Minds and Machines, A. R. Anderson, ed., Prentice-Hall, Englewood Cliffs, NJ, 1964,

Gunderson, K., *Mentality and Machines,* 2d ed., University of Minnesota Press, Minneapolis, 1985,

Webb, J., *Mechanism, Mentalism and Metamathematics,* Reidel, Dordrecht, 1980,

Minds, Machines and Evolution, C. Hookway, ed., Cambridge University Press, Cambridge, 1984,

Penrose, R., *The Emperor's New Mind,* Oxford University Press, Oxford, 1989.

For a summary of the current state of play in the AI game, see Chapter Five of the volume

Casti, J., *Paradigms Lost: Images of Man in the Mirror of Science,* Morrow, New York, 1989 (paperback edition: Avon Books, New York, 1990).

§9. Global modeling as a growth industry got its start with the well-chronicled works

Forrester, J. W., *World Dynamics,* Wright-Allen Press, Cambridge, MA, 1971,

Meadows, D., D. Meadows, J. Randers, and W. Behrens, *The Limits to Growth,* Universe Books, New York, 1972.

An excellent summary of the many global models that have been developed through the years, their strengths and weaknesses, uses in policymaking and future prospects is provided in

Meadows, D., J. Richardson, and G. Bruckmann, *Groping in the Dark,* Wiley, Chichester, UK, 1982.

From a system-theoretic perspective, the pathological instabilities inherent in the original Forrester-Meadows exercise was first pointed out in the paper

Vermuelen, P. J., and de Jongh, D. C. J., "Growth in a Finite World—A Comprehensive Sensitivity Analysis," *Automatica,* 13 (1977), 77–84.

Beginning with the work of Leontief on input/output models of national economies, there has been an ever-increasing interest in the problem of identification of input/output coefficients using tools and techniques from linear system theory and statistics. The ideas underlying these efforts are presented in the classic works

Gale, D., *The Theory of Linear Economic Models,* McGraw-Hill, New York, 1960,

Baumol, W., *Economic Theory and Operations Analysis,* 2d ed., Prentice-Hall, Englewood Cliffs, NJ, 1965.

Unfortunately, the past two decades of work on parameter identification in econometric models has not resulted in a completely satisfactory state of affairs. Some of the reasons why are detailed in

Kalman, R., "Identification from Real Data," in *Current Developments in the Interface: Economics, Econometrics, Mathematics,* M. Hazewinkel and A. H. G. Rinnooy Kan, eds., Reidel, Dordrecht, 1982, pp. 161–196,

Kalman, R., "Dynamic Econometric Models: A System-Theoretic Critique," in *New Quantitative Techniques for Economic Analysis,* G. Szegö, ed., Academic Press, New York, 1982, pp. 19–28.

The uses of modern tools from differential geometry, such as fiber bundles, in physics is of relatively recent vintage. For an account of how these ideas are employed, especially in the context of renormalization theory, see

Hermann, R., *Yang-Mills, Kaluza-Klein, and the Einstein Program,* Math Sci Press, Brookline, MA, 1978.

Figure 1.5 shows clearly the separation between the real world of natural systems on the left side of the diagram, and the purely imaginary world of mathematics on the right. In our view, the natural home of the system scientist, as opposed to the natural scientist or mathematician, is as the keeper of the encoding and decoding operations $\mathcal{E}$ and $\mathcal{D}$. Given the way in which universities and research institutes are administratively organized along traditional disciplinary lines, lines that place primary emphasis upon the two sides of Fig. 1.5, it's no surprise that the system scientist finds it difficult to find a natural professional home in modern institutions of higher learning.

To function effectively, the system scientist must know a considerable amount about the natural world **and** about mathematics, without being an expert in either field. This is clearly a prescription for career disaster in today's world of ultra-high specialization, as it almost invariably ensures comments like "This is interesting work, but it's not really mathematics," or "This is an intriguing idea, but it ignores the experimental literature." At the same time, we see an ever-increasing move toward inter- and trans-disciplinary attacks upon problems in the real world, resulting in the emergence of quasi-disciplines cognitive science, sociobiology, archaeoastronomy, and other fields that cannot be comfortably accommodated within the boundaries of traditional academic disciplines. Our feeling is that the system scientist has a central role to play in this new order, and that that role is to first of all understand the ways and means of how to encode the natural world into "good" formal structures. The system scientist must then see how to use these structures to interpret the mathematics in terms of the questions of interest to the experimental scientist. The sooner this role is explicitly recognized by the international academic establishment, the sooner system science will be able to claim a legitimate place in the coming intellectual order.

DQ #3. The thesis of *reductionism* so pervades every area of science as the tool of choice for investigation of the properties of systems that it's of interest to examine the reductionist program in somewhat greater detail. Basically, the reductionist creed rests upon the following principles:

A. There exists a *universal* way $\mathcal{U}$ of encoding any natural system Σ, and every other encoding of Σ is equivalent to $\mathcal{U}$.

B. The encoding $\mathcal{U}$ is canonically determined from an appropriate series of *fractions* (proper subsystems) of Σ.

C. The fractions of Σ may be isolated from each other (i.e., be made independent) by the imposition of appropriate dynamics on Σ.

Thus we see that reductionism is really an *algorithm,* i.e., a prescription for finding the universal encoding $\mathcal{U}$ by determining the fractions of Σ using imposed dynamics. Needless to say, no such algorithm has ever been offered, and it's difficult to understand the degree of tenacity with which the scientific community clings to the fiction of the reductionist program in the face of the unlikelihood of ever successfully carrying it through to completion. A simple counterexample to the possibility of successfully following the steps outlined above is provided by the results of Problem 2 of this chapter, in which we saw that, in general, we can know the properties of individual subsystems characterized by the observables f and g, but not be able to predict from this knowledge what the composite system $h = g \circ f$ will be doing.

Another way of seeing the difficulties involved is to note that the imposition of the dynamics on Σ involves a breaking of linkage relationships among the observables of Σ. These relationships are broken in order to simplifying the *state* description of the system, i.e., to "decompose" the original linkage relationships between the states. Unfortunately (for the reductionist), an operation geared to simplify state descriptions will usually do drastic and terrible things to the tangent vectors associated with the original system dynamics. This inability to simultaneously simplify *both* the state and velocity descriptions of Σ is the main reason why reductionism fails as a universal scheme for the analysis of natural systems. For a fuller account of these matters, see the Rosen book cited under §2 above.

CHAPTER TWO

Catastrophes, Dynamics and Life: The Singularities of Ecological and Natural Resource Systems

1. The Classification Problem

Most models of natural phenomena make use of the elementary functions of analysis such as $\exp x$, $\sin x$, $\cos x$ and x^k as the raw material from which to construct the equations of state. These are the smoothest of smooth functions (analytic, even) and, consequently, can display no discontinuous changes of any sort; what happens at a point $x + \Delta x$ is pretty much the same as what's happening at x, if Δx is small enough. Nevertheless, a substantial number of physical procsses *do* display discontinuities, with the onset of turbulence, cellular differentiation, beam buckling and stellar collapse being but a minuscule sample of such situations. So how can we capture such discontinuous behavior in mathematical models built out of smooth components? This is the question that modern bifurcation theory and its close relatives, singularity theory and catastrophe theory, strive to address.

We have already encountered the idea of a family of system descriptions in Chapter 1, where we considered the parameterized equation of state $\Phi_\alpha(x)$. Here α belongs to some space of parameters $\mathcal{A}$, usually R^k. Intuitively, we can see how discontinuous behavior can emerge out of gradual, smooth changes in the parameters by thinking of the following experiment: Let's start making our way through through the set $\mathcal{A}$ in a smooth fashion, beginning at some point $a \in \mathcal{A}$. The description $\Phi_{a'}$ looks and behaves much the same as the description Φ_a for all points a' near a, so we don't see any sharp behavioral discontinuities in passing from one description to the other. In short, the two descriptions are equivalent. However, at some time during the course of our tour through $\mathcal{A}$, we might come to a point a^* for which the description Φ_{a^*} is not equivalent to Φ_a for any a near a^*. As noted in Chapter 1, we call a^* a bifurcation point of the family of descriptions. The reason for this terminology is that the behavior of the description Φ_{a^*} diverges, or *bifurcates,* from that of the description Φ_a for all a sufficiently close to a^*.

Since the system behavior changes dramatically at every bifurcation point $a^* \in \mathcal{A}$ from the behavior seen at all nearby descriptions, it would be nice to be able to classify the various possible behaviors by dividing up the set $\mathcal{A}$ into different subsets in such a way that two points $a, \hat{a} \in \mathcal{A}$ will belong to the same subset if and only if their corresponding descriptions

Φ_a and $\Phi_{\hat{a}}$ yield the "same kind" of behavior. In order to make this idea mathematically precise, we have to formalize both the idea of "closeness" for two descriptions and what we mean when we speak of the behaviors of the corresponding systems being "the same."

Systems, as we shall see, can display rather complicated behaviors. But some of these behaviors are more common than others. System theorists have formalized this idea of what's "typical" using the concept of *genericity.* A property **P** of a system is generic if "almost all" systems display **P**. The hope is that generic systems are sufficiently well behaved that we can classify them in some useful sense. The problem of classification, often more colorfully described as the "Yin-Yang Problem," amounts to the following: Find systems that are simple enough to be classified (the *yin,*) but complicated enough to be generic (the *yang*). Since the overall idea of classification is so important—and not just for modeling, but for mathematics as a whole—let's take a few pages to talk about the classification problem in a bit more detail. For the sake of definiteness, we'll concentrate upon the most important case for modeling when the description $\Phi_a(x)$ is a differential equation.

A differential equation on a space X is expressed by $\dot{x} = f(x)$, where the function $x(t)$ has its values in X, the function $f(x)$ is a rule mapping X to itself, and $\dot{x}$ is the derivative of $x(t)$ with respect to time. The manifold X is often a "flat" space like R^n, although in some applications it may be a curved space like a circle or a torus. The rule $f(x)$ is often called a *vector field.* It is a prescription, or recipe, telling us, in effect, "if you're at the point x at time t, go to the point $x + f(x)\Delta t$ at time $t + \Delta t$," where Δt is an infinitesimal increment of time. To make things simple, let's assume that the vector field f is *smooth,* i.e., infinitely differentiable.

With the above setup in mind, it's clear that the problem of classifying differential equations is the same as the problem of classifying smooth vector fields on X. So let's take $\mathcal{V}$ to be the set of all smooth vector fields on X. Of course, $\mathcal{V}$ is an infinite-dimensional space. But while this makes it hard to visualize the space $\mathcal{V}$, it's still quite easy to work with it since $\mathcal{V}$ is a vector space with a topology. Consequently, we have some notion of what it means for two vector fields to be "close" to each other. So it makes good sense to speak about a neighborhood of any $f \in \mathcal{V}$.

The classification program for smooth vector fields on X has four steps:

A. Equivalence: Clearly, if we're going to classify differential equations into equivalence classes, we have to have an equivalence relation on $\mathcal{V}$. The choice we make for this relation will be dictated by several factors, including the utility of the equivalence relation for applications and its likelihood of leading to interesting theorems. If the relation is too coarse, then we will be lumping together vector fields that "should be different." On the other hand, if the relation is too fine, then it will separate things that ought to be "the same."

As we'll see throughout much of this book, the most important consideration in choosing the equivalence relation for differential equations is the idea of "stability." Note, however, that this does not refer to the stability of the trajectory arising from a *single* equation, but rather to the stability of the differential equation within a *family* of equations. This is the concept called *structural stability,* which we will consider in more detail a bit later. The basic idea is that in the absence of convincing reasons to the contrary, "good" models of physical processes should be robust with respect to small disturbances in the underlying dynamical equations.

B. Density: The second step in our classification program involves showing that the stable equations are dense in $\mathcal{V}$. What this means is that every neighborhood in $\mathcal{V}$, no matter how small, contains a structurally stable vector field. This condition is important, since it ensures that a stable model is just an arbitrarily small perturbation away from any vector field in $\mathcal{V}$.

C. Classification of the Stable Points of $\mathcal{V}$*:* At this step we try to list the stable classes in $\mathcal{V}$, giving a canonical example that's representative of each class. This step is analogous to the situation in matrix theory, where we classify the $n \times n$ real matrices under similarity. In that setting we use the following equivalence relation: the matrix A is equivalent to B if and only if there exists a nonsingular matrix T such that $A = TBT^{-1}$. In this situation, two matrices belong to the same class if and only if they have the same invariant factors. In other words, each class is characterized by matrices whose characteristic values and their multiplicities are the same. A canonical representative of a class is then given by the Jordan form of any matrix from that class. We'll see later how this classification step works in the case of vector fields.

D. Listing of the Unstable Points of $\mathcal{V}$*:* To see what's happening at this step, let's suppose for a moment that $\mathcal{V}$ is only 3-dimensional. We let $\mathcal{S}$ be the subset of $\mathcal{V}$ consisting of stable vector fields, while $\mathcal{T}$ denotes the complementary subset of structurally unstable points. It's clear that $\mathcal{T}$ will consist of a number of surfaces that intersect each other, dividing up $\mathcal{V}$ into several regions. For a visual image, think of the kind of tray you use for making ice cubes. The interior of the tray is divided up by flat pieces of metal or plastic that intersect each other at right angles. These flat pieces separate the interior of the tray into individual rectangular cells, each of which gives one ice cube. In this analogy, the complete tray constitutes the set $\mathcal{V}$. The interiors of the rectangular cells from which we make the ice cubes corresponds to the stable points $\mathcal{S}$, while surfaces of the flat separator pieces, together with all the edges and points of their various intersections, forms the set of unstable points $\mathcal{T}$.

In the foregoing setup, each of the surfaces in $\mathcal{T}$ represents an unstable class of *codimension 1*. Here codimension means the difference between the dimension of $\mathcal{V}$ (which we are assuming here is 3) and the dimension of the surface, which we pretend here is 2. What this means is the following: Suppose we are given a single vector field f. This will be a single point of $\mathcal{V}$ and, in general, will lie in the subset $\mathcal{S}$; if by chance it happens to lie in $\mathcal{T}$, then by a small perturbation we can move it into $\mathcal{S}$. However, if we are given a 1-parameter *family* of vector fields f_α, then this family will be a curve in $\mathcal{V}$. Although we can easily move such a curve off the edges and points of $\mathcal{T}$, it may not be possible to move the curve off the surfaces of $\mathcal{T}$. "Typically," the curve will intersect the surfaces of $\mathcal{T}$, which are the unstable classes of codimension 1. Therefore, the unstable classes of codimension 1 will classify the "typical" bifurcations of 1-parameter families of vector fields.

Of course, $\mathcal{V}$ is actually infinite-dimensional. And $\mathcal{T}$ consists of infinite-dimensional submanifolds that criss-cross each other in various ways. However, we can still decompose $\mathcal{T}$ into strata of codimensions 1, 2, and so on, with the property that an r-parameter family of vector fields will typically meet only those strata of codimension $\leq r$. Hence, the unstable classes of codimension $\leq r$ will classify the typical bifurcations of such r-parameter families of equations.

The simplest situation in which we can employ this classification program is for *gradient dynamical systems.* In this case, the vector field $f(x)$ is the gradient of a potential function $P(x): X \to R$, i.e., $f(x) = -\text{grad } P(x)$. Here we assume that $P(x)$ also has as many derivatives as we wish. In this gradient situation, the problem is reduced from the classification of vector-valued mappings in the space $\mathcal{V}$ to the much simpler problem of classifying *scalar-valued* functions in the space $\mathcal{P}$. Much of this chapter is devoted to an account of how steps A–D in the classification program outlined above are carried out in this crucially important setting of gradient systems. We shall find that the stable functions in $\mathcal{P}$ can be classified by their critical points (the points at which the gradient vanishes), while the unstable functions are classified by the elementary catastrophes. At the end of the chapter we shall give some consideration to the classification of more general types of differential systems, showing why a complete classification along the above lines seems, at present, to be but a dream.

Exercises

1. Describe explicitly the steps A–D above in the case of classification of real $n \times n$ matrices under the equivalence relations of (a) congruence and (b) similarity.

2. Consider the space of polynomials with complex coefficients in a single variable z. Regard two such polynomials $p(z), \hat{p}(z)$ as equivalent if p and $\hat{p}$

have the same number of real roots (counted according to their multiplicities). (a) Show that this relation is an equivalence relation. (b) Determine the equivalence classes. (c) Describe a canonical representative from each equivalence class. (d) Characterize the stable and unstable points in this space of polynomials.

3. Employ an analogous argument to that given above to conclude that the unstable classes of codimensions 1 and 2 classify the typical bifurcations of 2-parameter families of vector fields.

2. Smooth Functions and Critical Points

The key to making discontinuity emerge from smoothness is the observation that the overall behavior of both static and dynamical systems is governed by what's happening near the critical points. These are the points at which the gradient of the function vanishes. Away from the critical points, the Implicit Function Theorem tells us that the behavior is boring and predictable, linear, in fact. So it's only at the critical points that the system has the possibility of breaking out of this mold to enter a new mode of operation. It's at the critical points that we have the opportunity to effect dramatic shifts in the system's behavior by "nudging" lightly the system dynamics, one type of nudge leading to a limit cycle, another to a stable equilibrium, and yet a third type resulting in the system's moving into the domain of a "strange atttractor." It's by these nudges in the equations of motion that the germ of the idea of discontinuity from smoothness blossoms forth into the modern theory of singularities, catastrophes and bifurcations, wherein we see how to make discontinuous outputs emerge from smooth inputs.

We can formalize this idea of "nudging" the system by recalling that virtually all system descriptions—static or dynamic—contain constituent parameters whose values characterize the specific problem situation. Be they chemical reaction rates, technological coefficients, the Reynolds number of a fluid or the Federal Reserve discount rate, such parameters define a *family* of related systems. And we can regard a "small nudge" as being the same thing as a small change in these parameters, essentially a movement from one member of this family of systems to another one nearby. It then makes sense to ask the following question: If we make a small change in the input parameters, does this result in a *"large"* change in the system's steady-state behavior? It's in this way that we can see how it might be possible for smooth changes in the parameters defining a system to give rise to discontinuous changes in the long-run behavior—even for systems whose underlying equations of state are themselves smooth.

Example: A Family of Quartics

To fix the foregoing ideas, consider a system described by the one-parameter family of quartics

$$f(x, a) = \frac{x^4}{4} + \frac{ax^2}{2} + x.$$

The critical points are those x for which $\partial f/\partial x = 0$. These are given by the set

$$\mathcal{B} = \{x : x^3 + ax + 1 = 0\}.$$

From elementary algebra, it's easy to see that we have three real critical points when $|a| < 2/(3\sqrt{3})$, one real critical point when $|a| > 2/(3\sqrt{3})$ and a repeated (double) critical point when $|a| = 2/(3\sqrt{3})$. Thus, the set $\mathcal{B}$ has the structure shown in Fig. 2.1.

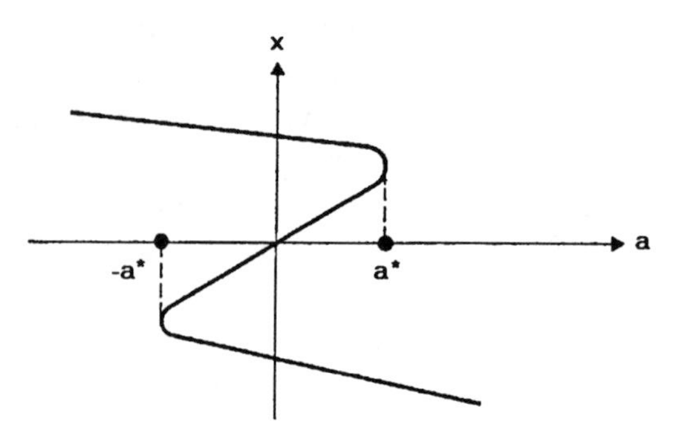

Figure 2.1. The Solution Set for $x^3 + ax + 1 = 0$

Here we have designated the roots as $x^*(a)$ to explicitly indicate their dependence upon the parameter a. In terms of the root structure, it's evident that the system undergoes a dramatic shift at the points $a^* = \pm 2/(3\sqrt{3})$, changing from having three real critical points to having just one. Thus, the map $a \mapsto x^*(a)$ is discontinuous (and multiple-valued) at $a = a^*$. So even though $f(x, a)$ is as smooth as it can be in both x and a, the behavior of the map from the parameter space to the set $\mathcal{B}$, which represents the observed behavior of the equation of state f, can be highly discontinuous—even when the parameter a is varied smoothly. It is this kind of discontinuity that catastrophe theory and its parents, singularity theory and bifurcation theory, have been created to study.

From now on we shall deal only with *smooth* maps and functions, i.e., maps $f\colon R^n \to R^m$ such that f has derivatives of all orders for all $x \in R^n$. Typical examples of such objects are $f(x_1, x_2) = x_1^2 + 2x_1x_2^3$, $f(x_1) = \exp x_1$ and $f(x_1, x_2) = \sin x_1 + \cos x_2$. By way of comparison, nonsmooth functions are objects like $f(x_1, x_2) = 1/\sqrt{x_1x_2}$ and $f(x_1) = \log x_1$. We denote the smooth maps by $C^\infty(R^n, R^m)$, calling them smooth *functions* when $m = 1$. As noted earlier, most of the relations, models and equations of state of mathematical physics, engineering and the life sciences are expressed using such functions, and our concern will be to look at the behavior of these smooth maps in the neighborhood of a *critical point* x^*, i.e., a point x^* at which $\partial f/\partial x = 0$. In component form, if $f(x_1, x_2, \ldots, x_n) = \{f_1(x_1, \ldots, x_n), \ldots, f_m(x_1, \ldots, x_n)\}$, then

$$\frac{\partial f}{\partial x} \doteq \begin{pmatrix} \frac{\partial f_1}{\partial x_1} & \frac{\partial f_1}{\partial x_2} & \cdots & \frac{\partial f_1}{\partial x_n} \\ \frac{\partial f_2}{\partial x_1} & \frac{\partial f_2}{\partial x_2} & \cdots & \frac{\partial f_2}{\partial x_n} \\ \vdots & & & \\ \frac{\partial f_n}{\partial x_1} & \frac{\partial f_n}{\partial x_2} & \cdots & \frac{\partial f_n}{\partial x_m} \end{pmatrix}.$$

The reader will recognize this as the usual *Jacobian* matrix of the map f. For notational simplicity, we shall always assume that a preliminary coordinate change in R^n has been carried out so that the critical point of interest lies at the origin, i.e., $x^* = 0$.

Consider the situation from elementary calculus, where we have f being a smooth function of a single variable ($n = m = 1$). If $f'(0) = 0$ and we scale the axes so that $f(0) = 0$, then the graph of f might look like the dotted curve in Fig. 2.2, while the Taylor series expansion near the origin would be

$$f(x) = f''(0)\frac{x^2}{2!} + f'''(0)\frac{x^3}{3!} + \cdots.$$

The sign of $f''(0)$ determines the character of $f(x)$ near the origin. Suppose $f''(0) > 0$. Then we have the situation of Fig. 2.2; if $f''(0) < 0$, then f has a local maximum at the origin. Finally, $f''(0) = 0$ means the origin is an inflection point. In this case, we need to consider higher-order derivatives to determine the local character of f near the origin.

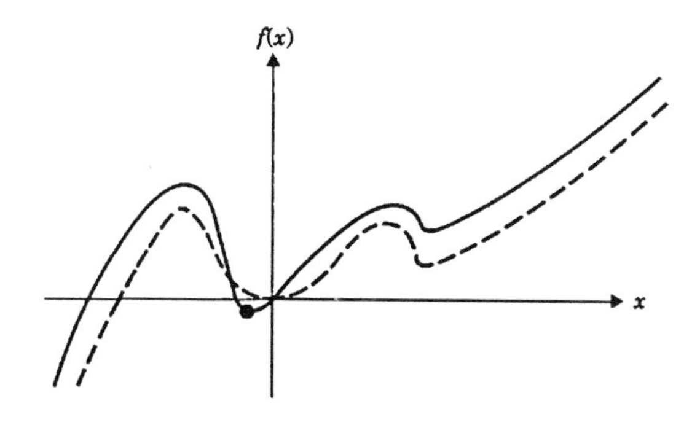

Figure 2.2. The Graph of f(x) and a Small Perturbation

What's important here is that the local behavior of f is entirely governed by the first nonvanishing term in the Taylor series expansion, in this case the term $\frac{1}{2}f''(0)x^2$ (assuming $f''(0) \neq 0$). Thus, locally f is a quadratic and, as we'll see later, there's a local coordinate change $x \to p(x) = x'$ such that in the x' variable we have the *exact* local representation $f(x') = \pm x'^2$ (Morse's Lemma). Furthermore, a small perturbation of f to $f + \epsilon g$ will preserve this quadratic structure, the effect of the perturbation g being only a small relocation of the critical point from the origin to a nearby point x^*_ϵ. This situation is also seen in Fig. 2.2. Singularity theory tries to generalize this type of result by addressing the following basic questions:

- What happens when n and/or $m > 1$?
- What about the case of degenerate critical points, i.e., those for which $f''(0) = 0$?
- What does a "typical" $f \in C^\infty(R^n, R^m)$ look like near the origin?
- If f is *not* typical, what is the simplest family that f can belong to such that the *family* is typical in the space of families of smooth functions?

We have fairly complete answers to these questions when $m = 1$ (functions) and some information when $m > 1$ (maps). A detailed mathematical account of the methods behind the answers would take us well beyond the scope of this book. But we shall try to convey as much of the underlying flavor of the results as we can during the course of our subsequent exposition. But first a few preliminaries pinning down more precisely what we mean when we speak of a function being "like" another function, and what properties qualify a function to be termed "typical."

Exercises

1. Compute the critical points of the following functions: (a) $f(x) = x^2 + 2x + 1$, (b) $f(x) = \sin 2x$, (c) $f(x, y) = x^2y + \frac{1}{3}y^3$, (d) $f(x) = x^{100}e^x$. Determine the nondegenerate critical point(s) for each function, i.e., those critical points x^* such that $f''(x^*) \neq 0$.

2. For those functions in Exercise 1 having degenerate critical points, try to find the "simplest" perturbation of f that removes the degeneracy. In other words, find simplest function g such that $f+g$ has only nondegenerate critical points.

3. Structural Stability and Genericity

Kepler's discovery of his famous laws of planetary motion are probably as good a place as any to mark the beginning of modern science. But, oddly enough, when you study celestial mechanics one of the first things you learn is that Kepler's laws are wrong. More precisely, we find that they are only good approximations, valid in a strict sense only for a solar system containing a single planet. But, of course, our solar system contains many planets, including several very large ones. As a result, the combined gravitational effects of all the planets act to pull each of them away from their perfectly elliptical Keplerian orbits. Fortunately, however, these departures from the Keplerian ideal can be easily calculated, enabling us to predict with great accuracy what the planets will be doing far into the future.

We regard Kepler's laws (or Newton's equations of motion) as good models of planetary behavior for the simple reason that small perturbations of these ideal, simplified equations yield predictions of planetary behavior that are in excellent agreement with what's actually observed. In more technical jargon, these models are *structurally stable,* which means that if we perturb Kepler's laws to some other rules of planetary motion that are "nearby," these new rules generate predictions that are very close to those of the Keplerian case. Since we can never make absolutely precise measurements, this kind of stability is crucial for any model that we're going to take seriously as a basis for predicting and/or explaining any natural phenomenon. So let's see what's involved in trying to dress-up this idea of "nearby" models in more formal mathematical clothes.

For our purposes, a function f is "like" another function $\hat{f}$ if the critical points of f and $\hat{f}$ share the same topological character (local min, max, saddle point, etc.). Looked at from another vantage point, we say that the two functions are alike if, by a coordinate change in the domain and range of f, we can transform f into $\hat{f}$, and conversely, using a coordinate change that preserves the nature of critical points. Thus, if $f, \hat{f} \in C^{\infty}(R^n, R^m)$, we say that f is *equivalent* to $\hat{f}$ if there exist coordinate changes

$$g \colon R^n \to R^n, \qquad h \colon R^m \to R^m,$$

such that the following diagram is commutative:

$$\begin{array}{ccc} R^n & \xrightarrow{f} & R^m \\ {\scriptstyle g}\downarrow & & \downarrow{\scriptstyle h} \\ R^n & \xrightarrow[\hat{f}]{} & R^m \end{array}$$

In this chapter we will always seek g and h as origin-preserving diffeomorphisms (i.e., one-to-one, onto, and smooth transformations, having the normalization $g(0) = h(0) = 0$).

The most natural way of obtaining a function $\hat{f}$ from a given function f comes about when f is a member of a *parameterized family* of functions. Let $\alpha \in R^k$ be a vector of parameters. Then we have $f_\alpha \colon R^n \to R^m$ for some fixed value of α. If we now make the perturbation $\alpha \to \hat{\alpha}$, we obtain a new member of the family $f_{\hat{\alpha}}$. In this situation we say that $f_{\hat{\alpha}}$ is equivalent to f_α if we can find coordinate changes g and h as above, together with a diffeomorphism $a\colon R^k \to R^k$, such that the above diagram commutes upon setting $f = f_\alpha$ and $\hat{f} = f_{\hat{\alpha}}$. In general, the critical points of f_α will also depend upon the parameter value α, so there is an induced map

$$\begin{aligned} \chi\colon R^k &\to R^n \\ \alpha &\mapsto x^*(\alpha), \end{aligned}$$

from the parameter space to the set of critical points of f. Of great theoretical and applied concern are those values of α where the map χ is discontinuous.

Example: A Cubic System

Consider the parameterized family of functions

$$f_\alpha(x) = \frac{x^3}{3} + \alpha x, \qquad \alpha \text{ real}.$$

The critical points $x^*(\alpha)$ are the real roots of the equation

$$f'_\alpha(x) = x^2 + \alpha = 0.$$

Consequently, the graph of the map χ is given by the diagram in Fig. 2.3. Here we see the discontinuity (nonexistence, actually) of the map χ as we pass smoothly through the point $\alpha = 0$. For $\alpha < 0$, the family has two real critical points of opposite sign (one a local max, the other a local min). These two points coalesce at $\alpha = 0$ to an inflection point, which then gives way to no real critical points for $\alpha > 0$.

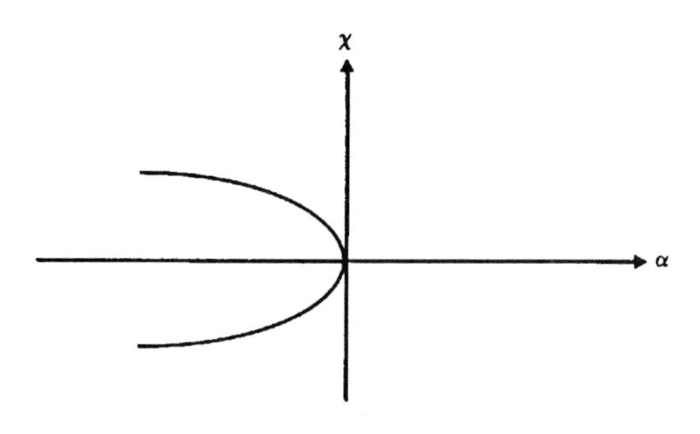

Figure 2.3. The Map χ

By the methods we will develop later on, it can also be shown that any function $f_{\hat{\alpha}}$ in this family is equivalent to f_α for $\hat{\alpha}$ in a local neighborhood of α—as long as that neighborhood does not include the point $\alpha = 0$. Thus, the bifurcation point $\alpha = 0$ splits the family $f_\alpha = x^3 + \alpha x$ into three disjoint classes—functions having two, one or no real critical points—and these classes are represented by the values $\alpha < 0$, $\alpha = 0$, and $\alpha > 0$, respectively.

The preceding considerations enable us to formulate precisely the notion of a function being "stable" relative to smooth perturbations. Let f and p be in $C^\infty(R^n, R^m)$. Then we say that f is *stable* with respect to smooth perturbations p if f and $f + p$ are equivalent, i.e., if there exist diffeomorphisms g and h such that the diagram

$$\begin{array}{ccc} R^n & \xrightarrow{\ f\ } & R^m \\ g\downarrow & & \downarrow h \\ R^n & \xrightarrow[f+p]{} & R^m \end{array}$$

commutes for all smooth p sufficiently close to f. For this definition to make sense, we need to impose a topology on the space $C^\infty(R^n, R^m)$ in order to be able to say when two maps are "close." The usual topology employed is the *Whitney topology,* which defines an ϵ-neighborhood of f by demanding that all functions in such a neighborhood agree with f and all of its derivatives up to accuracy $\epsilon > 0$. Details can be found in the references given in the chapter Notes and References.

If we pass to a parameterized family of functions, the preceding definition of stability must be modified to take into account variation not only in x, but also in the parameters α. Let $f_\alpha(x)$, $g_\alpha(x) \in C^\infty(R^n, R^m)$ be such smooth families. Then we say that $f_\alpha(x)$ is equivalent to $g_\alpha(x)$, $\alpha \in R^k$ if there exists a *family* of diffeomorphisms $q_\alpha \colon R^n \to R^n$, $\alpha \in R^k$, a diffeomorphism $e \colon R^k \to R^k$, and a smooth map $h \colon R^k \to R^m$ such that

$$g_\alpha(x) = f_{e(\alpha)}\big(q_\alpha(x)\big) + h(\alpha),$$

for all $(x, \alpha) \in R^n \times R^k$ in a neighborhood of $(0, 0)$.

The concept of structural stability now extends to families by saying that f_α is structurally stable *as a family* if f_α is equivalent to $f_\alpha + q_\alpha$, where q_α is a sufficiently small (i.e., nearby) *family* of functions.

Now let's turn to the question of what we mean, mathematically speaking, when we use the word "typical." Suppose we are given a set of objects S (any set in which a topology has been defined so that we can speak of one element being "close" to another). Further, let there be some property $\mathbf{P}$ that each of the elements of S may or may not possess. Then we say that the property $\mathbf{P}$ is *generic* relative to S under the following circumstances:

1) If $s \in S$ possesses property **P**, then there exists *some* neighborhood N of s such that every $s \in N$ also possesses property **P** (openness),

AND

2) *Every* neighborhood in S contains an element possessing property **P** (denseness).

In loose terms we can say that **P** is *generic* for the set S if "almost every" $s \in S$ possesses property **P**.

Example 1: Quadratic Polynomials

Let $S = \{$all quadratic polynomials in a single variable$\}$, i.e.,

$$S = \{f(x) : f = x^2 + bx + c,\ b \text{ and } c \text{ real}\}.$$

Let **P** be the property that the roots of f are distinct. Thus, $f \in S$ possesses property **P** if and only if $b^2 - 4c \neq 0$. To see that **P** is generic for the set S, we "coordinatize" S using the plane R^2. This is easily done by identifying each $f \in S$ with the point $(b, c) \in R^2$ formed from the coefficients of f. Those $f \in S$ not having distinct roots lie on the curve $b^2 = 4c$ (see Fig. 2.4), and it's evident that:

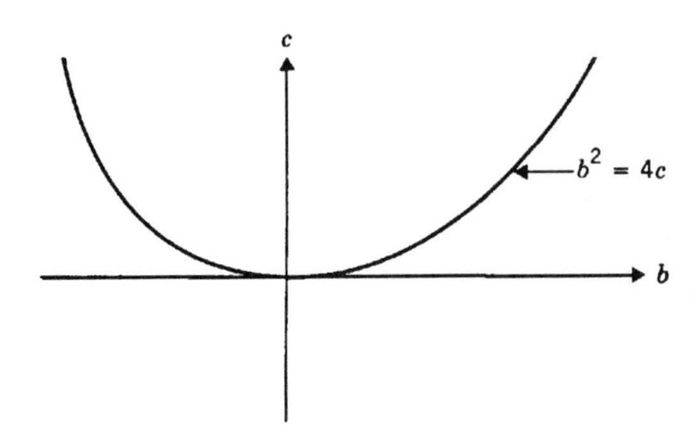

Figure 2.4. Genericity for Quadratic Polynomials

1) Every f such that $b^2 \neq 4c$ has some neighborhood in which all points (b', c') are such that $b'^2 \neq 4c'$, and

2) in every neighborhood of every point (b^*, c^*) such that $b^{*2} = 4c^*$, there exist points (b, c) such that $b^2 \neq 4c$.

Consequently, the property of having distinct roots is generic for quadratic polynomials.

Example 2: Nondegeneracy of Critical Points

Let $S = \{f \in C^\infty(R^n, R): f(0) = \text{grad } f(0) = 0\}$, and let the property **P** be that f has a nondegenerate critical point at the origin; i.e.,

$$\mathcal{H}(0) \doteq \det \left[\frac{\partial^2 f}{\partial x^2}\right](0) \neq 0.$$

We can see that nondegeneracy of critical points is a generic property by the following argument.

First of all, the characteristic polynomial of $\mathcal{H}$ is an nth-degree polynomial $p(\lambda)$. The condition that $\det \mathcal{H}(0) \doteq p(0) = 0$ means that $p(\lambda)$ must have no constant term, i.e.,

$$p(\lambda) = \lambda^n + a_{n-1}\lambda^{n-1} + \cdots + a_1\lambda.$$

But the set of all nth-degree polynomials can be identified with the points of R^n as in the previous example, and it is then easy to verify that points of the form $(a_{n-1}, a_{n-2}, \ldots, a_1, 0)$ correspond to nongeneric elements in $\mathcal{S}$. Hence, nondegeneracy of critical points is a generic property for smooth functions on R^n.

Exercises

1. Show that the functions $f(x, y) = x^2 + y^2 + x$ and $g(u, v) = u^2 + v^2$ are equivalent by finding a smooth change of variables $u = u(x, y)$, $v = v(x, y)$ that transforms f into g.

2. Prove that the function $f(x) = x^4$ is not structurally stable by finding an arbitrarily small perturbation of f that has a different set of critical points than f.

3. Determine whether the following parameterized families of functions are structurally stable: (a) $f(x, a, b) = \frac{1}{4}x^4 + \frac{1}{2}ax^2 + bx$, (b) $f(x, a) = e^x + ax^2 + x$, (c) $f(x, a, b) = \sin ax - \cos bx$.

4. Consider the set S consisting of all straight lines in the plane, i.e., $S = \{ax + by = k : a, b, k \in R\}$. Given a fixed line $s \in S$, prove that it's generic for all other lines in S to meet s in a single point.

5. Generalize the result of the preceding Exercise to the case when S consists of all smooth planar curves of degree less than or equal to 2. (Hint: parameterize all such curves using three real numbers).

6. Show that if the lines/curves of the previous Exercises are in three-dimensional space rather than in the plane, then it's generic for two elements to have **no** points of intersection.

4. Morse's Lemma and Theorem

The natural starting point for our study of the critical points of smooth functions is to consider functions that have a *nondegenerate* or *regular* critical point at the origin. Thus, we consider a smooth function $f\colon R^n \to R$, such that $f(0) = 0$, grad $f(0) = 0$ and $\det \mathcal{H}(f)(0) \neq 0$, where

$$\mathcal{H}(f) \doteq \left[\frac{\partial^2 f}{\partial x^2}\right]$$

is the Hessian matrix of f. Moreover, if the critical values are all distinct, i.e., $f(x^*) \neq f(\hat{x})$ for critical points $x^*, \hat{x}$, then f is called a *Morse function.*

If f is a Morse function, the Taylor series of f begins with quadratic terms:

$$f(x) = \sum_{i,j=1}^{n} \mathcal{H}_{ij}(f)(0)x_i x_j + \ldots .$$

We now investigate the degree to which the Hessian matrix $\mathcal{H}(f)$ determines the local structure of f near the origin. Clearly, if we can find a coordinate transformation $x \to y(x)$ such that in the y variables we have

$$f(y) = \sum_{i,j=1}^{n} \widehat{\mathcal{H}}_{ij}(0)y_i y_j,$$

where $\widehat{\mathcal{H}}$ is the transformed Hessian matrix, then the Hessian matrix would *completely* determine the behavior of f in a neighborhood of the critical point at the origin. Morse's Theorem asserts that this is indeed the case for Morse functions. Furthermore, the theorem states that Morse functions are both stable and generic in $C^\infty(R^n, R)$. As a result, the "typical" kind of smooth function we meet is a Morse function. So if we have a non-Morse function g at hand, an arbitrarily small perturbation of g will make it Morse. Furthermore, if f is Morse, then all functions sufficiently close to f are also Morse, where nearness is measured in the Whitney topology discussed earlier. Before sketching the proof of this key result, it's worthwhile reflecting for a moment upon its message.

At first glance, it might appear that Morse's Theorem covers virtually all cases of practical interest. After all, if we have the bad luck to be looking at a non-Morse function, we can always perturb it by a small amount and find ourselves back in the class of "good" functions. This is very reminiscent of the situation in matrix theory, where if we have a singular matrix A we can always make it nonsingular by perturbing some of its elements just a bit. But we know that in many important situations this procedure is not acceptable, and we must deal directly with the singular matrix A itself. So it is too with smooth functions. Sometimes the essence of the problem resides in the degeneracy of the critical point, as with phase transitions, and it's necessary to deal directly with non-Morse functions. Such situations arise naturally when we have a parameterized family of functions, since there will then almost always be some member of the family that is non-Morse. So it's important to know how the character of the critical point changes as we pass through these "atypical" functions in the family.

As an aside, it's tempting to speculate that the generic quadratic behavior of smooth functions ensured by Morse's Theorem accounts for the fact that almost all the basic laws of physics (conservation of energy, Fermat's principle, and the principle of least action, for example) are expressed as quadratic forms. Since it's both stable and "typical" to be locally quadratic, it's reasonable to expect that the local laws of classical physics will also be expressible as quadratics, as indeed they are.

As the first step toward Morse's Theorem, let us consider the following basic lemma:

MORSE'S LEMMA. *Let $f\colon R^n \to R$ be a smooth function with a nondegenerate critical point at the origin. Then there is a local coordinate system $(y_1, y_2, \ldots, y_n)$ in a neighborhood Y of the origin, with $y_i(0) = 0$, $i = 1, 2, \ldots, n$, such that*

$$f(y) = -y_1^2 - y_2^2 - \cdots - y_l^2 + y_{l+1}^2 + \cdots + y_n^2$$

for all $y \in Y$.

PROOF: We have

$$\begin{aligned} f(x_1, x_2, \ldots, x_n) &= \int_0^1 \frac{d}{dt} f(tx_1, \ldots, tx_n)\, dt \\ &= \int_0^1 \sum_{i=1}^n \frac{\partial f(tx_1, \ldots, tx_n)}{\partial x_i} x_i\, dt. \end{aligned}$$

Hence, we can write

$$f(x) = \sum_{j=1}^n x_j g_j(x),$$

in some neighborhood of the origin by taking

$$g_i(x) = \int_0^1 \frac{\partial f}{\partial x_i}(tx_1, \ldots, tx_n)\, dt.$$

Furthermore,

$$g_i(0) = \frac{\partial f(0)}{\partial x_i}.$$

Since the origin is a critical point, we have $g_i(0) = 0$. Hence, applying the above representation "trick" again, there exist smooth functions h_{ij} such that

$$g_j(x) = \sum_{i=1}^{n} x_i h_{ij}(x),$$

and we can write

$$f(x) = \sum_{i,j=1}^{n} x_i x_j h_{ij}(x).$$

If we symmetrize the h_{ij} by taking

$$\widehat{h}_{ij} = \frac{1}{2}(h_{ij} + h_{ji}),$$

the equation for $f(x)$ still holds, and we have $\widehat{h}_{ij} = \widehat{h}_{ji}$.

Differentiating the representation for $f(x)$ twice, we obtain

$$\frac{\partial^2 f(0)}{\partial x^2} = 2\widehat{h}_{ij}(0),$$

which shows that the matrix

$$\left[\widehat{h}_{ij}(0)\right] = \frac{1}{2}\left[\frac{\partial^2 f(0)}{\partial x^2}\right]$$

is nonsingular, by virtue of the fact that the origin is a nondegenerate critical point.

Now suppose that there exist local coordinates $(y_1, y_2, \ldots, y_n)$ in a neighborhood Y of the origin such that

$$f = \pm y_1^2 \pm y_2^2 \pm \cdots \pm y_{r-1}^2 + \sum_{i,j \geq r} y_i y_j H_{ij}(y_1, \ldots, y_n),$$

where $H_{ij} = H_{ji}$. By a relabeling of the last r coordinates, we may assume $H_{rr}(0) \neq 0$. Let

$$g(y_1, y_2, \ldots, y_n) = \sqrt{|H_{rr}(y_1, y_2, \ldots, y_n)|}\ .$$

By the Inverse Function Theorem, g is smooth in some neighborhood $V \subset Y$ of the origin. We again change coordinates to $(v_1, v_2, \ldots, v_n)$ defined by

$$v_i = y_i, \quad i \neq r,$$

$$v_r = g(y_1, y_2, \ldots, y_n)\left(y_r + \sum_{i>r} \frac{y_i H_{ir}(y_1, \ldots, y_n)}{H_{rr}(y_1, \ldots, y_n)}\right).$$

This is also a local diffeomorphism (again by the Inverse Function Theorem). Now we have

$$f(v_1, v_2, \ldots, v_n) = \pm v_1^2 \pm v_2^2 \pm \cdots \pm v_r^2 + \sum_{i,j \geq r+1} v_i v_j H'_{ij}(v_1, \ldots, v_n).$$

Hence, by induction, we have established that f is locally quadratic.

Remarks

1) The above proof should be compared with the procedure for reducing a symmetric matrix to diagonal form.

2) A function of the form

$$f(z) = z_1^2 + z_2^2 + \cdots + z_{n-l}^2 - z_{n-l+1}^2 - \cdots - z_n^2,$$

is called a *Morse l-saddle.* Thus, Morse's Lemma asserts that every smooth function f can be transformed by a smooth coordinate change to an l-saddle in the neighborhood of a nondegenerate critical point.

3) The number l is a topological invariant describing the *type* of the critical point at the origin, in the sense that l remains unchanged when we employ a smooth coordinate change in the original variables $x_1, x_2, \ldots, x_n$.

Now let's complete the discussion of Morse functions by giving a complete statement of Morse's Theorem.

MORSE'S THEOREM. *A smooth function $f\colon R^n \to R$ is stable if and only if the critical points of f are nondegenerate and the critical values are distinct (i.e., if x^* and $\hat{x}$ are distinct nondegenerate critical points of f, then $f(x^*) \neq f(\hat{x})$).*

Furthermore, the Morse functions form an open, dense set in the space of smooth functions $C^\infty(R^n, R)$.

Finally, in the neighborhood of any critical point of f there exists an integer l, $0 \leq l \leq n$, and a smooth coordinate change $g\colon R^n \to R^n$, such that

$$(f \circ g)(x_1, x_2, \ldots, x_n) = y_1^2 + y_2^2 + \cdots + y_l^2 - y_{l+1}^2 - \cdots - y_n^2.$$

We will see many examples of the use of Morse's Lemma and Theorem later on. So for the moment, let's pass directly to a consideration of what can happen if the critical point is *degenerate.*

Exercise

1. Determine which of the following functions are Morse functions: (a) $f(x) = \sin 2x$, (b) $f(x_1, x_2) = x_1x_2 + x_1^3 + x_1x_2^2$, (c) $f(x) = \log(1/x)$, (d) $f(x_1, x_2) = x_1^2x_2 - x_2^3$. Determine the canonical form in those cases where the function f is Morse.

5. *The Splitting Lemma and Corank*

We have already seen functions like $f(x) = x^3$ that have degenerate critical points. Even a cursory examination of the graph of such functions makes it clear that near a degenerate critical point, it's not possible to make the function "look quadratic" by any kind of smooth coordinate change. This raises the question: If these kinds of functions are not locally quadratic, what *do* they look like near a degenerate critical point? To address the matter, we need some additional concepts, starting with the development of a way to measure the degree of degeneracy of the critical point.

The most natural approach to measuring the degeneracy of a critical point is to recall that nondegeneracy is defined solely in terms of the rank of the Hessian matrix

$$H(f) = \left[\frac{\partial^2 f}{\partial x^2}\right],$$

evaluated at the critical point in question. Thus, if the critical point is at the origin and $\det H(f)(0) \neq 0$, then the origin is nondegnerate. This means that $H(f)(0)$ is of full rank n. In other words, rank $H(f)(0) = n$ if and only if the origin is a nondegnerate critical point. So one way to measure the degree of degeneracy is to take it to be the *rank deficiency* of H. This number r, termed the *corank* of f, is then defined to be

$$r \doteq \text{corank } f = n - \text{ rank } H(f)(0).$$

From what has gone before, it's evident that r is invariant under smooth coordinate changes.

Using the corank r and arguments similar to those employed in the preceding section, it's possible to prove the following key extension of Morse's Lemma:

THE SPLITTING LEMMA. *Let f have a degenerate critical point at the origin and assume corank $f = r$, i.e., $r = n -$ rank $H(f)(0)$. Then there exists a smooth coordinate change $x \to y$, such that near the origin f takes the form*

$$f(y) = g(y_1, \ldots, y_r) + Q(y_{r+1}, \ldots, y_n),$$

where g is a smooth function of order cubic and Q is a nondegenerate quadratic form.

Remarks

1) The terminology "Splitting Lemma" follows from the fact that we are able to "split" the new coordinate variables y into two *disjoint* classes: the subset $\{y_1, \ldots, y_r\}$ belonging to the higher-order function g, and the complementary subset $\{y_{r+1}, \ldots, y_n\}$, forming the nondegenerate quadratic form Q. The number of variables in each class is determined solely by the integer r, the corank of f.

2) In practice, the total number of independent variables in the function f may be very large. Yet the corank of f is usually still very small (often only 1 or 2). In such situations almost all of the variables of the problem are wrapped up in the quadratic form Q, an object with very nice analytic properties—especially for questions of optimization. Only a relatively small number of variables remain in the "bad" part of the decomposition of f. Consequently, the Splitting Lemma allows us to separate the analysis of f into a big "nice" part, and a small "bad" part.

3) The Splitting Lemma is a natural generalization of Morse's Lemma. This can be seen from the fact that if the origin is a nondegenerate critical point, we have $r = 0$. In this case the Splitting Lemma reduces to Morse's Lemma.

4) At this stage we can say nothing more specific about the function g, other than that it begins with cubic terms or higher. By imposing additional conditions upon f, we will later be able to assert that not only is $g \in O(|y|^3)$, but that g is actually a *polynomial.* The precise nature of these additional conditions is an integral part of the celebrated Thom Classification Theorem of elementary catastrophe theory.

Exercises

1. Determine the corank at the origin for each of the following functions: (a) $f(x_1, x_2) = x_1^3 + x_2^3$, (b) $f(x_1, x_2) = x_1^2 x_2$, (c) $f(x_1, x_2) = x_1 x_2 - \frac{1}{3}x_2^3$.

2. Let A be an $n \times m$ real matrix of rank $r \leq \min(n, m)$. Then the coranks of A in its domain R^m and range R^n are $m-r$ and $n-r$, respectively. If k is the dimension of the kernel of A, show that the coranks of A are related to k by the formula $k = m - r$ or, equivalently, $n - r = n - m + k$.

6. Determinacy and Codimension

Our main problem is to decide when two smooth functions f and g are locally equivalent, i.e., differ only by a smooth change of variables in some neighborhood of the origin. Let $j^k f$ denote the terms up to and including degree k in the Taylor series expansion of f about the origin. We call $j^k f$ the *k-jet* of f. The k-jet of f enables us to direct our attention to tests for the equivalence of f and g using finite segments of the Taylor series. These tests involve the concept of *determinacy.*

We say that f is *k-determined* if for all smooth g such that $j^k f = j^k g$, f and g are equivalent. In other words, f is *k-determinate* if f is equivalent to *every* g whose Taylor series agrees with that of f through terms of degree k. The smallest value of k for which this holds is called the *determinacy* of f, denoted $\sigma(f)$.

Note that to calculate $\sigma(f)$ by the foregoing definition involves an infinite computation, since we must test **every** smooth g whose k-jet agrees with that of f, at least in principle. For this reason we seek a test that implies finite determinacy, but that can be carried out in a finite number of steps. The basis of such a test is provided by the idea of *k-completeness.*

Let us define

$$m_n \doteq \{f \in C^\infty(R^n, R)\colon f(0) = 0\}.$$

So m_n is the set of smooth functions that vanish at the origin, i.e., functions having no constant term.

We say that a function f is $O(|x|^k)$ if the Taylor series expansion of f begins with terms of degree k. Further, let $f_{,i}$ denote the partial derivative of f with respect to x_i, i.e., $f_{,i} \doteq \partial f/\partial x_i$. Then we say that f is *k-complete* if every $\phi(x) \in O(|x|^k)$ can be written as

$$\phi(x) = \sum_{i=1}^{n} \psi_i(x) f_{,i}(x) + O(|x|^{k+1}),$$

where $\psi_i(x) \in m_n$.

Example: A Quartic Function

Let $f(x_1, x_2) = \frac{1}{4}(x_1^4 + x_2^4)$. The partial derivatives are

$$f_{,1} = x_1^3, \qquad f_{,2} = x_2^3.$$

Using these elements, we can generate any function of the form

$$\phi(x_1, x_2) = \psi_1(x_1, x_2)x_1^3 + \psi_2(x_1, x_2)x_2^3,$$

$\psi_1, \psi_2 \in m_n$. It's an easy exercise to see that no choice of ψ_1 and ψ_2 enables us to generate mixed fourth-order terms like $x_1^2 x_2^2$, although pure quartics like x_1^4 and x_2^4 are possible. However, all fifth-order terms—pure and mixed—can be generated through appropriate selection of $\psi_1, \psi_2 \in m_n$. Thus f is 5-complete, but not 4-complete.

The determination of exactly what kinds of functions are generated by the elements $\{f_{,i}\}$ leads to the final concept that we need, the notion of *codimension,* an idea we considered earlier in Section 1.

Note first that the space m_n is an infinite-dimensional real vector space (a ring, actually), having basis elements

$$\{x_1^{\alpha_1} x_2^{\alpha_2} \cdots x_n^{\alpha_n} \mid \textstyle\sum_i \alpha_i > 0\}.$$

Geometrically, any term in this list that *cannot* be generated using the elements $\{f_{,i}\}$ represents a "missing" direction in the space m_n, at least insofar as generation by the elements $f_{,i}$ is concerned. Loosely speaking, the *codimension* is the number of such missing directions. We denote it by codim f. More formally, if we define the *Jacobian ideal* of f as

$$\Delta(f) = \left\{ \phi \in m_n : \phi = \sum_{i=1}^{n} \psi_i(x) \frac{\partial f}{\partial x_i}, \quad \psi_i \in m_n \right\},$$

then

$$\text{codim } f = \dim_R [m_n / \Delta(f)] .$$

The importance of the codimension will emerge as we proceed. But for now it's sufficient just to note that even though both of the spaces m_n and $\Delta(f)$ are infinite-dimensional, the quotient space $m_n/\Delta(f)$ can be (and, in fact, usually is) finite-dimensional.

Example: A Quartic Function (cont'd.)

Returning to our earlier example with $f(x_1, x_2) = \frac{1}{4}(x_1^4 + x_2^4)$, we have already seen that all terms of order 5 and higher are generated by $f_{,1}$ and $f_{,2}$, with the only "bad" fourth-order term being $x_1^2 x_2^2$. However, no terms of order 3 or lower are contained in $\Delta(f)$. Thus, the "missing" directions in m_n are those generated by the following elements:

$$\{x_1, x_2, x_1^2, x_1x_2, x_2^2, x_1^3, x_1x_2^2, x_1^2x_2, x_2^3, x_1^2x_2^2\}.$$

There are ten elements in this set; hence, codim $f = 10$.

The connections between determinacy, completeness and codimension are given by the following theorem.

THEOREM 2.1.

i) f k-complete implies f is k-determinate.

ii) f k-determinate implies f is (k+1)-complete.

iii) codim $f < \infty$ if and only if f is finitely determined.

Remarks

1) The function f being finitely determined means that $f \sim j^k f$ (since we can always take $g = j^k f$). But $j^k f$ is just a polynomial of degree k. Thus, finite determinacy means that there is some local coordinate system in which f looks *exactly* like a polynomial.

2) The codimension of f and the determinacy $\sigma(f)$ are jointly infinite or finite, but they are not equal. It can be shown that codim $f \geq \sigma(f) - 2$. There are similar inequalities relating the corank r with both $\sigma(f)$ and codim f. These relations are of considerable applied significance, as we shall see later on.

3) It's possible for very simple looking functions to have infinite codimension (e.g., take $f(x_1, x_2) = x_1 x_2^2$). The moral here is that even simple models may have rather complicated behavior. We shall return to this point in the examples as well as in the Problems section.

4) Just like corank f, the codimension is a topological invariant. Thus, it provides a tool for classifying smooth functions near a critical point. We shall see that in some cases the corank and the codimension form a complete, independent set of *arithmetic* invariants for classifying the elements of m_n.

At this stage, it's natural to inquire to what degree a small perturbation of f introduces a change into the picture presented above. In other words, what kinds of perturbations can be neutralized by a suitable change of coordinate system? This question leads us directly to the idea of a *universal unfolding* of f.

Exercises

1. Compute the determinacy of the following functions: (a) $f(x) = x$, (b) $f(x) = x^n$, (c) $f(x_1, x_2, x_3) = x_1^2 + x_1x_2 + x_2^2 + x_2x_3 + x_3^2 + x_1x_3$, (d) $f(x_1, x_2, x_3) = x_1^2 + \frac{3}{2}x_2^2 + x_2^3 - 3x_2x_3^2$, (e) $f(x_1, x_2) = x_1^2(e^{x_2} - 1)$. (Answers: (a) 1, (b) n, (c) 2, (d) 4, (e) ∞.)

2. Compute the codimension of the following functions: (a) $f(x_1, x_2) = x_1^3 + x_2^3$, (b) $f(x_1, x_2) = x_1^3 - 3x_1x_2^2$, (c) $f(x_1, x_2) = x_1^4 + x_2^4 - 6x_1^2x_2^2$, (d) $f(x_1, x_2, x_3) = x_1^2 + \frac{3}{2}x_2^2 + x_2^3 - 3x_2x_3^2$, (e) $f(x_1, x_2) = x_1^2 + x_2^4$. (Answers: (a) 3, (b) 4, (c) 7, (d) 2, (e) 2.

3. Prove the following inequality relating the corank r and the codimension c: $c \geq \binom{r+1}{2}$.

4. Show that the codimension of the set of symmetric matrices of corank r in the space of all symmetric matrices of order n is $r(r + 1)/2$. (Remark: This result shows that the set of functions of corank r has codimension $r(r + 1)/2$, as well. Do you see why?).

5. Prove that $f(x_1, x_2, \ldots, x_n)$ has infinite codimension if and only if there exists an index i such that x_i^k in not generated for any value of k.

7. *Unfoldings*

We have seen that the basis elements in the factor space $m_n/\Delta(f)$ represent "directions" in m_n that cannot be generated using the first partial derivatives of f. Intuitively, we might suspect that if $p(x)$ is a smooth perturbation of f containing components in these "bad" directions, it would not be possible to remove these components from p by a smooth change of variables. On the other hand, if all the components of p in the "good" directions get intermixed with those same components of f, it's reasonable to suspect that these components of p can be transformed away by an adroit change of variables. The idea of an *unfolding* of f formalizes these speculations.

Roughly speaking, an unfolding of f is a k-parameter family of functions into which we can embed f, and which is stable *as a family.* If k is the smallest integer for which we can find such a family, then the unfolding is called *universal.* Let $\mathcal{F} = \{f_\alpha(x)\}$ be such a universal unfolding of f, with $f_0(x) = f(x)$. Then the family $\mathcal{F} - f$ represents all the perturbations of f that cannot be transformed to zero by means of a smooth coordinate change. Let's see how we can construct such a universal unfolding $\mathcal{F}$, making use of the codimension of f.

Let f be a function with codimension $c < \infty$, and let $\{u_i(x)\}$ be a basis for the space $m_n/\Delta(f)$, $i = 1, 2, \ldots, c$. Thus, the elements $\{u_i(x)\}$ are the directions in m_n that are **not** generated by $f_{,i} = \partial f/\partial x_i$. Then it can be shown that a universal unfolding of f is given by

$$f(x) + \sum_{i=1}^{c} \alpha_i u_i(x), \qquad \alpha_i \in R.$$

In other words, the smallest parameterized family that f can be embedded within in a stable way is a c-parameter family determined by a linear combination of the basis elements of $m_n/\Delta(f)$.

In terms of perturbations to f, this result means that if $\hat{f} = f + p$ is a perturbed version of f, with p expressed as $p = \sum_{i=1}^{\infty} \beta_i \phi_i(x)$, then we can write

$$\hat{f} = f + \sum_{i=1}^{c} \alpha_i u_i(x) + z(x),$$

where

$$z(x) = \sum_{i=1}^{\infty} \beta_i \phi_i(x) - \sum_{i=1}^{c} \alpha_i u_i(x).$$

The Universal Unfolding Theorem implies that there exists a smooth change of variables $x \to y(x)$, such that $z \equiv 0$ in the new coordinates. Consequently, the only part of the perturbation p that cannot be made to vanish by looking at $\hat{f}$ through the right pair of "glasses" is the part corresponding to the missing directions in m_n, i.e., those directions not generated by $\partial f/\partial x_i$, $i = 1, 2, \ldots, n$. These are the so-called essential perturbations; all other perturbations can be made to disappear if we express $\hat{f}$ in the right variables.

Often our function f already belongs to a k-parameter family of functions, and we would like to know how the parameterized family $\{f_\beta(x)\}$, $\beta = (\beta_1, \beta_2, \ldots, \beta_k)$ relates to the universal unfolding of f discussed above, an unfolding that depends upon the integer $c =$ codim f. Without loss of generality, we assume $k > c$, since otherwise $\{f_\beta\}$ would be equivalent to the universal unfolding $\{f_\alpha\}$. The Universal Unfolding Theorem guarantees the existence of the following:

1) A map

$$e\colon R^k \to R^c$$
$$\beta \mapsto \alpha,$$

where e is smooth (but, of course, not a diffeomorphism since $k > c$).

2) A mapping

$$y\colon R^{n+k} \to R^n,$$

such that for each β the map

$$y_\beta\colon R^n \to R^n$$
$$x \mapsto y(x, \beta),$$

is a local diffeomorphism at the origin.

3) A smooth map

$$\gamma\colon R^k \to R,$$

ensuring that the unfolding

$$\{f_\alpha(x)\} \doteq F(x, \alpha) = F(y_\beta(x), e(\beta)) + \gamma(\beta).$$

The maps e, y and γ smoothly reparameterize the unfolding $F(x, \beta)$ into the universal unfolding $F(x, \alpha)$. Note that if $k = c$ and e is a diffeomorphism, then the above situation is just the condition for equivalence of families of functions discussed in Section 3. To pull together all of the above material on codimension and unfoldings, let's look at an example showing how to compute all these various quantities and transformations.

Example: A Cubic Function

Consider the function

$$f(x_1, x_2) = x_1^2 x_2 + x_2^3.$$

The quantities $f_{,1}$ and $f_{,2}$ are given by

$$\frac{\partial f}{\partial x_1} = 2x_1 x_2, \qquad \frac{\partial f}{\partial x_2} = x_1^2 + 3x_2^2.$$

The space $\Delta(f)$ is

$$\Delta(f) = \{g\colon g = \phi_1(x_1, x_2)x_1x_2 + \phi(x_1, x_2)(x_1^2 + 3x_2^2),\ \phi_1, \phi_2 \in m_n\}.$$

After a bit of algebra, it's easy to see that no terms of the form

$$u_1 = x_1, \qquad u_2 = x_2, \qquad u_3 = x_1^2 + x_2^2,$$

belong to $\Delta(f)$, but all other quadratics as well as all higher-order terms are obtainable. Thus codim $f = c = 3$, and the elements $\{x_1, x_2, x_1^2 + x_2^2\}$ form a basis for $m_n/\Delta(f)$. Thus, a universal unfolding for f is given by

$$F(x, \alpha) = x_1^2 x_2 + x_2^3 + \alpha_1(x_1^2 + x_2^2) + \alpha_2 x_2 + \alpha_3 x_1.$$

Now consider the function $g(\hat{x}_1, \hat{x}_2) = \hat{x}_1^3 + \hat{x}_2^3$. It can be shown that g is also a cubic of codimension 3. Going through the above arguments for g, we are led to the unfolding

$$G(\hat{x}, \beta) = \hat{x}_1^3 + \hat{x}_2^3 + \beta_1 \hat{x}_1 + \beta_2 \hat{x}_2 + \beta_3 \hat{x}_1 \hat{x}_2.$$

Since f and g are equivalent via a linear change of coordinates, the two unfoldings $F(x, \alpha)$ and $G(\hat{x}, \beta)$ must also be equivalent.

To see this equivalence, set

$$e \colon R^3 \to R^3$$
$$(\beta_1, \beta_2, \beta_3) \mapsto (\alpha_1, \alpha_2, \alpha_3),$$

where

$$\beta_1 = -2^{2/3}\alpha_1,$$
$$\beta_2 = 2^{-2/3}(\alpha_2 + \alpha_3\sqrt{3}),$$
$$\beta_3 = 2^{-2/3}(\alpha_2 - \alpha_3\sqrt{3}).$$

Further, we choose the maps y and γ as

$$y \colon R^5 \to R^2$$
$$(x_1, x_2, \alpha_1, \alpha_2, \alpha_3) \mapsto (\hat{x}_1, \hat{x}_2),$$

with

$$\hat{x}_1 = 2^{-1/3}\left(\frac{x_2 + x_1}{\sqrt{3}} + \frac{2\alpha_1}{3}\right)$$
$$\hat{x}_2 = 2^{-1/3}\left(\frac{x_2 - x_1}{\sqrt{3}} + \frac{2\alpha_1}{3}\right)$$

and

$$\gamma \colon R^3 \to R$$
$$(\alpha_1, \alpha_2, \alpha_3) \mapsto \frac{-4\alpha_1^3}{27} + \frac{2\alpha_1\alpha_2}{3}.$$

Using these transformations and multiplying out, we find that

$$\hat{x}_1^3 + \hat{x}_2^3 + \beta_1\hat{x}_1\hat{x}_2 + \beta_2\hat{x}_1 + \beta_3\hat{x}_2 = x_1^2 x_2 + x_2^3 + \alpha_1(x_1^2 + x_2^2) + \alpha_2 x_2 + \alpha_3 x_1 + \gamma(\alpha_1, \alpha_2, \alpha_3),$$

as was to be shown.

The importance of this example is that it is far from immediately evident that the two functions f and g, together with their unfoldings F and G, are equivalent. The transformations e, y and γ given above are not difficult to find—once we know they exist—but it requires the machinery of singularity theory to decide the existence issue.

All the concepts presented thus far are drawn together in the proof of one of the major results of recent applied mathematics, the Thom Classification Theorem for elementary catastrophes. The next section takes up this famous result.

Exercises

1. Consider the function $f(x) = x^3$. Prove that the unfolding $F(x, a) = x^3 + ax$ is a universal unfolding of f by explicitly computing the various coordinate transformations as per the example in the text.

2. Write down the universal unfoldings for the functions of Exercise 2 of Section 6.

8. The Thom Classification Theorem

Equivalence under smooth coordinate changes defines an equivalence relation on the set of smooth functions $C^\infty(R^n, R)$: Specifically, two functions $f, g \in C^\infty(R^n, R)$ are equivalent if they can be transformed into each other (up to an irrelevant constant) by a locally smooth change of variables $y \colon R^n \to R^n$. As per the Yin-Yang Problem discussed in Section 1, the big questions now are: (1) How many equivalence *classes* are there? (2) What does a representative of each class look like? Under certain technical restrictions, these questions are answered by the Thom Classification Theorem.

It should be noted at the outset that the problem as stated is still not entirely settled. What the Classification Theorem does is show that the question is mathematically meaningful by imposing natural conditions on the elements of C^∞ that allow us to obtain a *finite* classification. This pivotal result was first conjectured by René Thom in the 1960s, a complete proof being given later by John Mather, based upon key work of Bernard Malgrange. V. I. Arnold has substantially extended the original results, and the implications of the Classification Theorem for applications have been extensively pursued by Christopher Zeeman, Tim Poston, Ian Stewart, Michael Berry and many others. So in a very real sense the Classification Theorem, although rightly attributed to Thom, is the product of a large group of mathematicians and represents a classic example of how fundamental mathematical advances are put together, piece by piece, in a collective international effort. Now let's turn to a statement of the Theorem itself.

Recall that Morse's Theorem told us that all functions were equivalent to quadratics near a nondegenerate critical point. Clearly, since quadratic functions themselves certainly fall into this category, only functions of order 3 and higher can possibly have degenerate critical points. Thom discovered that if we restrict attention to functions $f \in O(|x|^3)$ of codimension less than 6, then we can finitely classify this subset of $C^\infty(R^n, R)$. In addition, the codimension c and the corank r parametrize the equivalence classes, i.e., given a function f, the integers c and r determine the set of all functions that are locally equivalent to f at the origin—as long as $c \leq 5$. Here is a full statement of this fundamental result:

THOM CLASSIFICATION THEOREM. *Up to multiplication by a constant and addition of a nondegenerate quadratic form, every $f \in C^\infty(R^n, R)$ of codimension $c \leq 5$ is smoothly equivalent near the origin to one of the standard forms listed below in Table 2.1.*

Table 2.1. The Thom Classification of Smooth Functions

Corank/Codim	Function	Universal Unfolding
1/1	x^3	$x^3 + a_1 x$
1/2	x^4	$x^4 + a_1 x^2 + a_2 x$
1/3	x^5	$x^5 + a_1 x^3 + a_2 x^2 + a_3 x$
1/4	x^6	$x^6 + a_1 x^4 + a_2 x^3 + a_3 x^2 + a_4 x$
1/5	x^7	$x^7 + a_1 x^5 + a_2 x^4 + a_3 x^3 +$ $a_4 x^2 + a_5 x$
2/3	$x_1^3 - 3x_1 x_2^2$	$x_1^3 - 3x_1 x_2^2 + a_1(x_1^2 + x_2^2) +$ $a_2 x_1 + a_3 x_2$
2/3	$x_1^3 + x_2^3$	$x_1^3 + x_2^3 + a_1 x_1 x_2 + a_2 x_1 +$ $a_3 x_2$
2/4	$x_1^2 x_2 + x_2^4$	$x_1^2 x_2 + x_2^4 + a_1 x_1^2 + a_2 x_2^2 +$ $a_3 x_1 + a_4 x_2$
2/5	$x_1^2 x_2 + x_2^5$	$x_1^2 x_2 + x_2^5 + a_1 x_1^2 + a_2 x_2^2 +$ $a_3 x_1 + a_4 x_2 + a_5 x_2^3$
2/5	$x_1^2 x_2 - x_2^5$	$x_1^2 x_2 - x_2^5 + a_1 x_1^2 + a_2 x_2^2 +$ $a_3 x_1 + a_4 x_2 + a_5 x_2^3$
2/5	$x_1^3 + x_2^4$	$x_1^3 + x_2^4 + a_1 x_1 + a_2 x_2 +$ $a_3 x_1 x_2 + a_4 x_2^2 + a_5 x_1 x_2^2$

The Thom Theorem is of great theoretical and practical importance in just the same way that the Jordan Canonical Form Theorem is important in matrix theory. Both results consider a broad class of important mathematical objects, showing how we can constructively determine whether or not any two elements from this class can be transformed into each other by means of an appropriate change of coordinates. In the case of matrices, the coordinate changes are linear transformations themselves (i.e., matrices), while in the case of smooth functions the coordinate transformations are origin-preserving diffeomorphisms. Furthermore, each theorem provides us with a canonical representative from each class of equivalent objects, a representative containing all the structure present in every element in its class. Here by "structure" we mean all those properties that are left unchanged by coordinate transformations, i.e., the invariant—or coordinate-free—properties. From a system-theoretic point of view, these are the only properties that can rightly be regarded as properties of the system itself; hence, the importance of such classification theorems.

There are a number of amplifying remarks surrounding the meaning of Thom's fundamental result.

1) Table 2.1 omits functions that either have no critical point at the origin or have a nondegenerate critical point. In the first case, f is equivalent to any one of its linear coordinates. Morse's Theorem covers the second case.

2) We saw in the Splitting Lemma that every f of corank r could be written as $f = g(x_1, x_2, \ldots, x_r) + Q(x_{r+1}, \ldots, x_n)$, where $g \in O(|x|^3)$ and the function Q is a nondegenerate quadratic form. The column labeled "Function" in Table 2.1 shows only the first part of this decomposition and omits the quadratic form.

3) The terms in the universal unfolding are not unique; what is unique is the integer c, the codimension of f, which represents the *number* of such terms.

4) For functions of corank 1, the codimension suffices to determine the class of f; for corank 2, the codimension plus additional information is needed, the details of which would take us too far afield to go into here. The catastrophe theory literature cited in the Notes and References gives an account of how these corank 2 cases are obtained.

5) In codimensions greater than 5, the standard forms are no longer finite in number but may contain parameters (moduli). The classification in these cases bears analogy with the classification of matrices by similarity, where the characteristic values and the size of the Jordan blocks are the similarity invariants, implying an infinite classification. The situation is much the same in spirit, but far more difficult technically, for functions of codimension greater than 5.

For applications, the most important parts of the Classification Theorem are the canonical unfoldings and the behavior of the critical points of f as we vary the unfolding parameters $\{\alpha_i\}$. This circle of questions leads to what is usually termed (elementary) *catastrophe theory*, which we shall take up in more detail in later sections. For now, let's look at an example from the field of electrical power engineering illustrating many of the points we've developed thus far.

9. Electric Power Generation

Consider an electric power supply network with n generators. If we assume no transfer conductance between generators, the equations governing such a network are

$$M_i \frac{d\omega_i}{dt} + d_i\omega_i = \sum_{\substack{j=1 \\ j\neq i}}^{n} E_i E_j B_{ij}[\sin\delta_{ij}^* - \sin\delta_{ij}],$$

$$\frac{d\delta_i}{dt} = \omega_i, \qquad i = 1, 2, \ldots, n.$$

Here

$\omega_i =$ angular speed of rotor i,

$\delta_i =$ electrical torque angle of rotor i,

$M_i =$ angular momentum of rotor i,

$d_i =$ damping factor for rotor i,

$E_i =$ voltage of generator i,

$B_{ij} =$ short circuit admittance between generators i and j,

$\delta_{ij} = \delta_i - \delta_j$,

$\delta_{ij}^* =$ the stable steady-state value of δ_{ij},

$\omega_{ij} = \omega_i - \omega_j$.

Our interest is in studying the behavior of the equilibrium values of ω_i and δ_i as a function of the parameters M_i, E_i, B_{ij}, and d_i.

If we define $a_{ij} = d_i - d_j$, $b_{ij} = E_i E_j B_{ij}$, then it can be shown that the function

$$V(\omega_{ij}, \delta_{ij}) = \sum_{i=1}^{n-1} \sum_{k=i+1}^{n} \left[\tfrac{1}{2} M_i M_k \omega_{ik}^2 - a_{ik}\delta_{ik} - b_{ik}(M_i + M_k)\cos\delta_{ik}\right] - \sum_{i=1}^{n} \sum_{\substack{j=1 \\ j\neq i}}^{n-1} \sum_{\substack{k=j+1 \\ k\neq i}}^{n} M_i b_{jk} \cos\delta_{jk},$$

is a Lyapunov function for the above dynamics. We can use the function V to study the behavior of the network near an equilibrium employing the techniques of singularity theory introduced above.

To illustrate the basic ideas, consider the simplest case of $n = 2$ generators. In this case we have only two basic variables, $\omega_{12} \doteq x_1$, $\delta_{12} \doteq x_2$. The function V then reduces to

$$V(x_1, x_2) = \tfrac{1}{2} M_1 M_2 x_1^2 - a_{12} x_2 - b_{12}(M_1 + M_2)\cos x_2 + K,$$
$$\doteq \tfrac{1}{2}\alpha x_1^2 - \beta x_2 - \gamma \cos x_2 + K.$$

The local behavior of V near an equilibrium is not affected by the constant K, so we set $K = 0$. The critical points of V are at

$$x_1^* = 0, \qquad x_2^* = \arcsin\left(\frac{\beta}{\gamma}\right).$$

Using the rules of singularity theory, it can be shown that V is 2-determinate with codimension 0 if $\alpha \neq 0$ and $\gamma \neq \pm\beta$. The condition on α is necessary for the problem to make sense, so the only interesting possibility for a degeneracy in V occurs when we have $\gamma = \pm\beta$. If $\gamma \neq \pm\beta$, then V is equivalent to a Morse function in a neighborhood of its critical point and can be replaced by its 2-jet

$$\tfrac{1}{2}\alpha x_1^2 - \left(\sqrt{\gamma^2 - \beta^2}\right) x_2^2,$$

a simple Morse saddle. So let's assume $\gamma = \beta$. (The case $\gamma = -\beta$ is similar.)

The function V now assumes the form

$$V(x_1, x_2) = \tfrac{1}{2}\alpha x_1^2 - \beta(x_2 + \cos x_2).$$

Some additional calculations with V show that the corank and codimension are both 1, indicating that the canonical form for V is the so-called fold catastrophe. Thus, a universal unfolding of V in this critical case is

$$\tilde{V} = \frac{x_2^3}{3} + t x_2,$$

where t is the unfolding parameter.

From this analysis, we conclude that an abrupt change in the stability of the power network can only be expected when $\gamma = \beta$. In this case, the system's potential function is locally equivalent to $\tilde{V}$. The stability properties of the system will change abruptly when the mathematical parameter t, which is dependent upon the physical parameters α, β and γ, passes through the value $t = 0$. Results of this sort are routine to obtain by standard arguments in singularity theory, and will play an increasingly important role in engineering calculations in the future as the methods become more widespread.

10. Bifurcations and Catastrophes

Each of the unfoldings in Table 2.1 represents a c-parameter family of functions in the variables $x = (x_1, x_2)$. For most values of the parameters the corresponding unfolding $F(x, \alpha)$ has a nondegenerate critical point at the origin, so the local behavior is governed by Morse's Theorem. Of considerable theoretical and applied interest are those values of the parameters for which the origin is a degenerate critical point. The set of such parameter values is called the *bifurcation set* of the function $f(x)$, and it plays an important role in applications of singularity and catastrophe theory in physics, engineering and biology.

The bifurcation set $\mathcal{B}$ is given by

$$\mathcal{B} = \left\{ \alpha_k \colon \frac{\partial F}{\partial x} = 0, \quad \det \left[\frac{\partial^2 F}{\partial x^2} \right] = 0, \ k = 1, 2, \ldots, c \right\}.$$

Note that the definition of $\mathcal{B}$ involves two conditions on the function family $F(x, \alpha)$, which contains $r + c$ variables, where r = corank f, and c = codim f. As long as $r \leq 2$, we can always use these conditions to eliminate the x-dependence and express the bifurcation set $\mathcal{B}$ solely in terms of the parameters $\{\alpha_i\}$.

Example: The Cusp

Consider the famous "cusp" situation when $r = 1$, $c = 2$. In this case, a universal unfolding is given by

$$F(x, \alpha_1, \alpha_2) = x^4 + \alpha_1 x^2 + \alpha_2 x.$$

Consequently, the bifurcation conditions are

$$\frac{\partial F}{\partial x} = 0, \qquad \frac{\partial^2 F}{\partial x^2} = 0,$$

which lead to the equations

$$\begin{aligned} 4x^3 + 2\alpha_1 x + \alpha_2 &= 0, \\ 12x^2 + 2\alpha_1 &= 0. \end{aligned}$$

Eliminating x, we find that

$$\mathcal{B} = \left\{ (\alpha_1, \alpha_2) \colon 8\alpha_1^3 + 27\alpha_2^2 = 0 \right\}.$$

The geometrical picture of $\mathcal{B}$ in the (α_1, α_2)-plane is shown in Fig. 2.5. Here we see the two branches of the solution set of the equation $8\alpha_1^3 + 27\alpha_2^2 = 0$ coming together at the "cusp" point at the origin. This type

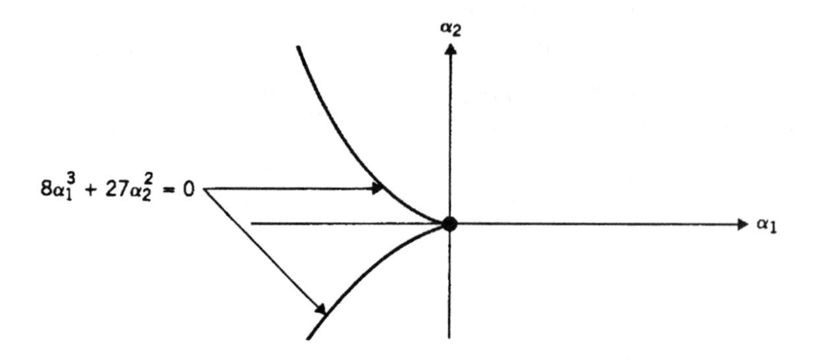

Figure 2.5. The Cusp Geometry

of geometrical structure in $\mathcal{B}$ serves to motivate the names like the "fold," "cusp," "butterfly" and "swallowtail" traditionally attached to the various classes in Table 2.1.

If we look at the solutions of the equation $\partial F/\partial x = 0$ as a function of the variable x *and* the parameters α, then we obtain a surface M in R^{r+c} that can be shown to be a smooth manifold. Each of the classes of functions in Table 2.1 has its own characteristic geometry for the surface M. But these geometries are related within the family of cuspoids or the family of umbilics in that as we pass from lower to higher codimension within a family, each manifold M is a proper submanifold of the higher-order singularity. The simplest version of this kind of dependence occurs for the fold and cusp geometries displayed in Figs. 2.6–2.7.

In the case of the fold, the equation for M is $3x^2 + \alpha = 0$, which leads to the solution $x^2 = \pm\sqrt{-\alpha/3}$. This is displayed as the dashed curve in Fig. 2.6. Passing to the cusp, the equation for M is $4x^3 + 2\alpha_1 x + \alpha_2 = 0$. The geometry is shown in Fig. 2.7, where we see how the fold geometry is embedded within the cusp by the fold lines. However, we see that the additional parameter in the cusp gives rise to a much richer geometry than for the fold, enabling us, for instance, to pass smoothly from one value of $x^*(\alpha_1, \alpha_2)$ to another on M when $\alpha_1 > 0$. Comparison with the fold geometry in Fig. 2.6 shows that such a passage is not possible when there is only a single parameter to vary.

In passing, we note that the bifurcation set $\mathcal{B}$ for the cusp as shown in Fig. 2.5 can be obtained by projecting the manifold M of Fig. 2.7 onto the (α_1, α_2)-plane. The bifurcation sets for the other singularity classes in Table 2.1 can also be obtained by projecting their characterizing manifold M onto the parameter space in the same manner.

A different way of viewing this setup is to regard the determination of

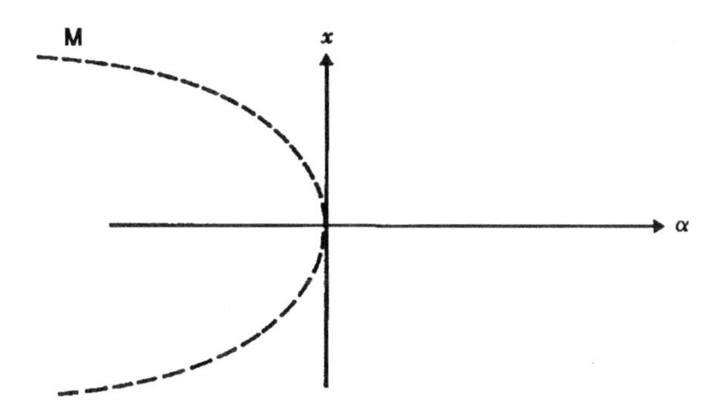

Figure 2.6. The Fold Manifold M

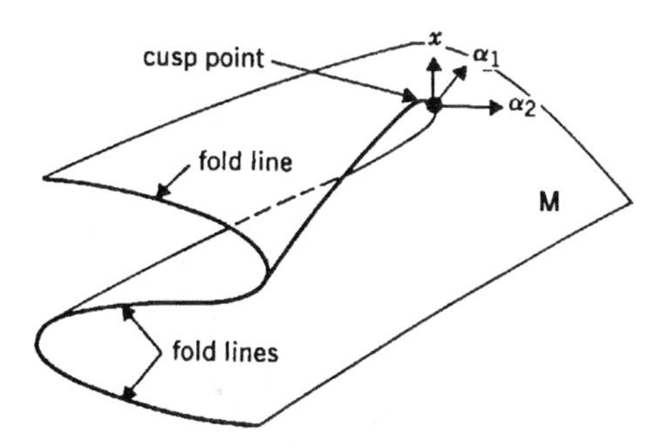

Figure 2.7. The Cusp Manifold M

the manifold $\mathcal{B}$ as the problem of finding a map

$$\chi \colon R^c \to R^r,$$
$$(\alpha_1, \alpha_2, \ldots, \alpha_c) \mapsto (x_1, x_2),$$

defined by the equation grad $F(x, \alpha) = 0$. The values of $\alpha \in R^c$ at which the map χ is discontinuous are exactly the points where the solution x^* acquires a new degree of degeneracy. Thom has termed these critical parameter values *catastrophes,* since they usually represent places where the physical process described by $F(x, \alpha)$ undergoes some sort of dramatic change in its behavior.

Great mathematics seldom comes from idle speculation about abstract spaces and symbols. More often than not it is motivated by definite questions arising in the worlds of nature and humans. And so it is with catastrophe theory, too, which was originally inspired by Thom's speculations about the developmental process of fertilized cells leading to the adult form of living things. In particular, Thom believed that under certain circumstances the shapes seen in the various catastrophe surfaces like the cusp and fold may manifest themselves in living organisms during the course of their embryological development. To illustrate these ideas, let's turn our attention from hard-core engineering and algebraic calculations to soft-core biology and geometry, showing how the pictorial representation of an unfolding of a singularity suggests a kind of language by which to describe the dynamical processes of embryology.

Thom's biological ideas are illustrated in Fig. 2.8, which shows two "unfoldings" side-by-side, one from mathematics, the other from biology. On the left side of the diagram, we see the unfolding of a sea urchin embryo as the fertilized zygote moves from the late blastula to the late gastrula stage in the developmental process termed gastrulation. The right side of the diagram shows various sections of the unfolding of the elliptic umbilic catastrophe (the first corank 2 catastrophe listed in Table 2.1) as the unfolding parameter a_1 moves from positive to negative values.

It's tempting to speculate along with Thom that there is some fundamental correspondence between these forms. On the other hand, most biologists are suspending judgment until the many experiments needed to correlate the control parameters in these catastrophe models with the spatiotemporal forms of living things have been carried out. Now let's leave the theory again and examine some illustrative uses of the preceding machinery for addressing questions of interest in natural resource systems.

Exercises

1. Consider the butterfly catastrophe associated with the unfolding $f(x, a, b, c, d) = x^6 + ax^4 + bx^3 + cx^2 + dx$. (a) Draw the sections of the bifurcation set corresponding to the cases a and b constants. (b) Show that this section has three cusps if the roots of the equation $20x^3 + 4ax + b = 0$ are all real, and that this will occur when $b^2 + 4(4a/3)^3 < 0$. (Remark: This condition can be satisfied only if a is negative. For this reason, the parameter a is called the "butterfly" factor, suggesting that it is by changes in a that we progress from a simple cusp to a surface in which an "intermediate" stable mode of behavior is possible.)

2. Use the geometrical picture of the bifurcation set to argue that, biologically speaking, the cusp can be interpreted spatially as a pleat or fault, while its temporal interpretation is to separate (or unite). Can you see why the butterfly is interpreted spatially as a pocket? What do you think its temporal interpretation might be?

3. In addition to biology, Thom has speculated about how the elementary catastrophes could be used to characterize features of language like verbs, nouns, subjects, objects and so forth. Think about how to interpret the cusp bifurcations in this context.

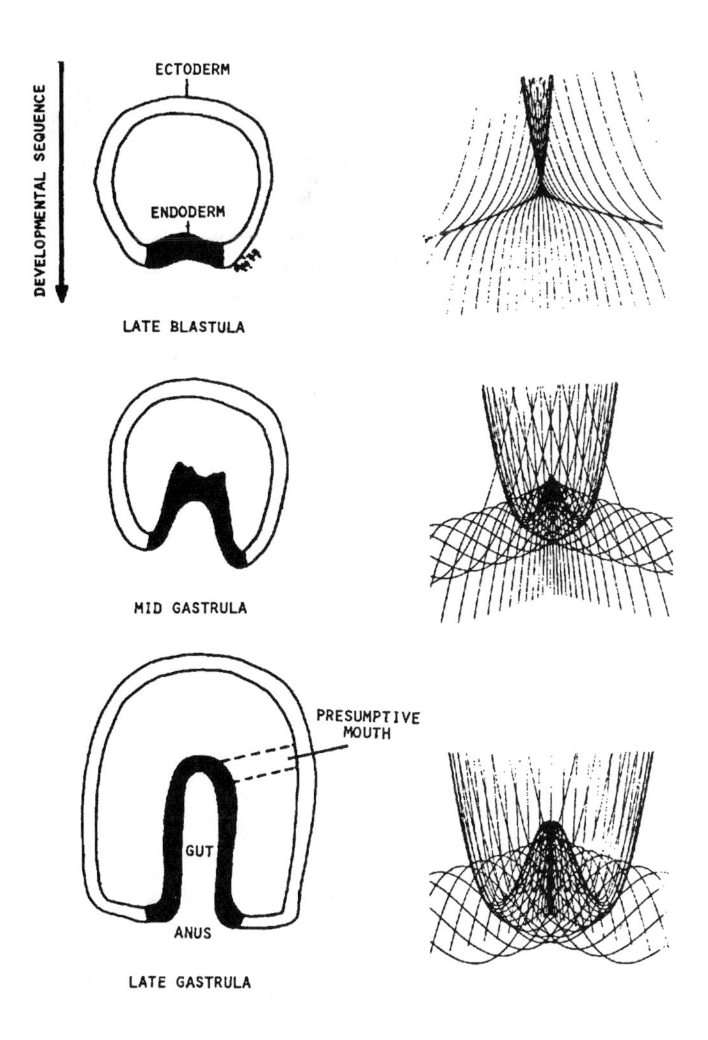

Figure 2.8. Unfoldings of a Sea Urchin Embryo and the Elliptic Umbilic Catastrophe

11. Harvesting Processes

There are two qualitatively distinct ways of making use of singularity and catastrophe theory in applications. These radically different ways depend upon the assumptions we can justify making regarding our knowledge of the equation of state governing the system's behavior. The first way, which we might term the *physical way,* assumes that a function $F(x, \alpha)$ is known, where $x \in R^n$ represents the system's observed behavioral states and $\alpha \in R^k$ is a vector of parameters (or inputs) determining the states. Using the known mathematical form of the function F, the machinery developed above can be employed to study the behavior of the critical points as functions of the inputs, as well as to identify those special values or regions in parameter space where the system undergoes various interesting types of behavioral discontinuties.

But very often the function F is not known, and we have knowledge only of the observed pattern of parameter values and the corresponding system behavior, i.e., we know the input/output behavior of the system but *not* the actual input/output map. Such situations are especially common in the social and behavioral sciences, where there are no laws of nature of the scope and majesty of those found in physics and engineering upon which to base an equation of state F. In these cases, we employ the *metaphysical way* of catastrophe theory, blithely **assuming** that the system variables x and α are given in the canonical coordinate system, with the number of x-variables equaling the corank of the unknown equation of state and the number of input parameters equaling the codimension. Furthermore, the metaphysical way also assumes that in applying the Splitting Lemma, we can ignore the nondegenerate quadratic form Q. This is tantamount to assuming that the essential system behavior is carried only by the higher-order, nonlinear part of the decomposition. The underlying justification for these assumptions is the invariance under smooth coordinate changes of most of the important properties associated with the system's critical points. Consequently, the behavior in the canonical model will be *qualitatively* the same as the behavior in whatever coordinate system is used to originally describe the problem. The example given earlier on electric power generation networks illustrates the physical way for applying the catastrophe ideas. In this section, we explore the metaphysical way.

A commonly occurring situation in the management of natural resource systems arises when we have a resource of some kind—fish or timber, for example—that if left to itself grows at some natural rate dependent upon the environmental carrying capacity and its normal fecundity. However, this resource has some economic value and is regularly harvested for human use. Due to technological advances, as well as laws of supply and demand, there are incentives for many harvesters to get into the business with high-tech equipment to reap the rewards of a high harvest. On the other hand, if too many harvesters (or harvesters that are too efficient) are allowed unrestricted access to the resource, the stock size declines to the point where the resource is unable to sustain itself and the whole industry becomes in danger of collapse. This situation is well documented, for example, in the harvesting of whales in the Pacific or anchovies off the coast of Peru. The first question of interest is how to balance the short- and long-term economics of the harvesting situation in order to construct a policy that serves both the economic needs of the harvesters and preserves the resource. A related question, and the one that we will address here, is how the equilibrium levels of the resource are affected by both the number of harvesters and the technical efficiency of their equipment. For the sake of definiteness, we formulate our discussion in terms of fishery harvests, although it should be clear that the arguments apply to many other types of natural resource processes.

Since we generally don't know the precise mathematical relationship between the equilibrium stock level and the levels of fishing effort and efficiency, we will invoke the metaphysical way of catastrophe theory and just assume that the fishing situation of interest can be described by the following input/output variables:

- *Input 1*: The level of fishing investment as measured by fleet size (the number of fishing boats or harvesting units).
- *Input 2*: The technological and economic efficiency measured in terms of the catchability coefficient and the capacity per boat, which are determined by the lowest stock size for which it is still economically attractive to continue fishing.
- *Output*: The stock size, which may be measured in many ways, and that may consist of several geographical substocks or even a mixture of many species.

Using the above variables and the metaphysical way, we assume that the unknown relationship between the stock size and the inputs is a function of codimension 2 and corank 1, i.e., if we let the inputs be denoted by α_1, α_2, while taking the output to be x, then there is a relationship $f(x, \alpha_1, \alpha_2) = 0$, such that codim $f = 2$ and corank $f = 1$. Since we are already being munificent with our assumptions, let's also assume that f is smooth. Then we are in the situation covered by the Thom Classification Theorem, and can assert that whatever form f may happen to have in the natural physical variables, there exists a smooth change of coordinates such that in the canonical mathematical variables, $f \to \hat{f} = \frac{1}{4}\hat{x}^4 + \frac{1}{2}\hat{\alpha}_1\hat{x}^2 + \hat{\alpha}_2\hat{x}$. Consequently, if our interest is only in the properties of f near the equilibrium stock levels, we can use the function $\hat{f}$. However, since we're making life easy anyway, we might as well assume that the natural coordinate system of the problem for x, α_1 and α_2 is also the canonical coordinate system, implying that we can just set $f = \hat{f}$.

In view of the above considerations, we know that the cusp geometry governs the steady-state stock size as a function of the harvesting effort and efficiency. Thus we can invoke the Classification Theorem to produce the geometrical depiction of this input/output relationship shown in Fig. 2.9, which is taken from real data on the antarctic fin and blue whale stocks during the period 1920–1975. The reader will recognize this figure as a relabeled version of the canonical geometry given earlier in Fig. 2.7. The arrows from the origin show the equilibrium stock level trajectory for different fleet sizes and catching efficiency levels. It is imperative to note that this trajectory is most definitely *not* the usual time-history of the stock levels that emerges as the inputs are varied. Rather, the trajectory shown in Fig. 2.9 should be interpreted in the following sense: First fix the fleet size and efficiency at given levels. Now observe the equilibrium stock level that results. This gives *one point* on the cusp manifold. We next change the inputs to new levels and wait for the stock level to again stabilize. This gives a second point on the cusp manifold. Continuing in this manner, we trace out the entire manifold shown in Fig. 2.9. Thus, the cusp manifold is the set of all possible *steady-state* levels that the stock can assume. So the curve shown is a curve of equilibrium positions for stock sizes, having nothing to do with the behavior of the system in its transient states as it moves toward equilibrium. To deal with the transient behavior, we need the much deeper theory of bifurcation of *vector fields,* bits and pieces of which we shall discuss later in the chapter.

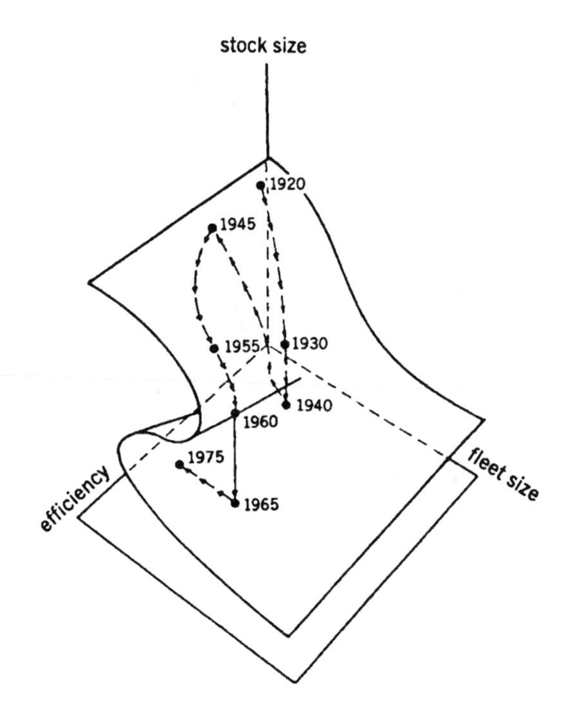

Figure 2.9. Stock Size Equilibria for the Antarctic Fin and Blue Whales

We can gain considerable understanding of fishery management practices and policies from an interpretation of the behavior shown in Fig. 2.9. The period prior to 1930 was characterized by expanding fishing effort, but with relatively low technological efficiency (e.g., shore-based processing plants). There is no clear evidence of catastrophic stock declines until the late 1930s. When the fishing industry came back to life after World War II, technological change was rapid, involving factory ships and fleets of highly mobile catcher boats. The catastrophic stock declines during the 1950s and 1960s were characterized by catches that remained very high until stocks became quite low. In this period the high catches were limited by quotas rather than vessel capacities.

The implications of this sort of analysis for fishery harvesting policy is quite clear. If catastrophic behavior is to be avoided in the face of increasingly sophisticated technological advances and continued government subsidies to fleets, then precise monitoring and control policies designed to avoid the fold lines in Fig. 2.9 are of vital importance. Since this seems to be difficult for a variety of technical and economic reasons, it may be necessary to create alternative schemes based upon a deeper consideration of the cusp geometry.

Such policies might involve avoidance of the cusp region by forcing a reduction in fishing efficiency, or broadening the cusp region by, for example, regulating the fleet size. A third possibility is to deliberately induce a sequence of small catastrophes by following a policy leading to many crossings of the fold lines, each crossing taking the system from a region of high stocks to low, or vice versa. Such a scheme would still preserve viability of the fishery system if the drops from the upper to the lower sheet were arranged to take place rather near the cusp point in the input space. One way to accomplish this would be to increase the harvesting capacity per vessel, while decreasing the total number of boats. Another would be to increase the stock size below which fishing is uneconomical. This is a very novel idea to most fishery managers, since it involves deliberately inducing a discontinuity in the equilibrium stock levels, a discontinuity that would oscillate between low and high populations. The political, social and psychological consequences of such actions go far beyond historical mangement wisdom; nevertheless, such deliberately induced "booms and busts" may ultimately turn out to be the only way to avoid the kind of catastrophic collapse of the fishing grounds that managers (and fishermen) want to avoid above all else. This situation is somewhat reminiscent of that encountered in earthquake prevention, in which it is often desirable to pump water into the fault lines to precipitate minor tremors and relieve the tension in the fault before it can build up to the point of generating a devastating quake.

Exercise

1. Comment upon the following research program aimed at developing system management procedures for coping with surprise in natural resource systems:

- Development of methods for stabilizing the system *outputs* without necessarily stabilizing all the state variables. This involves policies that decouple the outputs from some of the states.

• Design of "probing" policies involving variable control actions. Note that such policies are directly contrary to the objective of stabilizing system outputs.

• Identification of key variables, so that monitoring these variables provides the same information as monitoring the entire system state. This is a problem of observability of the sort we'll take up more fully in Chapter 6.

12. Estimation of Catastrophe Manifolds

Before catastrophe-theoretic models can be regarded as established tools of science, it's necessary to test them with real data. What this involves is the development of statistical procedures for estimating the values of the unfolding parameters and system outputs from observational data—under the assumption that the inputs and outputs do indeed satisfy one of the canonical catastrophe equations. In this section we present a statistical procedure due to Loren Cobb that enables us to estimate this relationship for cusp catastrophe models on the basis of the observed data.

Cobb's procedure starts with the following parameterization of the cusp manifold M:

$$(\alpha_0 + \alpha_1 z_1 + \alpha_2 z_2) + (\beta_0 + \beta_1 z_1 + \beta_2 z_2)(x - \lambda) + \delta(x - \lambda)^3 = 0. \qquad \text{(i)}$$

Here z_1 and z_2 are the *measured* input variables, while x is the measured output (behavioral) variable. The canonical control parameters for the cusp, a_1 and a_2, are given in terms of z_1 and z_2 by

$$\begin{aligned} a_1(z_1, z_2) &= \alpha_0 + \alpha_1 z_1 + \alpha_2 z_2, \\ a_2(z_1, z_2) &= \beta_0 + \beta_1 z_1 + \beta_2 z_2. \end{aligned} \qquad \text{(ii)}$$

The cusp point occurs when $a_1(z_1, z_2) = a_2(z_1, z_2) = 0$ and $x = \lambda$. Thus, the quantity λ is a "location" parameter, telling us where this singular point lies in the x-direction. The constant δ is a kind of "concentration" parameter, in the sense that the larger the value of δ, the more concentrated the data is in the vicinity of the cusp manifold. The bifurcation set $\mathcal{B}$ in this case is

$$\mathcal{B} = \{(z_1, z_2) \colon 27a_1^2(z_1, z_2) + 4a_2^3(z_1, z_2) = 0\}. \qquad \text{(iii)}$$

Let's emphasize that here we are assuming that the canonical control parameters a_1 and a_2 are linear functions of the measured control variables, and that the only transformation of the behavioral variable that's allowed is a simple translation of the zero point by an amount λ.

The statistical problem is now to estimate the eight parameters $\alpha_0, \alpha_1, \alpha_2, \beta_0, \beta_1, \beta_2, \delta, \lambda$ from measurements of the quantities z_1, z_2 and x. For this estimation, Cobb proposes the following algorithm:

1. Standardize each measured variable to zero mean and unit variance using standard statistical transformations of the measured values of each variable.

2. For each triple of observations (z_1, z_2, x), compute the following quantities:

$$\phi_1 = 1, \qquad \phi_2 = z_1, \qquad \phi_3 = z_2, \qquad \phi_4 = x,$$
$$\phi_5 = z_1 x, \qquad \phi_6 = z_2 x, \qquad \phi_7 = x^2, \qquad \phi_8 = x^3.$$

3. Compute the 8×8 matrix of sums

$$M_{ij} = \sum \phi_i \phi_j, \qquad i, j = 1, 2, \ldots, 8.$$

Here the summation is taken over **all** observed values of z_1, z_2 and x.

4. Compute the vector q whose components are given by:

$$q_1 = 0, \qquad q_2 = 0, \qquad q_3 = 0, \qquad q_4 = \sum \phi_1,$$
$$q_5 = \sum \phi_2, \qquad q_6 = \sum \phi_3, \qquad q_7 = 2\sum \phi_4, \qquad q_8 = 3\sum \phi_7.$$

5. Solve the linear algebraic equation $Mb = q$, obtaining the vector b.

6. We now take $\delta = b_8$ and $\lambda = -b_7/(3b_8)$ as our estimates of δ and λ. We also have

$$\begin{pmatrix} a_1(z_1, z_2) \\ a_2(z_1, z_2) \end{pmatrix} = \begin{pmatrix} b_1 + b_4\lambda + b_7\lambda^2 + b_8\lambda^3 \\ b_4 + b_7\lambda \end{pmatrix} + \begin{pmatrix} b_2 + b_5\lambda & b_3 + b_6\lambda \\ b_5 & b_6 \end{pmatrix} \begin{pmatrix} z_1 \\ z_2 \end{pmatrix}$$

7. We are now in a position to fix the directions that the axes point in the control parameter space. For the axis in the a_2-direction, we solve the equation $a_1(z_1, z_2) = 0$, while the solution of the equation $a_2(z_1, z_2) = 0$ sets the direction of the a_1-axis.

8. The bifurcation set $\mathcal{B}$ is obtained by solving Eq.(iii) for fixed values of z_1.

9. The estimated values of the behavioral variable x for given values of the measured controls z_1 and z_2 are found by solving Eq.(i).

10. Since all quantities were standardized in step 1, we destandardize the results obtained in steps 6–9.

There are a few points to note about the foregoing algorithm:

- The standardization in step 1 is important for computational accuracy. Do **not** omit it.
- The computations called for in steps 8 and 9 involve finding the roots of a cubic polynomial. There are standard algorithms available for doing this.
- If the estimated value of δ turns out to be negative or zero, then the cusp model is not appropriate for this set of data.

Example: Birth Rate Decline in Developing Nations

Cobb has applied the above algorithm to test whether or not the cusp can be used as an effective model for capturing the transition from high to low birth rates in developing countries. Historically, this transition has been preceded by declining death rates (from public health efforts) and increased literacy rates (from greater attention to public education). The question is see whether these two variables can serve as inputs for a cusp catastrophe model that explains and predicts transitions to lower birth rates.

To test this hypothesis, data were obtained for all 66 nations listed in the *World Almanac* having populations over one million people and an adult literacy rate of at least 50%. Birth and death rates were measured in units of live births or deaths per 1,000 inhabitants per year. Employing the above algorithm, the control space shown in Fig. 2.10 below was obtained. This calculation also led to the numerical results given in Table 2.2. On the basis of this exercise, one might speculate that North Korea is about to undergo a transition to dramatically lower birth rates.

Table 2.2. Predicted Birth Rates in Cusp Model

Country	Literacy	Death Rate	Birth Rate	(Predicted)
Argentina	93	9.4	22.9	(17.9)
Australia	98	8.7	18.4	(17.7)
Greece	84	8.9	15.6	(37.3, 20.4)*
Indonesia	60	16.9	42.9	(45.5, 16.3)*
Japan	99	6.5	18.6	(19.6)
Kuwait	55	5.3	47.1	(45.6)
North Korea	85	9.4	35.7	(19.5)
Mexico	76	8.6	42.0	(41.6)
Portugal	70	11.0	19.6	(43.1, 22.3)*

* (Two predicted values because there are two stable equilibria for these estimated control values)

The fish harvesting example of Section 11, coupled with the empirical techniques outlined in this section, show how we can use the metaphysical way of catastrophe theory to address some real-world policy issues in the natural resource management area. Now let's take a look at how actual knowledge of the system's underlying equation of state can enable us to employ the physical way to determine effective management policies for another kind of natural resource system, a spruce forest.

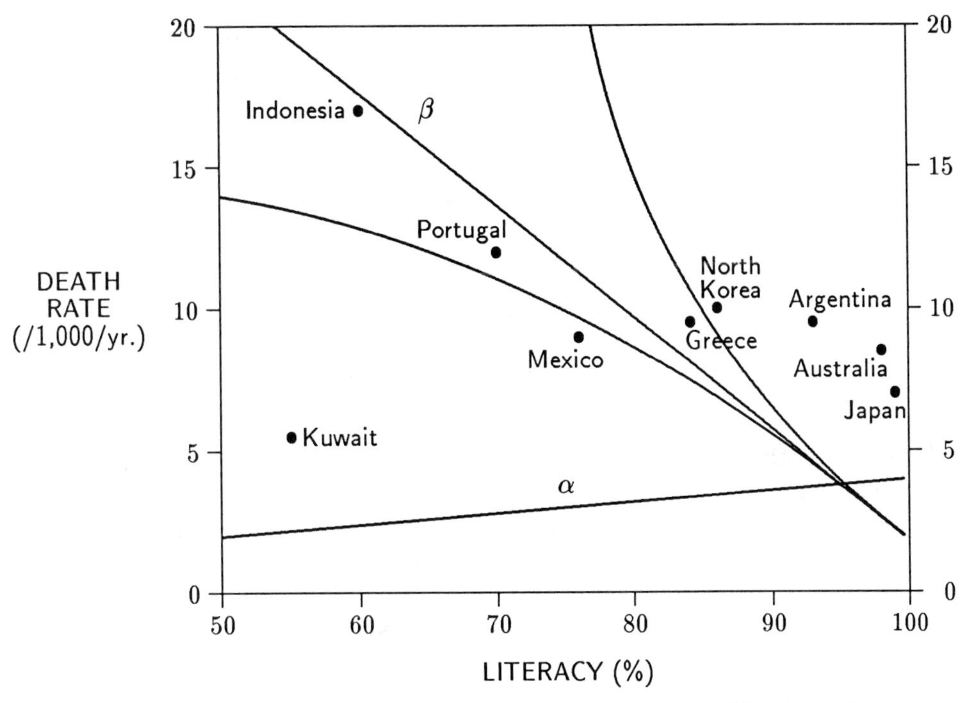

Figure 2.10. Cusp Parameter Space for Birth Rate Transition Model

Exercises

1. We know that as a and b range over the real numbers, the equilibria of the potential function $f(x, a, b) = \frac{1}{4}x^4 + \frac{1}{2}ax^2 + bx$ generate a surface having the cusp geometry. Verify this fact by first producing "pseudo" observations of a, b and x, and then using the numerical scheme of this section to estimate the cusp surface.

2. Consult a world almanac or other statistical database for data to test the hypothesis that the cusp geometry is a good way to model a nation's gross national product as a function of its population and number of schools.

13. Forest Insect Pest Control

To illustrate how much more detailed our analysis can be when we actually do know the relationship between the system control inputs and the outputs, let's examine a problem involving the growth and death of a forest insect pest, the spruce budworm, that infests the spruce forests of eastern North America. Budworm outbreaks occur irregularly at 40 to 80-year intervals, and may expand to cover many million hectares of forest in a few years. We will investigate the feasibility of generating controlling actions that could permanently suppress these outbreaks and their consequent ecological and economic disruptions.

Neglecting spatial dispersion factors, the dynamical equations governing the birth, growth and death of the spruce budworm at a given geographical site have been postulated to be

$$\frac{dB}{dt} = \alpha_1 B \left[1 - B\frac{(\alpha_3 + E^2)}{\alpha_2 S E^2}\right] - \frac{\alpha_4 B^2}{(\alpha_5 S^2 + B^2)},$$

$$\frac{dS}{dt} = \alpha_6 S \left[1 - \frac{\alpha_7 S}{\alpha_8 E}\right],$$

$$\frac{dE}{dt} = \alpha_9 E \left[1 - \frac{E}{\alpha_7}\right] - \frac{\alpha_{10} B E^2}{S(\alpha_3 + E^2)}.$$

Here $B(t)$ is the budworm density in the site, $S(t)$ is the total surface area of the branches in the stand of trees in the site, and $E(t)$ represents the stand "energy reserve," a measure of the stand foliage condition and health. These quantities satisfy the natural physical constraints

$$0 \leq E(t) \leq 1, \qquad B(t),\ S(t) \geq 0.$$

The parameters α_1–α_{10} represent various intrinsic growth rates, predation rates, and so forth. Our interest will be in the manner in which the equilibrium (steady-state) values of B, S and E vary with changes in these parameters.

Since our primary concern is with $\bar{B}$, the equilibrium value of the budworm population B, we set the right side of the above dynamical equations equal to zero. Performing the necessary algebra, the equation for $\bar{B}$ turns out to be

$$\alpha_1\alpha_7^3(\alpha_3 + \bar{E}^2)\bar{B}^3 - \alpha_1\alpha_2\alpha_7^2\alpha_8\bar{E}^3\bar{B}^2 + \alpha_1\alpha_5\alpha_7\alpha_8^2(\alpha_3 + \bar{E}^2)\bar{B} + \alpha_2\alpha_4\alpha_7^2\alpha_8\bar{E}^3\bar{B} - \alpha_1\alpha_2\alpha_5\alpha_8^3\bar{E}^5 = 0.$$

In obtaining this equation, we have assumed that the only physically interesting equilibrium points of the system are those with $\bar{S} \neq 0$.

In the equation for $\bar{B}$, the quantity $\bar{E}$ has been regarded as an 11th parameter in the problem. This approach can be justified by noting that the dynamical equations imply that the relationship between $\bar{E}$ and $\bar{B}$ is

$$\bar{B} = \frac{-\alpha_8\alpha_9}{\alpha_7^2\alpha_{10}(\bar{E}^3 - \alpha_7\bar{E}^2 + \alpha_3\bar{E} + \alpha_3\alpha_7)}.$$

Substituting this relation into the equation above for $\bar{B}$ yields an eleventh-degree polynomial equation for $\bar{E}$ alone. The only physically relevant values for $\bar{E}$ are in the range $0 \leq \bar{E} \leq 1$, and it can be shown that there is a *unique* root of this equation in the unit interval for each set of values for the parameters $\{\alpha_i\}$. Thus, in what follows we shall always use the symbol $\bar{E}$ to denote this unique root of the above equation for $\bar{E}$.

To reduce the equation for $\bar{B}$ to standard form (i.e., monic with no quadratic term), we introduce the new variable

$$y = \bar{B} - \frac{\alpha_2 \alpha_8 \bar{E}^3}{3\alpha_7(\alpha_3 + \bar{E}^2)} .$$

After a bit of manipulation, it turns out that y satisfies the cubic equation

$$-(y^3 + t_1 y + t_2) = 0,$$

where the parameters t_1 and t_2 are given in terms of the original system parameters as

$$t_1 = \frac{-\alpha_8 \bar{E}^2}{\alpha_7^2} \left[\frac{\alpha_2^2 \alpha_8 \bar{E}^4}{3(\alpha_3 + \bar{E}^2)^2} - \frac{\alpha_2 \alpha_4 \alpha_7 \bar{E}}{\alpha_1(\alpha_3 + \bar{E}^2)} - \alpha_5 \alpha_8 \right],$$

$$t_2 = \frac{-\alpha_2 \alpha_8^2 \bar{E}^5}{9\alpha_7^3(\alpha_3 + \bar{E}^2)} \left[\frac{2\alpha_2^2 \alpha_8 \bar{E}^4}{3(\alpha_3 + \bar{E}^2)^2} - \frac{3\alpha_2 \alpha_4 \alpha_7 \bar{E}}{\alpha_1(\alpha_3 + \bar{E}^2)} + 6\alpha_5 \alpha_8 \right].$$

Catastrophe theorists will recognize the y-equation as the equilibrium equation for the standard form of the cusp catastrophe as given earlier in Table 2.1, where now the two mathematically meaningful input parameters are t_1 and t_2. (In actuality, since further analysis shows that the equilibria on the upper and lower sheets of the cusp manifold for this problem are locally stable, this is called the *dual* cusp geometry rather than the cusp. This does not affect the geometry of the manifold but is of critical importance for applications, as we shall see shortly.) The geometry of the (y, t_1, t_2) interaction is shown in Fig. 2.11.

Examining Fig. 2.11, it's evident that if we want to manipulate the parameters α_1–α_{10} to stabilize the budworm density on the lower sheet of the manifold, then we're going to have to be able to select the parameter values so that they lie outside the shaded region, i.e., the region in control space where $4t_1^3 + 27t_2^2 \geq 0$. This is a necessary condition in order to be able to stabilize the budworm densities at a low level. If such a choice of parameters is not possible, then outbreaks will always occur whenever the equilibrium level is one of the points on the middle sheet of the manifold, i.e., when the equilibrium is a point corresponding to values of t_1 and t_2 for which $4t_1^3 + 27t_2^2 < 0$. Upon careful investigation of the physically realizable values of the parameters $\{\alpha_i\}$, it turns out that there is no magic range of values that will stabilize the budworm density on the lower sheet. Consequently, no amount of "knob-twisting" with the parameters $\{\alpha_i\}$ will suffice to control

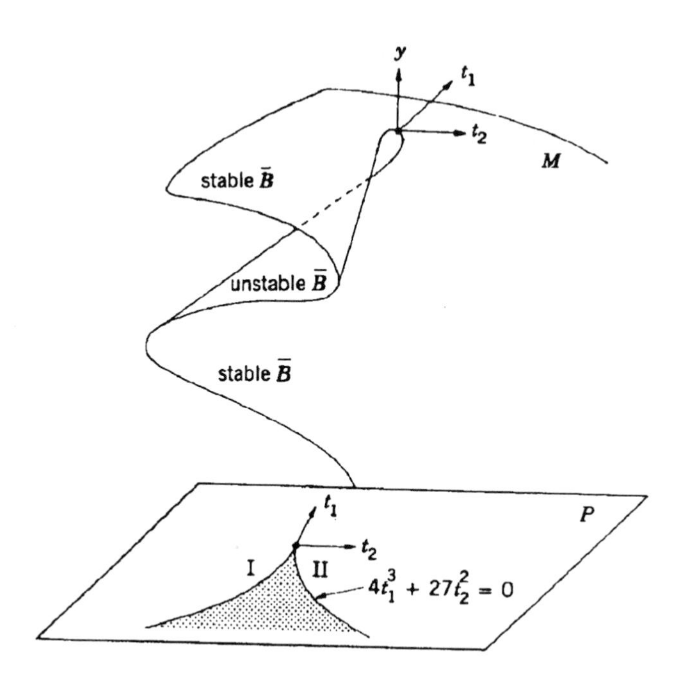

Figure 2.11. Equilibrium Budworm Densities

this system; any effective control scheme will have to be based upon more sophisticated methods of dynamic control theory of the sort we will consider in Chapters 6–7.

Although the budworm problem is interesting in its own right, by far the most significant aspect of this exercise is to show that the number of physically meaningful parameters in a problem may be very different from the number of *mathematical* parameters needed to address the questions of interest. Here we had 10 physically important parameters (the $\{\alpha_i\}$) given as part of the original problem statement; however, upon carrying out the elementary analysis of the equilibrium equation for $\bar{B}$, it turned out that the real question of interest regarding the possibility of regulating the budworm density by parametric variation came down to the interrelationship between the two mathematical parameters t_1 and t_2. Each of these parameters is a very complicated algebraic combination of all ten of the physical parameters. So it's very unlikely that any amount of a priori guesswork would find that this combination of the α-parameters—*and no other*—is the relevant combination for addressing the question of budworm outbreaks. This illustrates in the strongest possible fashion the advantage of the physical way of catastrophe theory in giving us a deeper understanding of what does and doesn't count in the analysis of a particular system.

Exercises

1. Suppose we have the scalar dynamics $\dot{x} = x[1 + P(x,a)/Q(x)]$, where P and Q are polynomials in x of degrees m and n, respectively, with $m < n$. Assume that a is a vector of parameters in R^k. Show that study of the equilibria surface $\dot{x} = 0$ as a function of the parameters a leads to consideration of the roots of a polynomial of degree $n+1$, whose coefficients are functions of the parameters a. Thus, conclude that the same trick that was used for the budworm problem to convert its study to that of the cusp catastrophe manifold can be generalized to higher-order catastrophes. At what point will this matchup between the equilibria surfaces and catastrophe manifolds break down?

2. In the budworm problem, we used a linear transformation to move from the physically-meaningful $\bar{B}$ variable to the more convenient mathematical variable y. This transformation enabled us to eliminate the cubic term in the quartic equation for $\bar{B}$. Show that such a transformation can always be used to eliminate the term of next-to-highest degree for any polynomial equation. That is, given the polynomial $p(x) = \sum_{i=0}^{n} a_i x^i$, show that we can always find a linear change of variables $x \rightarrow y$ such that in the y variables the coefficient of the next-to-highest order term is zero.

3. In the budworm equilibrium surface shown in Fig. 2.11, trace out a path through the control space that would take you *smoothly* from a region of high to low budworm population.

14. The Catastrophe Controversy

Catastrophe theory was announced to the general scientific community in 1972 with the publication of René Thom's remarkable book *Structural Stability and Morphogenesis.* The initial reviews in the most respectable scholarly journals were extremely positive, containing statements like "Both Newton's *Principia* and Thom's book lay out a new conceptual framework for the understanding of nature" [Clive Kilmister in *The London Times Higher Educational Supplement*] and "it [the book] gives me a sense of liberation and enlightenment akin to what I imagine astronomers must have felt when offered Copernican heliocentric geometry ... the sustained inspiration and the vast scope of the book put it firmly into the best tradition of natural philosophy" [Brian Goodwin in *Nature*]. But in less than five years, we find statements of the following sort appearing in equally prominent forums: "Exaggerated, not wholly honest ... the height of scientific irresponsibility" [Marc Kac in *Science*] and "Catastrophe theory actually provides no new information about anything. And ... it can lead to dangerously wrong conclusions" [James Croll in *New Scientist*]. What gives here? How can eminently respectable scientists come to such dramatically opposite opinions about a mere mathematical theory? What is it about catastrophe theory that led to statements of such lavish praise and heated outrage? Let's close our story about the classification program for gradient dynamical systems and its applications with a brief account of the pros and cons of this debate.

Happily, the controversy over catastrophe theory is a lot easier to understand than the theory itself—once we recognize that the controversy really separates into four quite distinct arguments. These involve: (1) the theory's mathematical and philosophical foundations, (2) the assumptions that are necessary to apply the theory, (3) the details of particular applications, and (4) the scientific attitudes, styles and even the personalities of the theory's supporters and opponents. We have already considered the second point within the context of the physical and metaphysical ways of catastrophe theory, so the balance of our discussion now will focus on the remaining elements of the debate.

The principal points raised against catastrophe theory from a mathematical point of view are that the theory is inherently local and that it applies only to a very restricted class of dynamical processes. Specifically, it applies to gradient (or gradient-like) systems, of which we shall say a bit more later on. The problem with locality is that the Thom Classification Theorem doesn't tell us what the system's behavior is like away from a critical point. Furthermore, since the relevant equivalence relation admits a stretching and bending of the coordinate system, it takes a certain act of faith to associate a mathematical jump on the catastrophe manifold with some observed discontinuity in a real-world process. Development of the tools needed to create a global theory of catastrophes that would justify this act of faith is a big conceptual and mathematical problem, one going far beyond the resolution of mere technical obstacles in the existing theory.

Looking at the problem of gradient systems, it turns out that there are many natural processes whose long-run behavior is not a point attractor; hence, such systems cannot be governed by gradient dynamics. For instance, many biological processes like the human heartbeat and the flowering of plants involve steady-state behaviors that are periodic. And as we shall see in Chapter 4, an even larger class of systems seem to settle into attractors that are "strange." In both cases, catastrophe analysis is inapplicable unless the problem has some kind of special structure allowing us to reduce the behavior of interest to an equivalent gradient form. Despite these incontrovertible mathematical facts, it is still the case that many interesting processes **do** fall into the domain of elementary catastrophe theory. Furthermore, we shall later see that a necessary and sufficient condition for a system $\dot{x} = f(x)$ to be gradient is that the system's Jacobian matrix $\mathcal{J} = [\partial f / \partial x]$ be symmetric. So it's often possible to closely approximate many non-gradient systems by a nearby gradient process, provided the symmetric part of the matrix $\mathcal{J}$ dominates the non-symmetric part. But these mathematical and philosophical difficulties are just the antipasto to the **real** catastrophe theory controversy. The main course revolves around the way the theory has been applied and the perceived motivations of its main practitioners, especially Christopher Zeeman.

In our discussion of fish harvesting, we saw that a number of assumptions have to be accepted in order to use catastrophe theory in situations where we do not know the explicit mathematical form of the underlying dynamics. We termed this approach the "metaphysical way" of catastrophe theory. And it's exactly this transition from "physics" to "metaphysics" that constitutes one of the main lines of attack against the application of catastrophe theory. In the words of Hector Sussman and Raphael Zahler, two of the leading anti-catastrophists, "catastrophe theory is one of many attempts that have been made to deduce the world by thought alone ... an appealing dream for mathematicians, but a dream that cannot come true." This sentiment is echoed by dynamical system theorist John Guckenheimer, who wrote in a review of Thom's book for the *Bulletin of the American Mathematical Society,* "they [Thom and Zeeman] show a real reluctance to get their hands dirty with the scientific details of the applications." But Guckenheimer concluded his review by saying that Thom might even be too cautious about the impact of his theory on biology, writing "Thom is pessimistic that this gap from singularity theory to experimental models can be bridged, but I am not so sure."

There are several aspects to these reservations about the applications of catastrophe theory, some centering on Thom's stated assumption that a good model for a physical process must be structurally stable, while others zero-in on the metaphysical assumptions that the system be governed by a potential function and that its behavior depends on only a small number of input parameters. Still another line of attack on the applications has been the claim that many of the published applications of catastrophe theory are just plain wrong, in the sense that they give predictions that are flatly contradicted by observational evidence. Let's look briefly at a couple of these claims.

Concerning the structural stability objection, John Guckenheimer has noted that "there are very reasonable models for occurrences in the real world that simply are not structurally stable, or even qualitatively predictable." In Chapter 4 we shall see evidence in support of this claim when we look at processes governed by chaotic dynamics. Such systems are manifestly unstable, as small shifts in the system parameters can (and do) give rise to transitions from simple steady-state motions like laminar flow in a fluid to complicated, inherently unpredictable behavior like fully-developed turbulence. The theory of elementary catastrophes as it currently exists cannot account for these kinds of behaviors, and it remains an open question as to whether the a suitable theory of "generalized" catastrophes can be developed to embrace such unstable processes.

As for catastrophe-theoretic models giving rise to bad models, one of the most flagrant examples of this sort is Zeeman's cusp model of the boom-and-bust behavior of prices on speculative markets. The reader should look at Discussion Question 15 for a fuller account of this model. Here it suffices just to note that the model implies that in a purely speculative market involving no investors who trade on the basis of economic fundamentals, there can never be a crash or a boom. In other words, if the only information investors use to make decisions is past price data, then there can never be a discontinuous change in prices. On the surface, such a conclusion seems patent nonsense. Real markets always experience booms and busts, and any model implying otherwise must be taken with several shakers full of salt. Other objections along similar lines have been leveled at many of Zeeman's models of things like prison riots and the aggressive behavior of animals. The degree to which these complaints hold water will ultimately turn on whether or not these preliminary models can be "souped up" to remove the offending pieces without at the same time having to throw out the catastrophe-theoretic baby with the bathwater. Now let's consider what is by far the most interesting (and probably most significant) aspect of the catastrophe controversy—the sociological side.

Public images notwithstanding, one should never forget that science, just like literature, music, dance and all the other arts, is a human activity. In fact, there's ample evidence to suggest that scientists are far from immune to the emotional outbursts and petty jealousies that are endemic and taken for granted in these more humanistic areas of intellectual pursuit. And many years of empirical evidence attests to the fact that mathematicians take a backseat to no one when it comes to hypersensitivity about the folkways and mores of their profession and the attention one of their number occasionally receives outside the cloistered walls of academia. The catastrophe theory controversy is about the best example I can think of in support of this thesis.

In its issue of January 19, 1976, *Newsweek* magazine ran a full-page article on catastrophe theory, the first story on mathematics they had published in over seven years. In this article the theory is described in terms rosy enough to emit heat, suggesting that Thom's ideas about discontinuous phenomena represent the most significant advance in applied mathematics since Newton's invention of the calculus. As one has come to expect whenever a laborer in the vineyards of mathematics receives even a smidgin of attention beyond the bounds of what the mathematical community feels is right and proper, the naysayers came crawling out from under their rocks. In this case, the charge was led by the aforementioned Sussman and Zahler, aided and abetted by a number of prominent members of the mathematical community. And in a 1977 article in *Science* magazine titled "The Emperor Has No Clothes," the battled was joined in earnest, with several pejorative quotes of the type given at the beginning of this section being uttered by prominent mathematicians like Stephen Smale of Berkeley and Joseph Keller of Stanford on the merits of catastrophe theory. In this notorious article by science writer Gina Kolata, Zeeman is described as a "publicist," and Thom is rightly quoted, but completely out of context, to the effect that in a world in which all concepts could be formulated mathematically, only the mathematician would have a right to be intelligent. What followed was a long bout of correspondence—pro and con—on the issues raised in the article, little of which had any bearing on anything other than the attention that Thom and Zeeman were receiving from the world outside science and, especially, mathematics.

In trying to summarize and evaluate the pluses and minuses of the many threads in the catastrophe theory debate, it's difficult for an uninvolved bystander not to wonder 'Why all the fuss?' As biologist Robert Rosen wisely counseled, "If an individual scientist finds such concepts uncongenial, let him not use them. There is no reason why he should take their existence as a personal affront." This is our view here, as well. Catastrophe theory will probably survive these broadsides, in much the same way and for much the same reasons that Darwin's theory of natural selection survived the bitter attacks mounted against it. Both theories are essentially explanatory rather than predictive, thereby failing to provide those who hunger for precise, quantitative predictions with the kind of numerology that has come to be synonymous with "science." But as René Thom so poignantly points out, "At a time when so many scholars in the world are calculating, is it not desirable that some, who can, dream?"

With this sociological intermezzo under our belts, let's turn our attention back to the mathematics and start looking at the question of how to extend the ideas of singularity and catastrophe theory to deal directly with the changes in *transient* behavior of a dynamical process. It turns out that this is an issue of far greater subtlety and depth than for the case of functions and maps, and we can only touch upon a few of the high points of this kind of dynamic bifurcation analysis in a book of this sort. We shall take up a consideration of these matters in Section 17. But as a stepping stone along the way, let's first look at the somewhat simpler extension involved in going from functions to maps.

15. Mappings

For the sake of exposition, we have concentrated our attention in this chapter on the case of bifurcations and singularities of smooth functions, i.e., transformations whose range is just the set of real numbers. Ironically, the setting that stimulated the development of the entire field of singularity theory involved the case of smooth *mappings* from the plane to the plane, i.e., mappings $f\colon R^2 \to R^2$. Allowing the range of our maps to be a space bigger than the real line opens up the possiblity of a host of additional types of behavior that the maps can display. And this added freedom greatly complicates the classification of behavioral modes. Nevertheless, a substantial body of results now exists for these more general situations. In this section we will give a brief taste of what can be done by considering the classical case of smooth maps of the plane to itself, as developed originally by Hassler V. Whitney in the 1950s.

Whitney's work on planar maps was focused on the following question: Can the results of Morse's Theorem for smooth functions be carried over to smooth *maps* of the euclidean plane to itself? Recall that the main features of Morse's Theorem are the genericity result, showing that Morse functions form an open, dense set in $C^\infty(R^n, R)$, and the explicit classification of smooth functions by the eigenvalue structure of the Hessian matrix of f. This classification also provided a detailed characterization of the canonical representative of each class as a nondegenerate quadratic form. It's these features of Morse's Theorem that Whitney was interested in trying to extend to the more general setting of planar maps. Perhaps surprisingly, Whitney was able to provide a total and complete analogue of Morse's results. Later we shall see that this kind of extension is not possible for general maps of $R^n \to R^m$, for the simple reason that there exist counterexamples to the genericity requirement, e.g., if $n = m = 9$, Thom has shown that the stable maps don't form a dense set. More generally, not all combinations of dimensions n, m are such that the smooth maps $f\colon R^n \to R^m$ are dense in $C^\infty(R^n, R^m)$. But before taking up these matters, let us give a complete statement of Whitney's classic result.

WHITNEY'S THEOREM. *Let M be a compact subset of R^2, and suppose that $\phi\colon M \to R^2$ with $\phi \in C^\infty(R^2, R^2)$. Denote ϕ componentwise as*

$$\phi(x, y) = \big(u(x, y), v(x, y)\big).$$

Then

i) ϕ is stable at $(x_0, y_0) \in M$ if and only if near (x_0, y_0), ϕ is equivalent to one of the three mappings

$$\begin{aligned} u &= x, & v &= y \quad \textit{(regular point)},\\ u &= x^2, & v &= y \quad \textit{(fold point)},\\ u &= xy - x^3, & v &= y \quad \textit{(cusp point)}. \end{aligned}$$

ii) The stable maps form an open and dense set in $C^\infty(R^2, R^2)$.

iii) ϕ is globally stable if and only if ϕ is stable at each point of M, and the images of folds intersect only pairwise and at nonzero angles, and the images of folds do not intersect images of cusps.

There are several comments in order regarding this basic result:

1) Compared with Morse's Theorem for smooth functions, Whitney's Theorem says that near a critical point maps of the plane look like either the fold or the cusp, whose canonical forms are described in the second part of the theorem. This result is to be compared with Morse's Theorem, which states that near a nondegenerate critical point a smooth function looks like either a "saddle" or a "bowl," i.e., it is quadratic.

2) The genericity conclusion is the same for both theorems: stable smooth maps of the plane are generic in $C^\infty(R^2, R^2)$, just as stable smooth functions are generic in $C^\infty(R^n, R)$.

3) The last part of Whitney's Theorem is a technical condition to ensure that we are dealing with planar maps that are "typical" in the space of such maps. The intersection condition given for global stability is the analogue for planar maps of the condition in Morse's Theorem requiring that the critical *values* be distinct.

The foregoing discussion has demonstrated the possibility of extending the fundamental aspects of Morse's Theorem to smooth planar maps. But what about *general* smooth maps of $R^n \to R^m$? As noted above, a general extension is not possible since there exist unstable maps from $R^9 \to R^9$ such that every map in an open neighborhood is also unstable. Thus, there is no possibility of extending the genericity result for smooth maps to this case, raising the question of whether there are certain "nice" combinations of n and m such that the stable maps from $R^n \to R^m$ do form an open and dense set. If you'll pardon the pun, openness remains an open question. But we do have a complete resolution of the matter with regard to density as the following theorem shows.

DENSITY THEOREM. *Let $q = n - m$. Then the stable smooth maps $f: R^n \to R^m$ are dense if and only if*

$$
\begin{array}{ll}
i)\ q \geq 4, & m < 7q + 8, \\
ii)\ q = 0, 1, 2, 3, & m < 7q + 9, \\
iii)\ q = -1, & m < 8, \\
iv)\ q = -2, & m < 6, \\
v)\ q \leq -3, & m < 7.
\end{array}
$$

COROLLARY. *Stable smooth maps are always dense if $n \leq 7$ or $m \leq 5$.*

Example: Urban Spatial Structure

To illustrate the Density Theorem, we consider an urban economic situation involving the flow of money from residents of one urban region to another. Let the zones of the region be labeled $i = 1, 2, \ldots, K$, and define

$$\begin{aligned} s_{ij} &= \text{flow of money from region } i \text{ to region } j, \\ e_i &= \text{per capita expenditure on shopping goods by residents of zone } i, \\ P_i &= \text{population of zone } i, \\ W_i &= \text{size of the "center" represented by zone } i, \\ c_{ij} &= \text{cost of travel from zone } i \text{ to zone } j. \end{aligned}$$

The standard aggregate model for s_{ij} is given by

$$s_{ij} = \frac{e_i P_i W_j^{\alpha} \exp(-\beta c_{ij})}{\sum_{k=1}^{K} W_k^{\alpha} \exp(-\beta c_{ik})}, \qquad i, j = 1, 2, \ldots, K,$$

where α and β are parameters representing consumer economies and ease of travel, respectively. Our interest here is in the map

$$\begin{gathered} S\colon R^K \to R^{K^2} \\ (W_1, W_2, \ldots, W_K) \mapsto (s_{11}, s_{12}, \ldots, s_{KK}). \end{gathered}$$

For clarity, consider the case of two regions ($K = 2$). The Jacobian matrix of the map S is given by

$$\mathcal{J}(W) = \alpha \begin{pmatrix} \frac{W_1^{\alpha-1} W_2^{\alpha}}{D_1^2} e^{-\beta(c_{11}+c_{12})} & \frac{-W_1^{\alpha} W_2^{\alpha-1}}{D_1^2} e^{-\beta(c_{11}+c_{12})} \\ \frac{-W_1^{\alpha-1} W_2^{\alpha}}{D_1^2} e^{-\beta(c_{11}+c_{12})} & \frac{W_1^{\alpha} W_2^{\alpha-1}}{D_1^2} e^{-\beta(c_{11}+c_{12})} \\ \frac{W_1^{\alpha-1} W_2^{\alpha}}{D_2^2} e^{-\beta(c_{21}+c_{22})} & \frac{-W_1^{\alpha} W_2^{\alpha-1}}{D_2^2} e^{-\beta(c_{21}+c_{22})} \\ \frac{-W_1^{\alpha-1} W_2^{\alpha}}{D_2^2} e^{-\beta(c_{21}+c_{22})} & \frac{W_1^{\alpha} W_2^{\alpha-1}}{D_2^2} e^{-\beta(c_{21}+c_{22})} \end{pmatrix},$$

where the D_i are distance parameters. To simplify matters, let's take $e_i = P_i = 1$ for this analysis. Then it's easy to see that rank $\mathcal{J} = 1$ for all values of α and β. This means that every point $W = (W_1, W_2)$ is a singular point for the map S. Consequently, S is an unstable map, implying that there exist arbitrarily close maps that are not equivalent to S. But the Density Theorem tells us that stable maps of $R^2 \to R^4$ are dense. So, although there are maps close to S that are not equivalent to it, there are other maps equally close to S that are stable; i.e., S can be arbitrarily closely approximated by a map that *is* stable.

Now let's finally turn our attention to the problem of dynamics. Up to now we have been using the ideas of dynamical systems and differential equations in a rather informal, taken-for-granted way. The next section formalizes these key concepts within the kind of mathematical framework we'll need in order to attack the Yin-Yang Problem of classifying vector fields.

Exercises

1. Consider the mapping of the complex plane into itself given by $z \to z^2$ (or, equivalently, $x \to x^2 - y^2$, $y \to 2xy$). Use Whitney's Theorem to prove that this map is not stable. (Hint: Show that it is not reducible by a smooth change of variables to one of the three canonical maps of part (i) of Whitney's Theorem.)

2. Find the critical points and the critical values of the Whitney map $u = xy - x^3$, $v = y$. (Answer: The critical points consist of the curve $y = 3x^2$. The critical values are found by substituting this expression into the equations for u and v. This leads to the parametric equations of the set of critical values as $u = 2x^3$, $v = 3x^2$. Thus, the set of critical values is a semicubical parabola in the (u, v)-plane.)

16. Dynamical Systems, Flows and Attractors

Intuitively speaking, we can think of a dynamical system as a formal way of describing how a point moves about on some surface M. For this description we need a rule telling us where the point should go next from its current location. Geometrically, we can think of the trajectory of the system as a curve drawn on M. Choosing one endpoint of the curve to represent the system's "initial state" at time $t = 0$, the other end of the curve then represents the system's final state at time $t = +\infty$. Let's now take a longer look at this intuitively appealing geometrical idea, formalizing it by what's called a *dynamical system.* To do this, we have to spell out precisely what we mean by a "surface," a "curve" and a "rule."

Mathematically, the notion of a surface is made precise by what's termed a *manifold.* Roughly speaking, an n-dimensional manifold M is a space in which it is possible to set up a coordinate system near each point such that locally the space looks like a subset of R^n. Of course, we may need to use a different local coordinate system around each point in M. So in order to ensure that the same point of M is described consistently using two different local coordinate systems, we need to have maps enabling us to translate back-and-forth between these different local descriptions. Depending upon the analytic properties of these "dictionary" maps, we speak of M as being a *differentiable*, C^r, *smooth*, *analytic*, *or topological* manifold. The general picture is shown in Fig. 2.12, where the U_i represent subsets of R^n, the maps φ_i are the local coordinate maps and the dictionary maps are given by φ_{ij}. Typical examples of smooth manifolds are R^n, an n-torus and the n-sphere S^{n-1}.

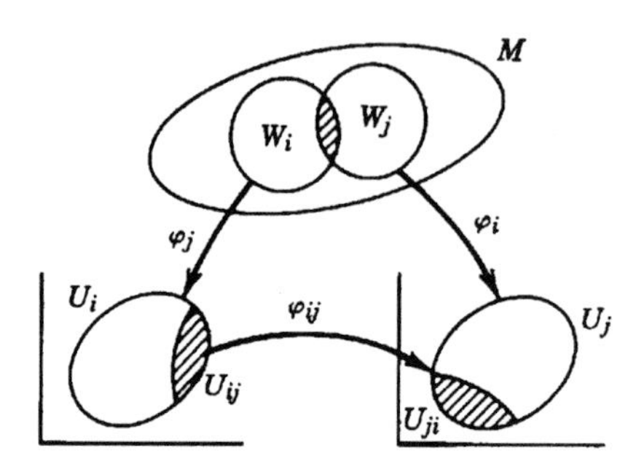

Figure 2.12. A Manifold M

Consider a point $x \in M$. The *tangent vector* to M at x consists of an equivalence class of all curves leaving x, two curves being considered equivalent if their images in R^n (with respect to any local coordinate system centered at x) can be transformed into each other by a differentiable transformation, i.e., they are equivalent in R^n. The set of all tangent vectors at x forms a vector space, termed the *tangent space* TM_x to M at x. The set $TM = \bigcup_{x \in M} TM_x$ forms the *tangent bundle* of M, i.e., the collection of all tangent planes to M (Fig. 2.13 illustrates these objects).

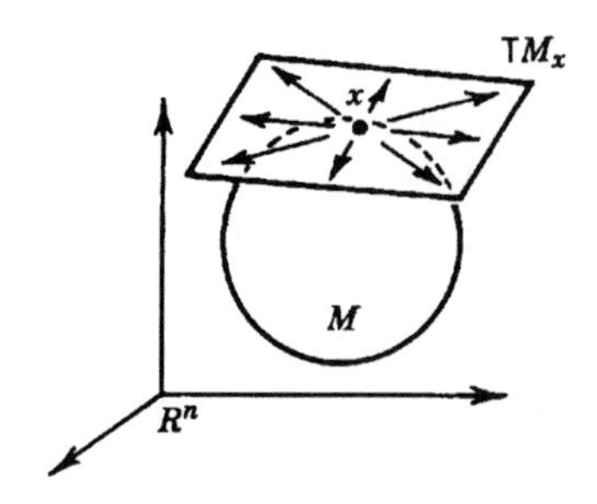

Figure 2.13. The Tangent Vectors and Tangent Space to M at x

The notion of a manifold with its tangent space and tangent bundle allows us to speak precisely about the idea of a surface and the concept of the tangent to a curve on this surface. In order to pin down what we mean by a "rule" specifying how a given point on a curve determines its successor, we need the idea of a *vector field* on M. Let $p: TM \to M$ be the mapping (projection) that assigns to each tangent plane $\tau \in TM$, the point $x \in M$ at which the plane τ is tangent to M. Then a vector field v on M is just a mapping $v: M \to TM$ such that the the composition $p \circ v: M \to M$ is the identity map on M. So, the idea is to start at a point $x_0 \in M$ and apply the map v to get a point in the tangent plane $TM_{x_0} \in TM$. The projection p then takes us back to the original point x_0. Consequently, the role of the vector field v is to *select* a tangent vector in TM_{x_0}. That vector then tells the curve which way to "move" away from the point x_0. If we have a local coordinate system $(x_1, x_2, \ldots, x_n)$ around the point $x_0 \in M$, the vector field v at x_0 can be written in the manner familiar from elementary differential equations as

$$\frac{dx}{dt} = v(x), \qquad x(0) = x_0.$$

It's rather straightforward to show that given a manifold M and a vector field v on a compact subset of M, there exists a one-parameter family of diffeomorphisms $g_t: M \to M$, such that

$$\frac{d}{dt} g_t x = v(g_t x).$$

Furthermore, the curve $g_t x_0$ is the unique solution of the equation

$$\frac{dx}{dt} = v(x), \qquad x(0) = x_0.$$

The mapping $\Phi(t, x_0) = g_t x_0$ is termed the *flow* of the vector field v at x_0, and formalizes our intuitive idea of a "curve" on the manifold M. The family of maps $\{g_t\}$ satisfies the semigroup property $g_{t+s} = g_t g_s$, which can be used to establish the uniqueness of the solution of the differential equation determined by v.

The main advantage of the formal set-up we have just described is that it is independent of the choice of any particular coordinate system at x. Thus, we can study the *intrinsic* properties of the vector field v without being distracted by artifacts introduced into the situation by a prejudicial choice of how to describe the system.

We are now in a position to define a *dynamical system* $\mathcal{D}$ as being simply a smooth manifold M, together with a vector field v defined on M. Compactly, $\mathcal{D} = (M, v)$. Our main interest for the remainder of this chapter will be to explore the question: Given a dynamical system $\mathcal{D}$ and a set of initial points U in M, what is the nature of the set of points $\lim_{t \to \infty} x_t$ for $x_0 \in U$? Here we will always take the time-parameter $t \in R$, the reals, or

$t \in \mathbb{Z}$, the integers, or possibly the subsets R_+ or $\mathbb{Z}_+$ consisting of nonnegative instants of time. To address this question, we need to consider the *stationary points* of the dynamical system $\mathcal{D}$.

Assume time is continuous. Then we call $x^* \in M$ a *stationary point* for the dynamical system $\mathcal{D}$ if $v(x^*) = 0$; in discrete time, the corresponding analogue is a *fixed point* of $\mathcal{D}$, which is defined to be a point $x^* \in M$ such that $v(x^*) = x^*$. Now let $\sigma > 0$. Then we say that a trajectory $\{x_t\}$ of $\mathcal{D}$ starting at the point x_0 is a *periodic orbit* of period σ if $x_t \neq x_0$ for $0 < t < \sigma$, and $x_\sigma = x_0$. A stationary point x^* is termed *stable* if for every $\epsilon > 0$, there exists a $\delta(\epsilon) > 0$ such that the distance $|x_t - x^*| < \epsilon$ for all $t > 0$, whenever $|x_0 - x^*| < \delta$. A stationary point is called *asymptotically stable* if it is stable and has a neighborhood U such that if $y \in U$, the distances between the points on the trajectory of $\mathcal{D}$ having any initial point $y \in U$ and the stationary point tend uniformly to zero as $t \to \infty$. Analogous definitions apply for the stability of periodic orbits. These stability concepts are illustrated in Fig. 2.14.

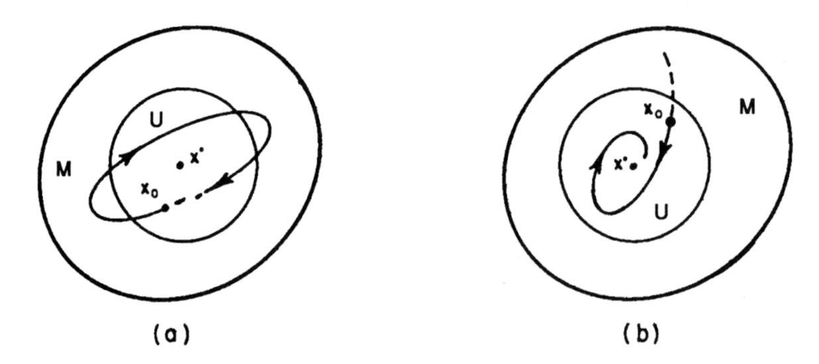

Figure 2.14. (a) Stable and (b) Asymptotically Stable Systems

Example: Linear Vector Fields

Consider a system of linear differential equations

$$\frac{dx}{dt} = Ax, \qquad x \in R^n,$$

where A is an $n \times n$ matrix. The solution curves of the system are given by $x_t = e^{At}x_0$. If the coordinates in R^n are chosen so that the matrix A is in Jordan form, then each solution curve $\{x_t\} \neq 0$ has one of the following kinds of behaviors:

(a) $\|x_t\| \to \infty$ as $t \to \pm\infty$.

(b) $\|x_t\| \to 0$ as $t \to +\infty$, $\|x_t\| \to -\infty$ as $t \to -\infty$.

(c) $\|x_t\| \to 0$ as $t \to -\infty$, $\|x_t\| \to +\infty$ as $t \to \infty$.

(d) $\|x_t\|$ and $\|x_t\|^{-1}$ are bounded.

The last case occurs only if A has a nonrepeated characteristic value lying on the imaginary axis. In particular, if 0 is a characteristic value of A there is a linear subspace of stationary points. If there is a pair of nonzero, pure imaginary characteristic values of A, then there are periodic orbits, and more than one such pair may lead to almost periodic orbits, depending upon whether the ratios of these characteristic values are rationally independent or not.

Since a matrix will not generically have only purely imaginary characteristic values, we can expect one of the first three cases to occur. Assume now that A is generic. Those solutions having behavior of type (b) span a linear subspace of R^n. This subspace is generated by the characteristic vectors of A associated with the characteristic values having negative real parts. We call this subspace the *stable manifold* of the stationary point at the origin. Similarly, the solutions displaying behavior of type (c) span a linear subspace called the *unstable manifold* of the origin. Since we have assumed that A has no purely imaginary characteristic values, the stable and unstable manifolds are complementary subspaces. In such situations we say that the stationary point at the origin is *hyperbolic,* since all solutions not on the stable or unstable manifold follow trajectories that look like hyperbolas twisted in space.

The situation just described for linear systems carries over to "generic" nonlinear dynamical systems as the following result shows.

STABLE MANIFOLD THEOREM. *Let v be a smooth vector field having a stationary point p on an n-dimensional manifold M. Assume that the flow of v is defined for all t, and that the derivative of v at p has s characteristic values with negative real parts and $n - s$ characteristic values with positive real parts, counted according to their multiplicities. Then there are submanifolds $W^s(p)$ and $W^u(p)$ of M such that:*

i) $p \in W^s(p) \bigcap W^u(p)$.

ii) If $\{g_t\}$ represents the flow of v, then

$$W^s(p) = \{x : g_t x \to p \text{ as } t \to \infty\},$$

$$W^u(p) = \{x : g_t x \to p \text{ as } t \to -\infty\}.$$

iii) The tangent space to W^s at p is spanned by the characteristic vectors of the derivative of v at p corresponding to the characteristic values having negative real parts. Similarly, the tangent space to W^u at p is spanned by the characteristic vectors corresponding to the characteristic values of the derivative of v at p having positive real parts.

There is a corresponding result for the fixed points of invertible maps, in which the conditions on the derivative of the map at the fixed point p

involve the characteristic values being inside or outside the unit circle rather than lying in either the right- or the left-half of the complex plane. Note that the characteristic values in this discrete-time case correspond to the exponentials of those in the continuous-time situation.

One of the most important features of the stable and unstable manifolds is that they intersect *transversally* at p. This means that their tangent spaces taken together span the tangent space to M at p. Such intersections are important because they persist under perturbations of the submanifolds. This notion of transversality, in turn, forms the basis for the idea of the "structural stability" of the vector field v under perturbations of various sorts.

In many physical situations the time constants are such that the transient motion of the system is quite rapid, and the real questions of interest center about the limiting behavior. Furthermore, even if the transient motion is important for the functioning of the system, it is the behavior near the stationary points that ultimately "organizes" the way in which the transient motion takes place. Thus, we need to have some tools for working with the sets describing the way the system trajectories behave as $t \to \infty$.

Suppose we are given the dynamical system $\mathcal{D}$, with $\{\Phi(x, t): t \in R\}$ being the trajectory through the point $x \in M$. We call x a *wandering point* if there is a neighborhood U of x such that $\Phi(U, t) \bigcap U = \emptyset$ whenever $|t| > t_0$, for some $t_0 > 0$. Points that are not wandering are called *nonwandering* points. Intuitively, a wandering point x has the property that after some initial time, no point near x returns to a point near x. The set of nonwandering points of the flow Φ is called the *nonwandering set* of Φ, and is denoted by Ω. The basic idea in dynamics is to first describe the nonwandering set Ω, together with its dynamics, and to then describe the way in which trajectories flow from one piece of Ω to another. To do this, we need the concept of an *attractor* of the flow Φ.

A closed, invariant set S for the flow Φ is called *topologically transitive* if there is a trajectory γ inside S such that every point of S is the limit of a sequence of points taken from γ. Thus S is the closure of γ, and γ is dense in S. Then a topologically transitive subset S of Ω is called an *attractor* for the flow Φ if S has a neighborhood U such that Φ carries U into itself as $t \to \infty$, and every point near S approaches S asymptotically as $t \to \infty$. Thus, small changes in initial conditions from those in S don't send the trajectory far away from S, and all such trajectories ultimately return to S. Finally, we call U the *domain of attraction* of S. As a technical point, we note that an attractor is something more than just an *attracting set,* i.e., a set A such that all points nearby to A ultimately go to A as $t \to \infty$. To be an attractor, a set must be an attracting set that also contains a dense orbit. So, for example, an attracting set that contains only a stable periodic orbit cannot be an attractor.

Example 1: A Cubic System

To fix some of the foregoing ideas, consider the system

$$\dot{x} = x - x^3, \qquad \dot{y} = -y, \qquad (x, y) \in M = R^2.$$

The set $A = \{(x, y)\colon -1 \leq x \leq 1,\ y = 0\}$ is an attracting set for the system, although most of the points in this set are wandering. The attractors of the flow are the two points $(\pm 1, 0)$.

Example 2: A Circle Map

$$\dot{\theta} = 1, \qquad \dot{\phi} = \pi, \qquad (\theta, \phi) \in M = T^2 \text{ (the 2-torus)}.$$

The irrational linear flow has a dense orbit for this system, and as a result, every point on T^2 is nonwandering. The attractor for this flow is the entire manifold T^2.

Exercises

1. Show that all of the following objects are manifolds: (a) any open subset of R^n, (b) the circle S^1 defined by the equation $x_1^2 + x_2^2 = 1$ in the plane, (c) the torus T^2 defined as the cartesian product $S^1 \times S^1$.

2. Prove that the set of all nonsingular $n \times n$ matrices has the structure of a differentiable manifold. (Hint: Think of a matrix as a point in R^{n^2}.)

3. The set $SO(3)$ of orthogonal matrices of order 3 and determinant $+1$ can be regarded as a subset of R^9. Prove that this subset is a submanifold.

4. Show that if a vector field v is defined on a manifold M, the trajectory of the vector field on M can be extended from time $t = 0$ to times $t = \pm\infty$ if the manifold M is compact. (Note: the compactness condition cannot be dropped, as illustrated by the equation $\dot{x} = x^2$ on the manifold $M = R$.)

5. Consider the dynamical system $\ddot{x} = 1 + 2\sin x$, describing the motion of a pendulum subject to constant forcing. A natural state manifold for this system is the surface of a cylinder. (a) Can you see why? (b) Draw the phase portrait of this system on the surface of the cylinder.

6. Find the fixed points and determine their stability or instability for the following linear dynamical systems: (a) $\dot{x} = -y,\ \dot{y} = x$, (b) $\dot{x} = y,\ \dot{y} = -(x + y)$.

7. Consider the system $\dot{x} = x,\ \dot{y} = -y + x^2$, which has a single equilibrium at the origin. (a) By eliminating the time variable t, show that this system can be written as a linear equation and integrated directly to yield

$$y(x) = \frac{x^2}{3} + \frac{c}{x},$$

where c is a constant of integration. (b) Show that the stable and unstable manifolds for the origin are given by

$$W^s(0) = \{(x,y)\colon x = 0\},$$
$$W^u(0) = \{(x,y)\colon y = x^2/3\}.$$

(c) Draw a graph of these two manifolds, and compare it with the corresponding graph of the stable and unstable manifolds for same system linearized about the origin.

8. Prove that the origin is the only nonwandering point for the damped harmonic oscillator $\ddot{x} + a\dot{x} + x = 0$, but that for the damped oscillator ($a = 0$) all points in R^2 are nonwandering. (Hint: Consider this problem in the $(x, \dot{x})$-phase space.)

9. Consider the system $\dot{x} = x - x^3$, $\dot{y} = -y$. (a) Show that the closed interval $-1 \leq x \leq 1$ is an atttracting set for this system, even though most points in it are wandering. (b) Prove that the attractors for this system are the points $(x,y) = (\pm 1, 0)$.

17. Bifurcation of Vector Fields

In the opening section of the chapter we posed the general classification problem for vector fields, i.e., for dynamical systems. But up to now we have focused our attention almost exclusively on the singularities of smooth *functions* and *mappings.* So it's high time we turned to our primary goal and looked at what can be done by way of classifying differential equations.

The natural bridge between classifying singularities of functions and mappings and classifying the bifurcations of dynamical systems is provided by the so-called gradient or gradient-like processes. We call a dynamical system

$$\frac{dx}{dt} = f(x, \mu), \qquad x \in R^n, \qquad \mu \in R^k, \tag{Σ}$$

a *gradient* system if there exists a function $V\colon R^n \to R$, such that $f(x, \mu) = -\text{grad}\, V$ for all $\mu \in R^k$. It's easy to check that a necessary and sufficient condition for (Σ) to be a gradient system is that the Jacobian matrix $\mathcal{J}$ be symmetric, i.e.,

$$[\mathcal{J}]_{ij} \doteq \left[\frac{\partial f_i}{\partial x_j}\right] = [\mathcal{J}]_{ji}\,.$$

In this case, a suitable function V is given by

$$V = \sum_{i=1}^{n} \int_0^1 f_i(tx_1, tx_2, \ldots, tx_n)x_i\, dt.$$

We call f a *gradient-like* system provided there exists a smooth function $V\colon R^n \to R$, such that

$$\text{grad } V(x) \cdot f(x, \mu) \geq 0 \text{ for all } x \in R^n.$$

In either of the above cases, the equilibrium states of the system Σ coincide with the critical points of the function V. As a result, we can directly employ the arguments outlined earlier to study the critical points of V in order to shed light upon the behavior of the dynamical process Σ in a neighborhood of an equilibrium point. Gradient and gradient-like systems form the basis for the theory of *elementary* catastrophes, providing the mathematical starting point for most of the applications that have been presented in the literature. Dynamical systems having equilibrium sets more complicated than simple point equilibria or periodic trajectories involve analyses that go beyond the elementary catastrophes discussed thus far. We shall give indications of where some of the difficulties lie in Chapter 4. But for now, let's introduce the approach to linking catastrophes and dynamics by taking a look at an example of how to use gradient behavior to categorize the behavioral modes for the growth of blue-green algae in a shallow water pond.

Example: Phytoplankton Dynamics

Many shallow water ponds experience the phenomenon of phytoplankton "bloom," in which there is a rapid increase (or die-off) of blue-green algae, such as *Anabaena,* during certain seasons of the year. Typically, this bloom process occurs according to the following sequence:

1) A superphosphate fertilizer is put into the pond while the algae concentration is low.

2) An algal bloom occurs during which *Anabaena* increase rapidly. The maximal algal concentration depends upon the time of the year when the pond is fertilized.

3) An algal bloom die-off takes place without immediate remineralization of phosphate.

The above scheme indicates that the die-off is part of the overall dynamics of the system and is not triggered by other environmental factors. Our goal is to model this process using elementary catastrophe theory.

Denote the total algal concentration by $x(t)$, with the concentration of the blue-green nitrogen fixer *Anabaena* represented by $b(t)$. Since the algal bloom seems to be triggered by the amount of phosphate available, we let $a(t)$ represent the phosphate concentration of the pond. Using arguments based upon properties of the logistic equation, it's been argued that these quantities are governed by the following dynamics:

$$\frac{dx}{dt} = -(C_1 x^3 - C_2 ax + C_3 b),$$

$$\frac{da}{dt} = -C_4 x(a - a_0),$$

$$\frac{db}{dt} = C_5 ab - C_6 bx,$$

where the parameters C_1–C_6 represent various conversion factors and birth and death rates, while a_0 is the equilibrium concentration of phosphate for the pond.

The equation for x can be thought of as a gradient system for the function

$$V(x, a, b) = \tfrac{1}{4}C_1 x^4 + \tfrac{1}{2}C_2 ax^2 + C_3 bx,$$

with a and b considered as time-varying parameters. To bring this function into the canonical form of the cusp geometry, we introduce the coordinate change

$$x = C_1^{-1/4} z, \qquad a = \left(C_1^{1/2}/C_2\right) p_1, \qquad b = \left(C_1^{1/4}/C_3\right) p_2,$$

so that the dynamics for x are governed by the scalar potential function

$$V(z, p_1, p_2) = \tfrac{1}{4}z^4 - \tfrac{1}{2}p_1 z^2 + p_2 z.$$

Using typical values of the parameters C_1–C_6 and a_0, trajectories of the total algae, *Anabaena* and phosphate concentrations from the model and from experimental observations are depicted in Fig. 2.15. In general the agreement is good, with the discrepancy in the phosphate concentration explainable by a variety of factors that are discussed in the paper cited in the chapter Notes and References. The main point of this example is not so much to show the agreement with the actual experimental data, but rather to demonstrate how a gradient dynamical system can naturally arise in the process of investigating a realistic natural resource situation. It should be noted, of course, that every scalar dynamical process is a gradient system. Consequently, whenever we are interested in the behavior of a process whose output is a scalar quantity, as in this lake pollution example, it's possible to employ elementary catastrophe theory analysis along the lines outlined above.

From the symmetry condition required for gradient systems, it's evident that gradient, and even gradient-like, systems do not form a dense set in the space of all smooth dynamical processes. It is exactly at this point that

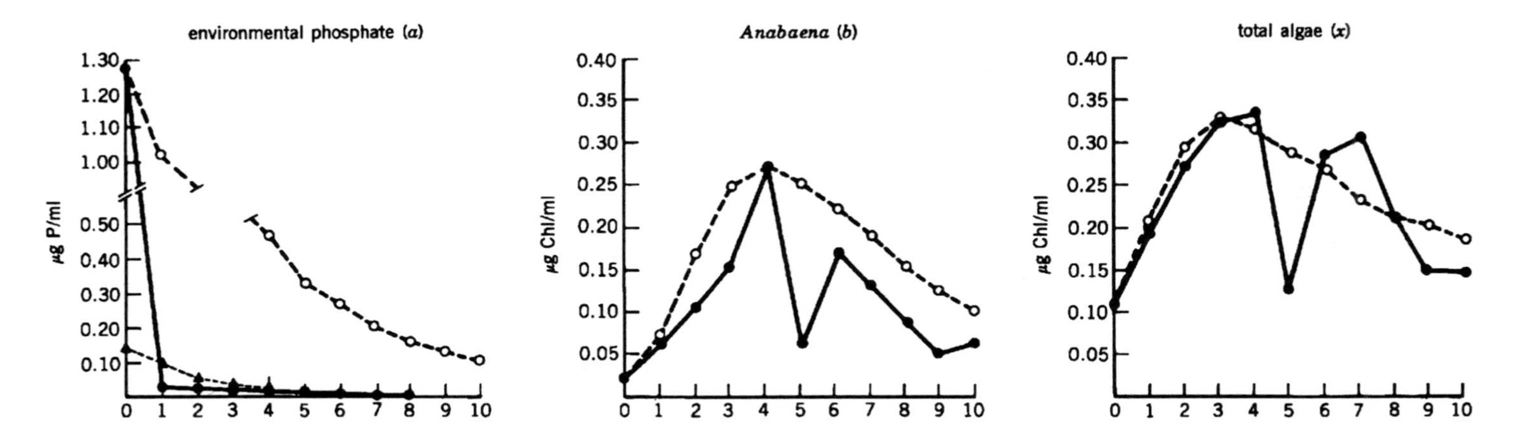

Figure 2.15. Simulated and Experimental Values of Pond Variables

the real troubles begin when we attempt to classify the types of behaviors that can arise in the structure of the attractor set of the system Σ. Even in the simplest-looking one- and two-dimensional systems, a bewildering array of behavioral modes can arise as we change system parameters, and only the most elementary types of dynamics can really be said to be understood. Since Chapter 4 considers the subclass of such systems displaying the kind of chaotic behavior associated with "strange" attractors, here we only look at the simplest type of nongradient system involving *codimension 1* bifurcations as an indication of the kind of results obtainable when one starts trying to classify vector fields.

Recall from our discussion in the opening section of this chapter that a 1-parameter family of vector fields is a curve in the space $\mathcal{V}$ of all smooth vector fields. Furthermore, such a family intersects the unstable classes generically in a point lying on a surface of dimension one less than the dimension of $\mathcal{V}$. That is, the typical, or "expected," way for such a family to undergo a qualitative change in behavior is by going through a codimension 1 bifurcation. Generically, there are two ways in which this can happen.

The most important cases to consider in the single-parameter case are when for a certain value of the system parameter μ, say $\mu = \mu_0$, the system Jacobian matrix $\mathcal{J}$ has either a pair of characteristic values on the imaginary axis (the Hopf bifurcation), or $\mathcal{J}$ has 0 as a simple characteristic value (the saddle-node bifurcation). It can be shown that if the Jacobian matrix $\mathcal{J}$ at $x = 0$, $\mu = \mu_0$ has k, ℓ and m characteristic values with positive, negative and zero real parts, respectively, then locally (in (x, μ)-space) near $(0, \mu_0)$, the canonical saddle-node dynamics assume the form

$$\frac{dx}{dt} = \mu^* - x^2, \qquad \frac{dy}{dt} = A_+ y, \qquad \frac{dz}{dt} = A_- z,$$

where $A_\pm$ are matrices having k and ℓ roots in the right and left half-planes, respectively, and where $\mu^* = \mu - \mu_0$. Thus, the only locally nonlinear behavior comes from the one-dimensional "x-part" of the dynamics.

Similarly, if $\mathcal{J}$ has a complex conjugate pair of roots cross the imaginary axis at $(0, \mu_0)$, then a smooth coordinate change produces the standard form

$$\frac{dx}{dt} = -y + x\left(\mu^* - (x^2 + y^2)\right),$$

$$\frac{dy}{dt} = x + y\left(\mu^* - (x^2 + y^2)\right),$$

$$\frac{dw}{dt} = B_+ w,$$

$$\frac{dv}{dt} = B_- v,$$

where again the matrices $B_{\pm}$ represent the linear parts of the dynamics coming from the subspaces corresponding to the roots of $\mathcal{J}$ in the left and right half-planes.

The importance of the saddle-node bifurcation is that all bifurcations of one-parameter families at an equilibrium with a zero characteristic value can be perturbed to saddle-node bifurcations. So we can expect that the zero characteristic value bifurcations that we meet in practice will be saddle-nodes. And if they are not, then there is probably something special about the problem that restricts the context to prevent the saddle-node from occurring. A typical type of constraint of this sort arises when there is some sort of symmetry in the problem that forces the roots of $\mathcal{J}$ to be symmetric with respect to the imaginary axis. If the entries of $\mathcal{J}$ are not constrained by such factors, then the *generic* way that a root can cross the imaginary axis is either for a real root to go through the origin, or for a complex conjugate pair to cross the imaginary axis. The first case leads to the saddle-node; the second to the Hopf bifurcation.

As far as the qualitative structure of the trajectories goes, in the case of the Hopf bifurcation the stable equilibrium at the origin becomes unstable at $\mu = \mu_0$, and a stable limit cycle is born whose radius is proportional to $\sqrt{\mu - \mu_0}$. For the saddle-node, we have either the appearance or disappearance of a stable and unstable equilibrium, depending upon the direction in which the parameter μ passes through the critical value μ_0. The bifurcation diagrams for these two cases are displayed in Fig. 2.16.

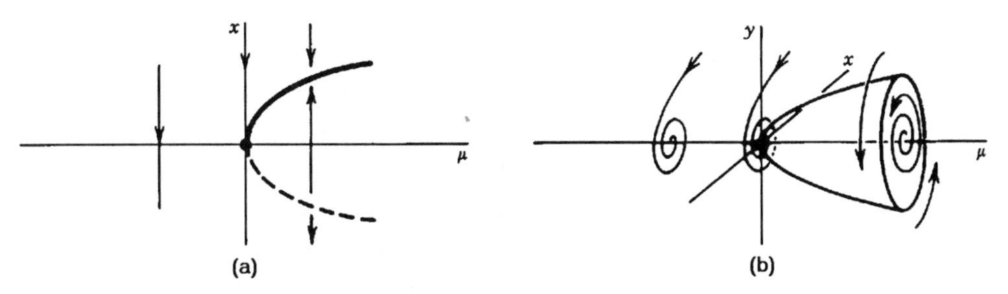

Figure 2.16. (a) Saddle-Node and (b) Hopf Bifurcations

Exercises

1. Prove that the system $\ddot{x} + a\dot{x} + bx + x^2 x + x^3 = 0$ undergoes a Hopf bifurcation on the lines $a = 0,\ b > 0$ and $a = b,\ a, b < 0$.

2. Show that the quadratic scalar map $x \rightarrow \mu - x^2$ undergoes a saddle-node bifurcation at $(x, \mu) = (-\frac{1}{2}, -\frac{1}{4})$. (Note: Here we are dealing with a *map* rather than a flow. Thus, a fixed point of the map may have fail to be hyperbolic by having a characteristic value $+1$, -1, or a complex pair of characteristic values $\lambda, \bar{\lambda}$, with $|\lambda| = 1$. The bifurcation theory

for fixed points with characteristic value $+1$ is analogous to the case for equilibria of flows with characteristic value 0, i.e., the saddle-node. The cases involving characteristic value -1 are called *flip* bifurcations, while the complex conjugate case leads to a bifurcation theory somewhat more subtle than, but similar to, the Hopf situation.)

18. Stochastic Stability and the Classification of Vector Fields

Following Thom's success in classifying gradient vector fields, mathematicians have been eagerly trying to extend the classification to more general types of systems. As we saw in Section 1, this means that some sort of equivalence relation has to be developed for the space of smooth vector fields $\mathcal{V}$. The favored relation has been that of topological equivalence. This relation leads to the view of two vector fields $f, \hat{f}$ on a manifold X being *topologically equivalent* if there is a homeomorphism of X onto itself that maps the trajectories of f onto the trajectories of $\hat{f}$. Note here that the appropriate coordinate transformation on X is now required to be only continuous, rather than infinitely differentiable as has been our standing assumption to this point. With the relation of topological equivalence, a vector field f is called structurally stable if every vector field close to f is equivalent to it.

Unfortunately, after several decades of trying mathematicians are now more-or-less in agreement that the steps of the classification program cannot be successfully carried out using the notion of toplogical equivalence. Here are but a few of the reasons:

i. Structurally stable systems are not dense in $\mathcal{V}$. So a "typical" vector field in $\mathcal{V}$ will not be structurally stable. As a result, the kind of structural stability arising from topological equivalence is of very limited use in applications, where we want to have models that are robust to small perturbations.

ii. Topological equivalence emphasizes the details of the trajectories of the system, which takes attention away from what may well be more important aspects of the overall behavior like the long-run attractors.

iii. The concept of topological equivalence is at variance with the category of smooth vector fields. So, since a homeomorphism may not be differentiable, to classify systems by this notion of equivalence leads to a situation in which two smooth vector fields could be equivalent, but there might not be any smooth change of variables that would transform one into the other. So, again, topological equivalence is of limited use in applications.

On balance, then, the failure of the classification program resides in the decision to focus attention on the trajectories of the vector fields. Recently,

Christopher Zeeman proposed a new line of attack on the problem, introducing a concept of equivalence that involves only the long-term attractor of the vector field, leaving aside all aspects of the transient motion. This idea, which leads to a kind of structural stability that Zeeman terms *ϵ-stability,* enables us to successfully negotiate all of the steps in the classification program for smooth vector fields. The remainder of this section is devoted to giving a brief introduction to Zeeman's important results.

First, a little intuition. Suppose we have a vector field $f: X \to X$ and an initial point $x_0 \in X$. Geometrically, the trajectory of x_0 under f can be thought of as a flow moving through the state-space X like a streamline of some hypothetical fluid. Now suppose that the fluid is noisy, so that the flow is disturbed by random perturbations. In this case, we have a stochastic dynamical system $dx = f(x)\,dt + \epsilon\,dw$, with noise level ϵ and an explicit random term dw whose statistics can be taken as we wish.

In such a system the trajectories from different initial points are highly irregular, and it makes more sense to try to look at their average behavior than to focus on an individual trajectory. Think of a cluster of points colored with ink. The density of the ink may not be uniform, since some parts of X may be visited more often than other regions. But if the entire cluster obeys the noisy dynamic, the averaged effect of the stochastic noise acts like diffusion. Therefore, the cluster combines a deterministic motion given by the original vector field f with a diffusion process driven by the noise. So the ink smears out into a fuzzy but structured pattern. The evolution of the density distribution of the ink is governed by a deterministic partial differential equation called the Fokker-Planck equation.

Suppose now that the state-space X is compact—a smooth space of bounded extent. Then an initial distribution cannot diffuse away to nothing, but must settle into some more interesting long-run behavior. Often it moves to a stationary distribution, representing the steady-state probability of finding the system trajectory in a given region of X.

Zeeman's idea is to note that this steady-state distribution is a fundamental invariant of the original vector field f. So, associated with every such vector field in $\mathcal{V}$, there is a unique family of density distributions—one for each value of the noise ϵ. To find this family, add a noise term $\epsilon\,dw$ to the original dynamics $f(x)\,dt$ and solve the resulting Fokker-Planck equation. What Zeeman shows is that there is a natural generic property associated with the steady-state distribution. And since each vector field has a unique family of distributions associated with it, it is then possible to solve the Yin-Yang Problem for smooth vector fields by transferring it to the setting of these smooth families of distribution functions. In this sense, what Zeeman has accomplished is a direct generalization for all smooth vector fields of Thom's classification of gradient fields by smooth potential functions. Now let's look at some of the details.

Let U denote the space of smooth probability functions on X, i.e., functions $u\colon X \to R$ such that $u > 0$ and $\int u = 1$. If we are given a vector field $v \in \mathcal{V}$ and a noise level ϵ, the Fokker-Planck equation for u is

$$\frac{\partial u}{\partial t} = \epsilon \Delta u - \Delta \cdot (uv).$$

The solution $u(t)$ of this equation represents a "blob" of ink being driven along by the vector field v, while at the same time being subject to diffusion of a level ϵ. Eventually, the blob homes in on a steady state u^*, which is independent of the initial distribution. This fact leads to the

STEADY-STATE THEOREM. *If X is compact, then the steady-state u^* exists and is unique.*

With this crucial fact in hand, we are now in a position to use the steady-state distribution u^* as a tool with which to study the vector field v. In fact, we now have a map $\pi_\epsilon\colon \mathcal{V} \to U$ that takes the vector field v to the unique steady-state element u^*. Here we have explicitly indicated that this map depends on the noise level ϵ that's been selected. The properties of the map π_ϵ are given by the

PROJECTION THEOREM. *The map $\pi_\epsilon\colon \mathcal{V} \to U$ is smooth, open and onto for each $\epsilon > 0$.*

This result enables us to use the map π_ϵ to transfer the whole of Thom's classification of smooth *functions* (the probability functions in U) to a classification of vector fields in $\mathcal{V}$ satisfying all of the steps A–D of the classification program.

The first step of the program, the selection of the equivalence relation, goes as follows. We define two vector fields $v, \hat{v}$ to be *ϵ-equivalent* if the steady states $u^*, \hat{u}^*$ are equivalent as smooth functions (i.e., are diffeomorphic to each other). We then say that v is *ϵ-stable* if there is a neighborhood of v in $\mathcal{V}$ such that all vector fields in this neighborhood are *ϵ-equivalent* to the vector field v. The Projection Theorem tells us that v can be ϵ-stable if and only if u^* is a Morse function. This fact, coupled with our results in Section 4 on Morse functions, immediately fills in the remaining steps of the classification program. The final result is the

EPSILON CLASSIFICATION THEOREM.

(a) ϵ-stable vector fields form an open, dense set in $\mathcal{V}$, i.e., ϵ-stability is a generic property in $\mathcal{V}$.

(b) The ϵ-stable classes are classified by Morse functions.

(c) The ϵ-unstable classes are classified by the elementary catastrophes using the Maxwell convention.

Now let's close this discussion with a few examples illustrating Zeeman's classification results.

Example 1: A Linear System

Let $X = R$, $v(x) = -x$. The relevant Fokker-Planck equation here is

$$\frac{\partial u}{\partial t} = \epsilon \frac{\partial^2 u}{\partial x^2} + \frac{\partial}{\partial x^2}(ux).$$

The steady-state distribution is found by setting the right-hand side of the above equation equal to zero. This yields the equation for u^* as

$$\frac{d}{dx}\left(\frac{du^*}{dx} + \frac{u^* x}{\epsilon}\right) = 0.$$

Integrating this equation, we obtain

$$\frac{du^*}{dx} + \frac{u^* x}{\epsilon} = A, \text{ a constant.}$$

Thus,

$$\frac{d}{dx}(u^* e^{x^2/2\epsilon}) = A e^{x^2/2\epsilon}.$$

Integrating again, we obtain

$$u^* = \left(A \int_0^x e^{y^2/2\epsilon}\, dy + B\right) e^{-x^2/2\epsilon},$$

where B is a constant. A little algebra shows that if A is nonzero, then for sufficiently large values of x we will have u^* negative, contradicting the requirement that u^* is a probability density function. Therefore, $A = 0$. Moreover, the condition $\int u^* = 1$ fixes the value $B = 2\pi\epsilon^{-\frac{1}{2}}$. Thus, we obtain the final result

$$u^* = \frac{1}{\sqrt{2\pi\epsilon}} e^{-x^2/2\epsilon} \doteq N_\epsilon.$$

So we see that in this case the steady-state solution of the Fokker-Planck equation is just the density function for the normal distribution with mean 0 and standard deviation ϵ.

The original vector field $v(x) = -x$ had a point attractor at the origin. So if there were no diffusion term ($\epsilon = 0$), the entire population of trajectories would move toward the origin, tending to the Dirac δ function as $t \to \infty$. However, when $\epsilon > 0$ the diffusion term pushes the cluster of

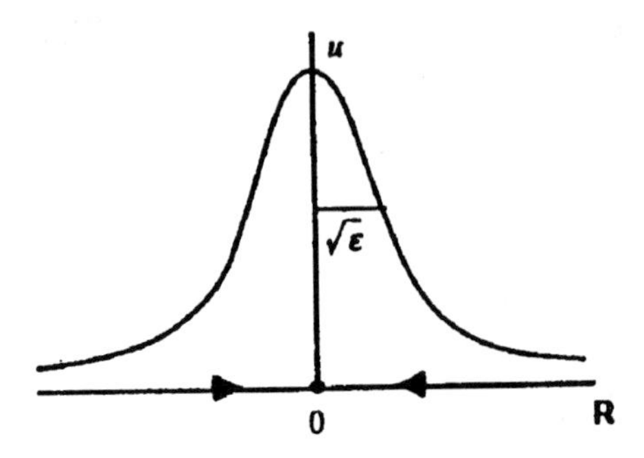

Figure 2.17. The Steady-State Distribution for the Vector Field v(x) = − x

trajectories away from the origin until they distribute themselves in accordance with the normal density function. The overall situation is shown in Fig. 2.17 above.

We note here that the vector field $v(x) = -x$ is ϵ-stable because the steady-state distribution function u^* is a Morse function. Consequently, any small perturbation of v will have a similar steady-state, again with a single maximum. Now let's look at a case where the field is ϵ-unstable.

Example 2: The Van der Pol Equation

Consider the equation

$$\frac{d^2x}{dt^2} + k(3x^2 - 1)\frac{dx}{dt} + x = 0,$$

where k is a large positive constant. The flow of the trajectories of this system has a limit cycle A and a point repellor at the origin. All other trajectories move toward A. On the limit cycle A itself, there are two fast vertical legs separated by two slow horizontal legs. Consequently, the steady state u^* represents two parallel mountain ridges with two maxima above the two slow legs, two saddles above the two fast legs, and a minimum above the origin. The overall situation is shown in Fig. 2.18.

The vector field for the van der Pol equation is ϵ-unstable because by symmetry the two maxima are at the same height. Thus, the steady-state distribution u^* cannot be a Morse function. But a small perturbation to the vector field, such as adding a small constant to the original equation, will stabilize it.

We shall see more of these classifications in the Problems for this chapter, as well as in Chapter 4, where the ideas will be used to advantage to stabilize vector fields whose flows are chaotic.

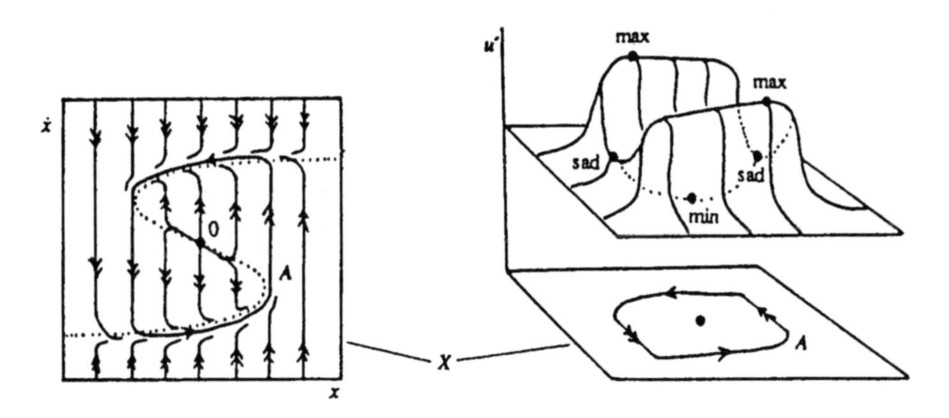

Figure 2.18. The Steady State for the Van der Pol Equation

Exercises

1. In the example of the text, suppose the vector field is $v(x) = -\alpha x$, $\alpha > 0$. Here the parameter α is the curvature of v at the minimum. Show that the steady-state solution of the Fokker-Planck equation is now $N_{\epsilon/\alpha}$. That is, the greater the curvature, the stronger the attraction and the higher and narrower the peak of the normal distribution.

2. Prove that if the initial condition for the Fokker-Planck equation is $u(x,0) = N_{\sigma^0}(x - \mu_0)$, then the time-dependent solution of the equation is given by $u(x,t) = N_\sigma(x-\mu)$, where $\mu = \mu_0 e^{-t}$, $\sigma = \epsilon+(\sigma_0-\epsilon)e^{-2t}$. In short, if the initial function is a normal distribution with mean μ_0 and standard deviation σ_0, then the entire evolution of the system is a sequence of normal distributions with the mean and variance "homing-in" exponentially to those of the steady state.

3. Consider the case when $X = R$ and the vector field $v(x) = -(x^3 - x + a)$ is a section of the usual cusp catastrophe. (a) Show that in this case the steady-state of the Fokker-Planck equation is $u^* = ke^{-f/\epsilon}$, where $f = \frac{1}{4}x^4 - \frac{1}{2}x^2 + ax$. (b) Confirm that just like the general cusp geometry, this steady-state will also be bimodal, having two peaks at x_1 and x_3 separated by a minimum at x_2. (c) Show that if $a > 0$, the peak at x_1 is higher. Moreover, as $\epsilon \to 0$, the right peak disappears and u^* tends to the Dirac delta function at x_1. What happens if ϵ always stays positive?

Discussion Questions

1. Starting with observations by Jacques Hadamard regarding criteria for "good" mathematical models of physical phenomena, it has often been claimed that in order for a model to be a credible representation of the

physical phenomenon, the behavior of the model should not change dramatically if we perturb the mathematical description a little bit. In other words, only descriptions that are stable with respect to small perturbations in either their dynamics or their initial conditions can serve as candidates for the representation of real-world processes. Discuss the pros and cons of this contention. Can you give examples of *unstable* descriptions that nevertheless serve as "good" representations of physical processes?

2. Certain mathematical relationships between variables describing physical situations have been elevated to the status of a "natural law," while other seemingly similar types of dependencies are termed merely "empirical relationships." The law of Conservation of Energy is an example of the former, while Ohm's "Law" illustrates the latter. Can you identify any criteria by which we might be able to distinguish a natural law from an empirical relationship? Should invariance under coordinate transformations enter into the requirements for a natural law? If so, what does such a requirement have to do with the classification of smooth maps?

3. It has been suggested in Chapter 1 that the *complexity* of a system Σ can be measured by the number of inequivalent descriptions of Σ that can be constructed by a given observer $\mathcal{O}$. Clearly, this is a "relativistic" measure, dependent upon the particular observer. If we take "equivalence" to mean that the descriptions differ only by a smooth change of variables, this view of complexity leads to the conclusion that the maximum complexity of Σ as seen by *any* observer would be the number of equivalence classes that that observer could distinguish in the space of smooth descriptions of Σ. By the Thom Classification Theorem, this number is finite only for descriptions of codimension five or less. In higher codimensions, each class is characterized by a canonical description containing parameters called *moduli;* hence, there are an infinite number of classes.

a) Discuss this relativistic view of system complexity. How would you use it to measure the complexity of a stone? An electrical circuit? A living organism? An automobile? A corporation? A political system?

b) Consider how you might compare the complexities of two systems even though their codimensions may both be greater than five.

c) We could term the above measure of complexity, the complexity of Σ as seen by $\mathcal{O}$, and denote it as $\mathcal{C}_{\mathcal{O}}(\Sigma)$. However, the observer $\mathcal{O}$ is also a system that is in interaction with Σ. Hence, the system Σ can itself form a measure of complexity for $\mathcal{O}$. Call this the complexity of the observer as seen by the system, denoting it by $\mathcal{C}_{\Sigma}(\mathcal{O})$. In the physical sciences, it's usually tacitly assumed that the interaction between the system and the observer is highly asymmetric, with the system's impact upon the observer being negligible. Hence, $\mathcal{C}_{\Sigma}(\mathcal{O}) \approx 0$. Discuss this assumption within the context of social and behavioral systems. In cases where the system's impact upon the observer is not negligible, consider how you might use the difference $\mathcal{C}_{\mathcal{O}}(\Sigma) - \mathcal{C}_{\Sigma}(\mathcal{O})$ as a tool for policymaking on the part of the observer and/or the system.

4. Intuitively, our notion of "surprise" involves the discrepancy between the predictions we make on the basis of models (mathematical, mental or otherwise) and the actual observed behaviors of these systems. How could you use the notions of system description and bifurcation in order to develop a mathematical theory of surprises? Consider the connection between the concept of complexity given in the preceding Discussion Question and the problem of measuring the possibility of surprising behavior of the system Σ.

Suppose you were faced with the spruce budworm system of the text, in which the equilibrium levels of budworm density are governed by the cusp geometry shown in Fig. 2.5. If you were on the lower sheet of the cusp manifold, but near the fold line leading to a jump to the upper sheet, would you consider the magnitude of the jump from lower to upper sheet as being a satisfactory measure of the surprise value of the situation? Why? If your objective as the manager of the forest is to avoid this kind of unpleasant surprise, how would you arrange harvesting, tree planting, spraying of insecticide, and so forth in order to avoid the surprise entirely? Can you imagine situations in which it might be better to allow many small jumps (surprises) in budworm population as opposed to trying to avoid any kind of discontinuous jumps in the population? (See Chapter 8 for more details on the concept of surprise.)

5. The principal objective of the managers of most natural resource systems is to minimize the variability of the economic resource represented by the resource, while at the same time maximizing the economic yield from harvesting. For example, fishery managers want to take action to generate maximum sustainable yields, as do forestry managers, mink farmers and citrus-fruit farmers. Basically, this philosophy involves trying to manipulate controllable parameters in the system so that the stability boundary of the desired equilibrium is very large and the speed of return to that equilibrium is very fast, the ideal situation being a controlled system having a single equilibrium with an infinitely large domain of attraction. Elementary facts from dynamical system theory demonstrate the unattainability of this ideal. Consider how close you could come to achieving this managerial nirvanna when the system equilibria structure is given by one of the elementary catastrophe geometries.

6. A variety of "myths" have arisen regarding the stability properties of natural resource systems. C. S. Holling has classified these myths into four categories according to the way nature has arranged the long-term behavior of the system:

• *Global Stability*—A benign and infinitely forgiving nature, in which the system is able to absorb any type of perturbation and return to its original operating point.

• *Small Is Beautiful*—The system consists of a number of decentralized islands of stability and displays a high level of heterogeneity in its behavior.

• *Nature, the Practical Joker*—Multiple equilibria exist of both good and bad types, and they impose restrictions on behavioral variability and possible movement of the stability boundaries.

• *Nature Resilient*—The system is designed to absorb and *benefit* from change rather than trying to control it.

Give examples of natural processes displaying each of these kinds of stability properties. Why do we call these stability properties "myths"?

7. In Charles Hermann's theory of international conflict, he identifies three major components contributing to the level of intensity of a crisis: *magnitude* of the threat, *decision time* available and degree of *surprise.* These factors are displayed in the "crisis cube" shown below, together with the positions of several important twentieth-century crises.

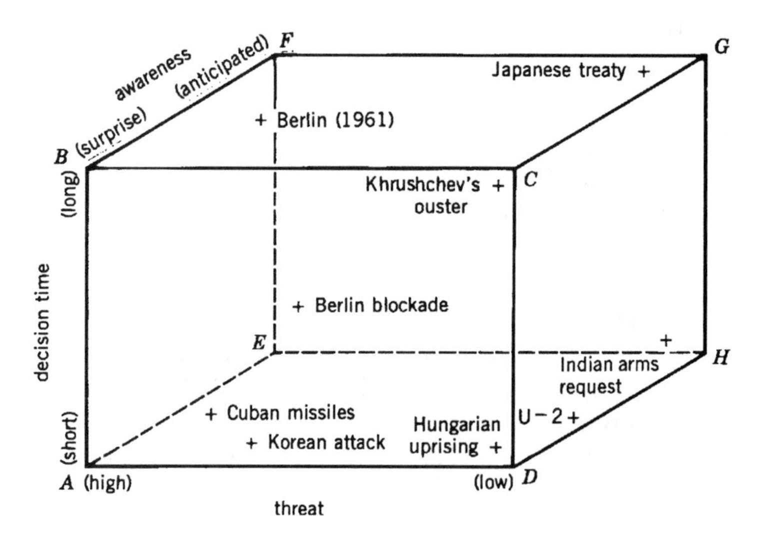

Hermann's Crisis Cube

Using Hermann's quantities as control variables, how would you use the "metaphysical way" of catastrophe theory to develop a model for international crises? What output variable(s) would you use and what catastrophe geometry would then be generated? How would you suggest calibrating your model using actual data from real conflicts?

8. Arnold Toynbee's *A Study of History* emphasizes the rise and fall of great civilizations as being a process of challenge and response. In Toynbee's view, an external threat can either strengthen or reduce a civilization's integrity. As a consequence, that civilization can either dominate nearby cultures or be subjugated by them. The following diagram shows a "cusp eye's" view of the growth and decline of the Roman Empire, in which *challenge* and *political and economic integration* are taken as control variables.

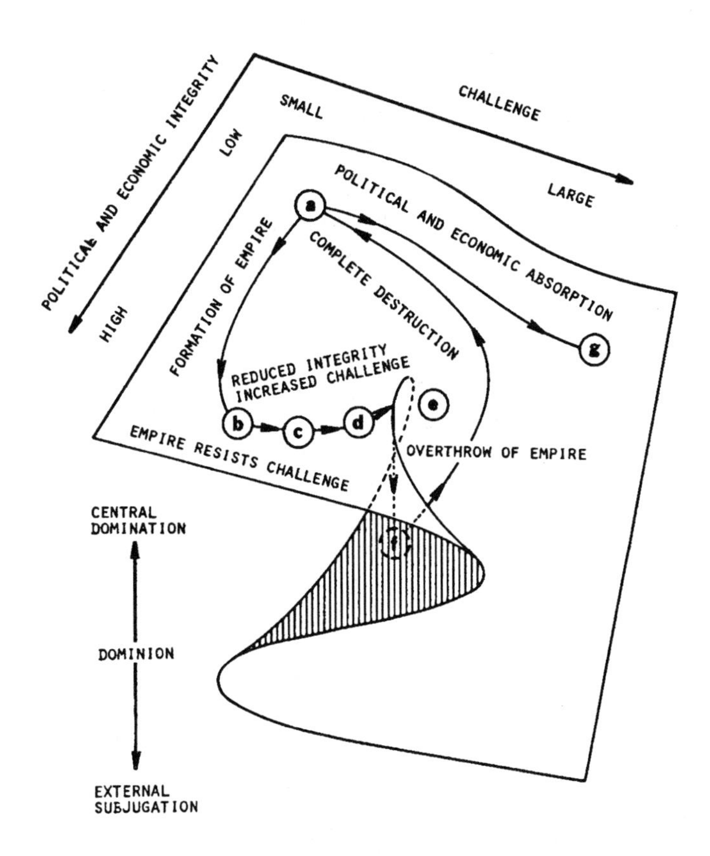

The Rise and Fall of the Roman Empire

a) Most historians agree that in the Third and Fourth centuries, Rome was able to meet its external challenges from the Persian rulers by granting its eastern client states more autonomy. Which part of the path on the diagram is most likely to correspond to this easing of central control?

b) On the basis of the cusp picture given here, what do you think was the primary cause of the fall of the Roman Empire?

c) Can you give a plausible interpretation of the transition from complete collapse at *(f)* to political and economic absorption at *(a)?*

9. Thom's original motivation for the development of many of the ideas and techniques of catastrophe theory was to provide a language with which to discuss the biological processes of morphogenesis, i.e., the emergence of physical form. Roughly speaking, his idea is that the distinctive geometries of the various catastrophe manifolds like the cusp, butterfly, and swallowtail serve as geometric "archetypes" for the type of local physical structure that could possibly unfold from a single fertilized cell. The ways in which these pure archetypes can interact then provide insight into processes such as cellular differentiation, constraining the various phenotypic forms that might result.

Discuss the pros and cons of basing a theory of embryological development upon this idea. What are the types of issues that would have to be addressed in order to carry out such a program? Could the same set of concepts be employed to create a program for classifying the developmental patterns and resulting forms of other types of life-like objects, such as languages, societies or corporations? How would the phenomena of heredity, variation and natural selection be incorporated into such a research program?

10. Many empirical investigations lead to S-shaped curves describing the manner in which one measured variable changes as a function of another. For example, a worker's salary as a function of time spent on the job, or the growth rate of bacteria on a petri dish. Mathematically, these relationships are usually described by some variant of the classical logistic equation

$$\frac{dx}{dt} = rx\left(1 - \frac{x}{K}\right),$$

where r represents a growth-rate parameter, while K involves the environmental "carrying capacity." What is the connection between this logistic growth dynamic and the geometry of the fold and cusp catastrophes? How would the higher-order cuspoid catastrophes be used to extend the above logistic dynamics to generate "S-curves" with more bends and levels? What types of physical situations might lead naturally to these "generalized" logistic curves?

As an example of the kind of thing one gets when you start using the logistic curve, Cesaré Marchetti has found a large number of very dissimilar physical and social processes, **all** of which seem to obey this kind of logistic law. The figure on the next page shows one of the more amusing of his many examples (in the figure the population level x is normalized to its maximum level K, with the dependent variable then being expressed in terms of the quantity $F = x/K$, which renders the logistic law linear.) The interested reader should consult Marchetti's work cited in the Notes and References for a much more detailed account of this logistic phenomenon.

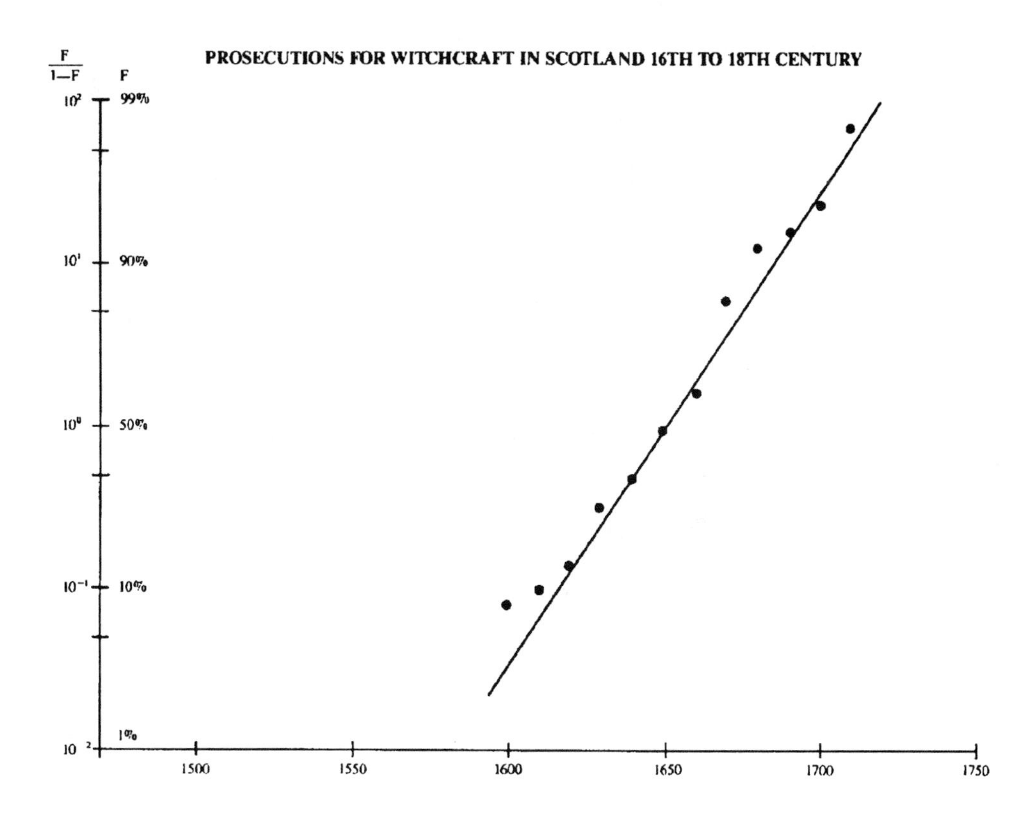

Scottish Witchcraft Trials and the Logistic Law

11. In the standard set-up from classical von Neumann-Morgenstern game theory, there are two players, call them Al and Bob. They simultaneously and independently take actions a_i and b_j. These actions result in a payoff $V(a_i, b_j)$ to, say, Al (in zero-sum games, this means that Bob's payoff is $-V(a_i, b_j)$). Al tries to maximize his payoff, while Bob attempts to minimize. Von Neumann showed that there exist optimal strategies for Al and Bob, telling them how to make their choices in accordance with a probability distribution determined by the payoff function V.

Consider the situation in which Al is unaware of Bob's existence and only feeds his decision a_i into a "black box," observing the output $V(a_i, b_j)$. Note here that Bob's choice b_j is incorporated into the interior of the box and is not seen by Al. So from Al's point of view, he's playing a game against the box, trying to maximize his payoff over some sequence of decisions. Thus, the original game can now be cast in the form of an input/output pattern of a black box. How would Al interpret the "spirit" of Bob as embodied in the black box?

Now consider the inverse problem: Given a black box with a particular input/output pattern, under what circumstances can we associate a zero-sum, two-person game with it? How does this situation relate to elementary catastrophe theory? (Hint: Consider when the internal dynamics of the black box can be characterized as the gradient of a potential function.)

12. Suppose we have a physical process whose equilibrium state is given by the minimum of the parameterized potential function $V(x, \alpha)$. Let the function V have a global minimum at the point (x^*, α^*). Then as we vary α, there will be some value $\hat{\alpha}$ at which the minimum at x^* either ceases to be a global minimum or disappears entirely. There are now two possibilities for a change of state of the system:

- *Delay Rule*—The system waits until the local minimum at x^* disappears entirely, moving then to the nearest local minimum associated with the parameter value $\hat{\alpha}$.

- *Maxwell Convention*—The system immediately moves to the new global minimum of V that's associated with the parameter value $\hat{\alpha}$ as soon as the minimum at x^* ceases to be a global minimum. The figure below shows the geometry of the cusp catastrophe under this convention.

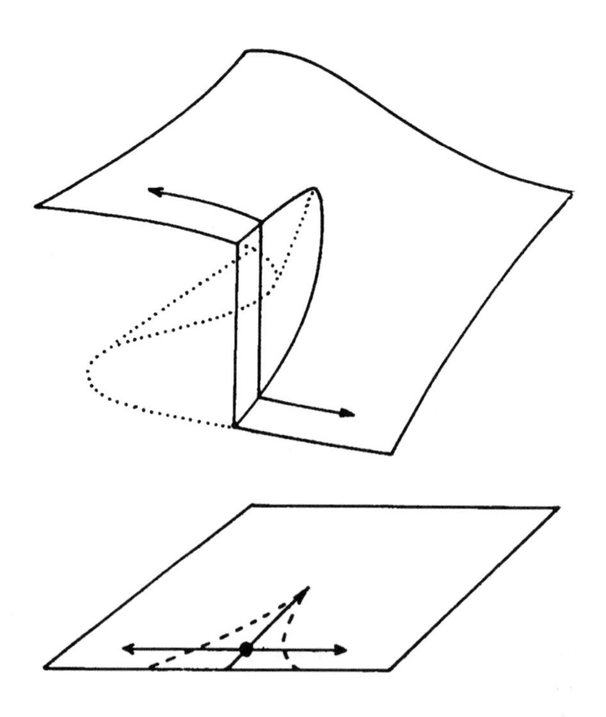

The Maxwell Convention for the Cusp Catastrophe

Give examples in the physical, social and behavioral sciences in which the system operates according to one or the other of the foregoing rules. In our examples in the text, we have implicitly assumed the Delay Convention. How would it change the cusp geometry for the budworm example if we used the Maxwell Convention instead? What difference would this make to the interpretation of the results? Can you think of any other "conventions" that might be used to govern the change of state?

13. In linguistics and biology we see the following type of hierarchical morphological patterns:

$$\text{Level I—}\{\text{Sentence}\} \longleftrightarrow \{\text{Egg}\}$$

$$\text{Level II—}\{\text{Noun and Verb Phrases}, \ldots\} \longleftrightarrow \{\text{Ectoderm, Mesoderm}, \ldots\}$$

$$\text{Level III—}\{\text{Noun, Verb, Article}, \ldots\} \longleftrightarrow \{\text{Bone, Skin}, \ldots\}$$

$$\vdots \qquad\qquad \vdots$$

Discuss the similarities and essential differences between these two morphological sequences. In particular, comment upon the essential linearity of linguistics versus the intrinsic three-dimensional geometry of biological forms.

14. In applications of singularity and catastrophe theory, the variables appearing in the nondegenerate quadratic form in the Splitting Lemma are often neglected in the problem analysis. Discuss the role that these "hidden variables" might play in characterizing the total behavior of the system. As a specific example, consider the case of a simple harmonic oscillator whose dynamics are described by the equation $\ddot{q} = -\omega^2 q$. If all we can observe is the position q, then we see the system behavior as being a linear motion back-and-forth along the q-axis from $+A$ to $-A$, where A is the amplitude of the oscillation. However, if we include the velocity in our observed output, then the motion becomes the transit of a circle with radius A^2 in the $(q, \dot{q})$-plane. This kind of situation recalls Plato's famous metaphor that what we see of nature is only a projection of the true reality, as with a shadow cast against the wall of a cave. Consider the possibility that the unseen variables in the quadratic form are those that have been "projected away" in Plato's Cave.

15. In a well-known (and strongly criticized) attempt to use catastrophe theory in the social sciences, Christopher Zeeman suggested the cusp geometry to model the "boom-and-bust" behavior of speculative markets. Taking the control variables to be excess demand for stock by investors and the percentage of speculators in the market, with the observed output being the rate of change of an index of the market's state (e.g., the Dow

Jones Industrial Average), the cusp geometry depicted below was used to qualitatively model the process of stock market crashes.

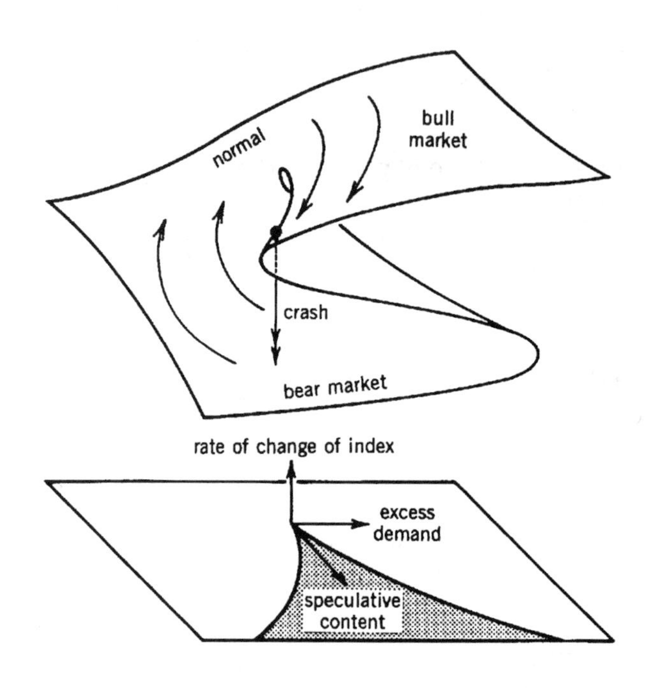

Stock Market Geometry

a) Describe in words the dynamical behavior of a market governed by this geometry.

b) The above model seems to suggest that a purely speculative market (i.e., one with no fundamentalist investors, but only speculators) could never experience a crash. How could you answer this objection?

c) Could a model of the above type be used to describe other types of "crashes" like political revolutions, corporate takeovers and urban decay? How would you define the variables in such situations?

16. Instead of displaying equilibrium behavior consisting of a single *point*, many natural systems have a steady-state behavior that is oscillatory (e.g., the human heartbeat, planetary orbits and business cycles). How could you use the ideas of catastrophe theory to account for such cyclic equilibrium behavior?

17. Consider the following experiment: We are given N points in R^n and must separate them into two disjoint subsets, X and Y. We now want

to find a polynomial function p of degree m such that

$$p(x) = \begin{cases} > 0, & \text{for } x \in X, \\ < 0, & \text{for } x \in Y. \end{cases}$$

It can be shown that as N becomes larger than $2m$, the chance of success in finding such a classifier drops dramatically. In other words, if a model classifies more than twice as many points as it has adjustable parameters, it is unusually good in a very strong sense. Discuss the implications of this type of result for assessing the credibility of a catastrophe theory model based upon one of the polynomial canonical forms; i.e., how could you develop a test for the goodness of fit of a catastrophe theory model to observed data?

18. In structural mechanics, the functions needed for a universal unfolding correspond to failure modes of things like beams, struts and other mechanical objects. These different failure modes may be interpreted as indicating that the structure has buckled, broken, collapsed or whatever. The universal unfolding of the system's potential function is then an enumeration of all possible failure modes.

Discuss the fact that, in general, mechanical failure modes are irreversible, since otherwise we could rebuild or repair a failed structure simply by reversing the path through the parameter space taking us from the original structure to the failed one. Why can't we use bifurcation theory to represent system *generation*, as well as system failure? Consider the possibility that construction modes are somehow more special, or less generic, than failure modes.

19. Recently, Alexander Woodcock has proposed a butterfly catastrophe model for the political structure of a society. In this model the society can be anything from a complete autocracy to a complete democracy, depending upon political and legislative strength, judicial bias and the influence of the media. Woodcock uses these four variables as inputs. A typical section of this model involving low judicial bias and high media influence is shown in the diagram on the next page.

a) Draw corresponding diagrams for the situations when there is high or low judicial bias and/or media influence.

b) Are there other input parameters you can think of that could be used instead of those chosen by Woodcock?

c) An alternate view of the same kind of question has been studied by Christopher Zeeman, who uses the two control parameters *political power* of the population and *economic equality*. Considering the political spectrum ranging from authoritarian left to authoritarian right with liberal left/right in between, draw a plausible catastrophe manifold for this situation.

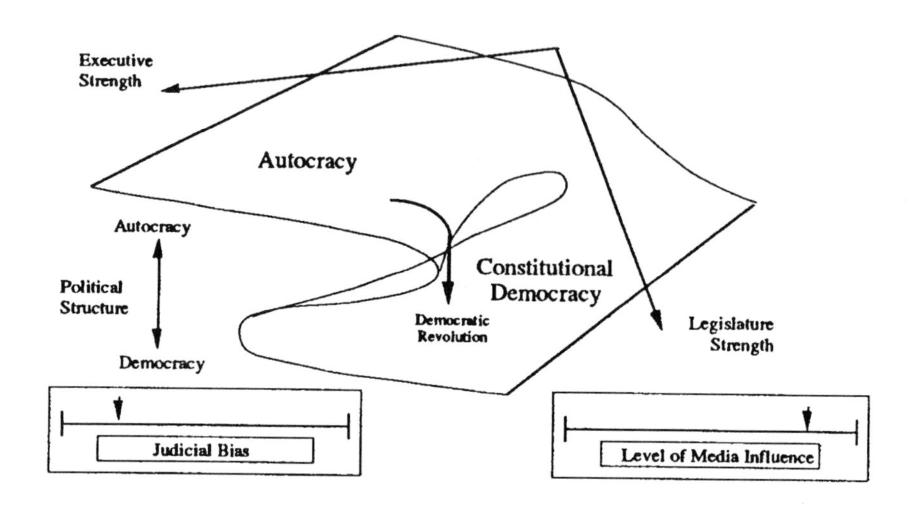

Butterfly Model of Societal Political Structure

20. A function f that is not k-determinate for any k requires an infinite number of unfolding parameters for a full description of the possibilities for what the function $f + \epsilon g(x)$ looks like around the origin for arbitrarily small ϵ. Consider a mechanical object described by such a function f. What kind of characteristics would you expect it to have?

Problems

1. Consider the function $f(x, y) = x^2y + \frac{1}{3}y^3 + \frac{1}{2}y^2$.

a) Compute the determinacy of f.

b) Calculate the corank and codimension of f for the critical point at the origin.

c) Determine a universal unfolding for f in a neighborhood of the origin.

2. Given the smooth dynamical system

$$\dot{x} = f(x),$$

where $f(0) = 0$, it's customary to linearize the system near the origin as a way of locally approximating the original process. This linearization yields the system

$$\dot{z} = Fz,$$

where $F = \left[\partial f(0)/\partial x\right]$. Under what conditions will the trajectory of the linear approximation remain qualitatively the same as that of the original system near the origin? What conditions will cause such a linearization to break down, even locally?

3. Consider the following system of equations sometimes used to describe the behavior of a single species of microorganism growing in a chemostat,

$$\begin{aligned}
\dot{x} &= (\mu - D)x, \\
\dot{s} &= D(s_r - s) - \frac{\mu x}{Y}, \\
\mu &= \frac{\mu_m K_i s}{s^2 + K_i s + K_i K_s} .
\end{aligned}$$

Here x is the biomass, μ the specific growth rate and s the nutrient concentration. All other quantities are constants.

Let the system be brought to equilibrium, and then slowly increase the dilution rate D. Interpret the subsequent behavior of s as a "fold" catastrophe. Find the smooth coordinate transformation that takes the system in the physical variables x and s to the canonical variables associated with the fold geometry.

4. The ideal gas law is given by

$$\left(P + \frac{a}{V^2}\right)(V - b) = RT,$$

where P is the pressure, V the volume, T the temperature, R the gas constant, and a, b constants representing the particular gas. If we regard V as the dependent variable, this law can be written as

$$V^3 - \left(b + \frac{RT}{P}\right)V^2 + \left(\frac{a}{P}\right)V - \frac{ab}{P} = 0.$$

This form strongly suggests that the surface it represents is smoothly equivalent to that of the canonical cusp manifold. So it should be possible to interpret the behavior of the system in catastrophe theory terms, taking V as the output variable and P and T as inputs. When we attempt to do this, however, we find that the predictions do not agree with observations. For example, there is no hysteresis: water generally boils at the same temperature as steam condenses. Nor is there bimodality, since we can predict V uniquely if we know the temperature and pressure. What is the problem here and how can you fix it?

5. Consider the mappings $f_{\pm}$ of the plane defined by

$$(x, y) \mapsto (x^2 \pm y^2 + ax + by, xy),$$

where a and b are real parameters. Show that the maps f_+ and f_- are not equivalent for any values of the parameters, i.e., there do **not** exist smooth coordinate changes g and h with $g(0) = h(0) = 0$, such that the following diagram commutes:

$$\begin{array}{ccc} R^2 & \xrightarrow{f_+} & R^2 \\ {\scriptstyle g}\downarrow & & \downarrow{\scriptstyle h} \\ R^2 & \xrightarrow[f_-]{} & R^2 \end{array}$$

6. Let $R^{n\times m}$ be the set of all real $n \times m$ matrices, and let M_r be the subset of $R^{n\times m}$ consisting of those matrices of rank r.

a) Show that M_r is a smooth submanifold of $R^{n\times m}$ and that the codimension of M_r equals the product of the coranks $(m - r)$ and $(n - r)$. That is,

$$\dim M_r = mn - (m - r)(n - r).$$

b) Specialize the above result to show that square matrices of corank k have codimension k^2 in $R^{n\times n}$.

7. The following model has been suggested to explain drug response in a human when a drug is repeatedly given under the condition that the drug D first binds with the target T according the law of mass-action, and the drug target complex DT then acts as an inducer to depress the binding of further drug administrations:

$$\dot{x} = R(y) - k_1x(d - y) + k_{-1}y - g(x),$$
$$\dot{y} = k_1x(d - y) - k_{-1}y.$$

Here $k_{\pm 1}$ are mass-action coefficients, x and y are the concentrations of the target complex and target, respectively, $g(x)$ is the degradation rate of x, d is the drug concentration at the biophase (= constant) and $R(y)$ is a sigmoidal function. For definiteness, assume

$$R(y) = \alpha_1 \frac{1 + K_1y^p}{K + K_1y^p}, \qquad g(x) = k_{10}x,$$

where α_1 = a production constant, k_{10} is the removal rate constant and K_1 represents the measure of tightness of the binding of the target complex to the repressor, while $K > 1$ is a constant. Define the quantities

$$A = \frac{k_1\alpha_1(K_1)^{1/p}d}{k_1\alpha_1 + k_{-1}k_{10}}, \qquad B = \frac{Kk_{10}k_{-1} + k_1\alpha_1}{k_1\alpha_1 + k_{-1}k_{10}}.$$

a) Show that for $A \ll B$ (or $A \gg B$) the system has only a single equilibrium with $y > 0$. What does this inequality imply about the size of the constant d?

b) If A and B are in the range $2\sqrt{B} < A < B/2$, show that there are three equilibria such that $y > 0$. Interpret this physically in terms of the amount of drug d that can be administered and still achieve a multiple steady-state concentration in the body.

c) Develop the cusp geometry governing this situation making use of A and B as control parameters, y being the behavioral variable.

8. A version of the Lotka-Volterra equations describing the population levels of n interacting species is given by

$$\frac{dN_i}{dt} = N_i \left[k_i + \sum_{j=1}^{n} a_{ij} \frac{N_j}{N} \right], \qquad \sum_{i=1}^{n} N_i = N, \quad i = 1, 2, \ldots, n.$$

Here N_i represents the population of the ith species, while k_i and a_{ij} are natural growth and interaction rates, respectively.

a) Show that under the transformation $x_i \to N_i/N$, the above system becomes a set of cubic equations for the quantities x_i.

b) Consider the case of $n = 2$ species. Show that if

$$A = k_1 - k_2 + a_{12} - a_{22} < 0, \qquad B = a_{11} + a_{22} - a_{12} - a_{21} > 0,$$

with $|A| < |B|$, the system has two stable equilibria at $(1, 0)$ and $(0, 1)$ and one unstable equilibrium at $N_1 = -A/B$, $N_2 = 1 + A/B$. Determine what initial conditions lead to each of the stable equilibria.

c) Using the initial population level of the first species as a control parameter, show how the shift of equilibrium from one stable point to the other can be modeled via the "cusp" catastrophe geometry.

9. Suppose we are given the independent, identically distributed random variables $\{X_i\}_{i=1}^{n}$, each having normal distribution with mean θ and variance τ^2. Assume that τ is known and we want to estimate θ. Our beliefs before the experiment began are that $\theta = \mu$ with probability α. For the *a priori* distribution function of θ, we take

$$g(\theta|\beta, \mu, \sigma) = \alpha(\sigma) f(\theta|\mu, \sigma^2) + (1 - \alpha(\sigma)) f(\theta|\mu, \sigma^{-2}),$$

where $0 \leq \sigma < 1$, μ and β real, with $\beta > 0$. Here we assume

$$f(\theta \mid \mu, \sigma^2) = (2\pi)^{-\frac{1}{2}} \sigma^{-1} \exp\left\{-\tfrac{1}{2}\sigma^{-2}(\theta - \mu)^2\right\},$$

with the multiplier being given by

$$\alpha(\sigma) = \frac{\beta\sigma}{1 + \beta\sigma}\,.$$

Thus, $g(\theta)$ is a mixture of two normal distributions with the same mean μ. (Usually we will choose σ small to model the type of *a priori* distribution described above.)

After observing $x = \{X_1, X_2, \ldots, X_n\}$, the *a posteriori* density $g(\theta \mid x)$ has the form

$$g(\theta \mid x) = \alpha^* f(\theta \mid \mu_1^*, V_1^*) + (1 - \alpha^*) f(\theta \mid \mu_2^*, V_2^*),$$

where

$$\mu_1^* = \frac{\sigma^{-2}\mu + n\tau^{-2}\bar{x}}{\sigma^{-2} + n\tau^{-2}}\,, \qquad \mu_2^* = \frac{\sigma^2\mu + n\tau^{-2}\bar{x}}{\sigma^2 + n\tau^{-2}}\,,$$

with

$$V_1^* = (\sigma^{-2} + n\tau^{-2})^{-1}, \qquad V_2^* = (\sigma^2 + n\tau^{-2})^{-1}.$$

The quantity α^* is given by the relation

$$\frac{\alpha^*}{1 - \alpha^*} = \beta\sigma \left(\frac{\sigma^{-2} + n\tau^{-2}}{\sigma^2 + n\tau^{-2}}\right)^{\frac{1}{2}} \exp\left\{-\tfrac{1}{2}(\mu - \bar{x})^2 \times \left[(\sigma^2 + n^{-1}\tau^2)^{-1} - (\sigma^{-2} + n^{-1}\tau^2)^{-1}\right]\right\}.$$

This *a posteriori* density will be bimodal provided that $(\mu - \bar{x})^2$ is much greater than zero.

Now suppose we want to estimate θ by minimizing the conjgate loss function

$$L(\delta, \theta) = h[1 - \exp\{-\tfrac{1}{2}k^{-1}(\theta - \delta)^2\}].$$

a) Show that the expected loss function $E(\delta)$ satisfies

$$1 - E(\delta) = \hat{\alpha} f(\delta|\mu_1^*, V_1^* + k) + (1 - \hat{\alpha}) f(\delta|\mu_2^*, V_2^* + k),$$

where

$$\hat{\alpha} = \frac{\alpha^*(1 + V_1^* k^{-1})^{-\frac{1}{2}}}{\alpha^*(1 + V_1^* k^{-1})^{-\frac{1}{2}} + (1 - \alpha^*)(1 + V_2^* k^{-1})^{-\frac{1}{2}}}\,,$$

with $f(\delta)$ as defined above.

b) Suppose that either k or n is large, so that

$$\frac{V_1^* + k}{V_2^* + k} \approx 1.$$

Show that $E(\delta)$ has a cusp point at

$$(\alpha^*, (\bar{x} - \mu)^2) \approx (\tfrac{1}{2}, 4(V_1^* + k))\,.$$

c) Using the approximate symmetry of $E(\delta)$, show that the lowest minimum will be the one nearest μ if $\alpha^* > \frac{1}{2}$, and the one nearest $\bar{x}$ if $\alpha^* < \frac{1}{2}$.

d) What happens if k and n are both small?

10. Assume that the dynamics of development in an urban housing area can be described by the quantities

$\dot{N}$ = the rate of growth of housing units in the area at time t,

a = the *excess* number of vacant units in the area relative to the norm,

b = the *relative* accessibility of the area to the regional population base.

Under the assumption that $\dot{N}(t)$ moves so as to maximize the potential

$$V = \pm(\tfrac{1}{4}\dot{N}^4 + \tfrac{1}{2}a\dot{N}^2 + b\dot{N}),$$

show how to characterize the behavior of $\dot{N}$ as a function of a and b using elementary catastrophe theory. How do you know which of the signs, plus or minus, to use for the potential V?

11. In studies of the collapse of ancient civilizations such as the Classic Maya, Mycenaean, Hittite and others, Colin Renfrew has identified the following characteristic features in the "collapse" phase:

- Collapse of the central administrative organization.
- Disappearance of the traditional elite class.
- Collapse of the centralized economy.
- Settlement shift and population decline.

During the "Aftermath" period, we observe a transition to a lower level of sociopolitical integration and the development of a romantic Dark Age myth.

In addition, such collapses display the following temporal features:

- The collapse may take on the order of one hundred years for completion, although in the provinces of an empire the withdrawal of central authority can occur more rapidly.
- Dislocations are more evident early in the collapse period, and show up as human conflicts like wars, destruction, and so on.
- Border maintenance declines during the period so that outside pressures can be seen in the historical record.
- The growth of many variables like population, exchange and agricultural activity follow a truncated S-form.
- There is no obvious single "cause" of the collapse.

a) Let the observed output variable for such collapses be taken to be D, the degree of centrality or control of the governing authority, while the control quantities are the accumulated investment in charismatic authority I and the economic balance for the rural population N. Using these quantities, develop a cusp catastrophe model for the decline and fall of the society. Using the geometry developed, trace out the rise and fall of the civilization as a sequence of changes of I and N.

b) Explain why at least two control variables are needed to explain the observed dynamical behavior of these types of collapses.

c) The cusp model describes a bimodal polity involving a rapid transition from a centered to a noncentered society, as measured by the degree of central authority. Suppose you wanted to include a third type of social structure corresponding to, say, a tribe or chiefdom, as opposed to the extreme of an egalitarian society. The simplest catastrophe geometry that allows for this type of intermediate behavior is the *butterfly,* which involves four control parameters. In addition to the quantities I and M used above, there are now two additional factors, call them T and K, which determine whether or not a tribal structure is possible. What social interpretation can you attach to these quantities?

12. Consider the simple Euler arch depicted below.

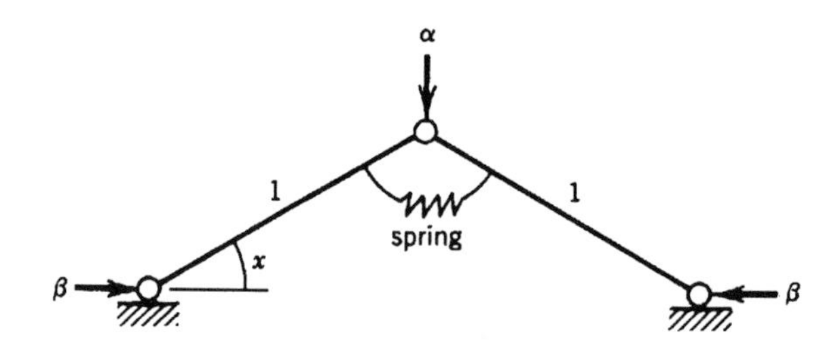

The Euler Arch

a) Considering the energy in the spring, the energy gained by the load and the energy lost by compression, show that the total energy in the system is given by

$$V(x, \alpha, \beta) = 2\mu x^2 + \alpha \sin x - 2\beta(1 - \cos x),$$

where μ is the modulus of elasticity for the spring. Using V, calculate the surface of equilibria, the fold lines and the cusp point for the system.

b) Prove that the arch buckles when $\beta = 2\mu$.

c) Show that there exists a coordinate change in x, α and β, such that in the new coordinates V has the form

$$V \sim \frac{1}{6}\mu x^4 + ax - bx^2.$$

Hence, V is the potential for the cusp catastrophe.

13. *Duffing's equation,* given by

$$\ddot{x} + \epsilon k \dot{x} + x + \epsilon \alpha x^2 = \epsilon F \cos \Omega t,$$

is often used to describe nonlinear oscillations in a variety of physical situations. Here ϵ, k and F are positive constants, with ϵ small. The quantities $\Omega = 1 + \epsilon\omega$, ω and α are real parameters. For sufficiently small values of α and ω, the attractors of this system consist of limit cycles with amplitude A and phase ϕ.

a) Use the substitution

$$x = A \cos (\Omega t - \phi)$$

in the original equation to obtain the following estimate (up to order ϵ) for the amplitude and phase of the limit cycles:

$$A^2(\tfrac{3}{4}\alpha A^2 - 2\omega)^2 = F^2 - k^2 A^2,$$

$$\tan \phi = \frac{4k}{3\alpha A^2 - 8\omega} .$$

b) The first of the above equations gives the amplitude A as a function of the parameters α and ω. Show that the graph of A has two cusp points located at

$$(\alpha, \omega) = \pm \left(\frac{\sqrt{3}k}{2}, \frac{32k^3}{9\sqrt{3}F^2} \right) .$$

c) Assume $\alpha > \sqrt{3}k/2$ (a hard spring). Now slowly increase ω from negative to positive values. Show that A smoothly increases to a maximum value $A^* = F/k$ at the point $\omega^* = 3\alpha F^2/8k^2$. What happens now if ω is further increased?

d) Draw a graph of the function $A(\alpha, \omega)$.

14. What are the critical points and critical values of the mapping $\chi\colon S^2 \to R^2$, which projects the sphere (tangent to the plane at the origin) onto the horizontal plane?

15. Show that the above projection map of the sphere to the plane is stable, but that the analogous map of the circle to the line given by $x \to \sin 2x$ is unstable.

16. Consider the *complex* map $w(z) = z^2 + 2\bar{z}$. Show that the set of critical points of this map consists of the entire unit circle $|z| = 1$. What is the geometric structure of the set of critical values?

17. Show that the two functions $f(x) = x^2$ and $g(x) = x^2 - x^4$ can be transformed into each other by means of the coordinate change

$$y(x) = \frac{x}{|x|}\sqrt{\frac{1-\sqrt{1-4x^2}}{2}}\,.$$

Over what range of x-values is this change valid? Why can't the change work globally?

18. Consider the function $f(x, y) = x^4 + y^4 - 6x^2y^2$.

a) Prove that f is 4–determinate.

b) Show that codim $f = 9$.

c) Construct a universal unfolding of f. What terms form a basis for the ideal $m_n/\Delta(f)$?

19. The scalar dynamical system

$$\dot{\theta} = a - \cos\theta, \qquad a \text{ real},$$

with state-space $U = \{\theta\colon 0 \leq \theta < 2\pi\}$ has several different types of equilibria, depending upon the value of the parameter a. Classify them.

20. The following equations describe the flow of power at each node in an electric power transmission network:

$$P_i = \alpha_{in}\sin\theta_i + \sum_{j=1}^{n-1}\alpha_{ij}\sin(\theta_i - \theta_j),$$

where P_i is the power injected at the ith node, θ_i is the voltage angle at the ith node and the quantities α_{ij} represent the manner in which the power at each node is transmitted to other nodes in the network. Assume that $\alpha_{ij} = \alpha_{ji} > 0$ if $i \neq j$ and $\alpha_{ii} = 0$, $i = 1, 2, \ldots, n$.

a) Show that the trigonometric substitution

$$x_i = \sin(\theta_i - \theta_j), \qquad y_i = \cos(\theta_i - \theta_j),$$

transforms the above equations into the set of nonlinear algebraic equations

$$\sum_{j=2}^{n}\alpha_{1j}x_j = P_1,$$

$$\sum_{\substack{j=2\\ j\neq i}}^{n}\alpha_{ij}(x_jy_i - x_iy_j) - \alpha_{1i}x_i = P_i, \quad i = 2, 3, \ldots, n-1,$$

$$x_i^2 + y_i^2 = 1, \quad i = 1, 2, \ldots, n.$$

b) Show that there exists a scalar function V such that the system

$$\begin{pmatrix} \sum \alpha_{1j} \sin(\theta_1 - \theta_j) - P_1 \\ \sum \alpha_{n-1,j} \sin(\theta_{n-1} - \theta_j) - P_{n-1} \end{pmatrix} = \begin{pmatrix} \frac{\partial V}{\partial \theta_1} \\ \frac{\partial V}{\partial \theta_{n-1}} \end{pmatrix},$$

if and only if $P_i = 0$ for all i. That is, the right side is the gradient of a potential field if and only if the power injected at each node is zero.

c) Prove that for $n = 3$ and $\alpha_{ij} = 1$ there are exactly six solutions to the system of part (a), and that of these solutions only one is stable, i.e., the linearized approximation in the neighborhood of this solution has a coefficient matrix whose characteristic values all lie in the left half-plane.

21. The equation describing the temperature at which paper ignites is given by

$$mc^0 \frac{dT}{dt} = q - h_0(T - T_a)^{\frac{4}{3}} - KT^4 + r_2 a_2 m \exp\left(\frac{-e_2}{RT}\right), \qquad T(0) = T_a,$$

where

$T_a =$ the ambient temperature,

$m =$ surface density of the paper,

$c^0 =$ specific heat,

$q =$ the rate at which heat is applied,

$h_0 =$ temperature independent constant,

$K =$ radiation factor,

$r_2 =$ heat of the reaction,

$a_2 =$ preexponential factor,

$e_2 =$ activation energy,

$R =$ universal gas constant.

a) Show that the equation describing the equilibrium temperature T_e is

$$q - h_0(T_e - T_a)^{\frac{4}{3}} - KT_e^4 + r_2 a_2 m \exp\left(\frac{-e_2}{RT_e}\right) = 0.$$

b) Prove that the above relation for T_e is equivalent to the fold catastrophe geometry. Plot the quantity T_e as a function of q, keeping all other quantities constant.

22. Suppose we are given the dynamical process

$$\dot{x}_i = f_i(x_1, x_2, \ldots, x_n), \qquad i = 1, 2, \ldots, n.$$

Define the elements

$$u_{ij}(x) \doteq \frac{\partial}{\partial x_j}\left(\frac{dx_i}{dt}\right), \qquad i, j = 1, 2, \ldots, n.$$

The quantity $u_{ij}(x)$ measures the effect of a change in x_j on the *rate of change* of the quantity x_i when the system is in the state x. If $u_{ij}(x) > 0$, we call u_{ij} an *activator* of x_i, whereas if $u_{ij}(x) < 0$, it is an *inhibitor.* It's clear that if we *know* the dynamics, then we can compute the *activation-inhibition pattern* $\{u_{ij}(x)\}$ from the functions $\{f_i\}$. Moreover, in that case we have the relation

$$df_i = \sum_{j=1}^{n} u_{ij}\, dx_j, \qquad i = 1, 2, \ldots, n.$$

a) Suppose the dynamics are *unknown,* but that the activation-inhibition pattern is given. Show that under these circumstances it's possible to construct the dynamics if and only if

$$\frac{\partial}{\partial x_k} u_{ij} = \frac{\partial}{\partial x_j} u_{ik},$$

for all $i, j, k = 1, 2, \ldots, n$. (This means that the differential forms $\{df_i\}$ are what's termed *exact.*)

b) Show that the condition for exactness of a differential form is (highly) nongeneric, i.e., in the space of differential forms, the exact forms do *not* form an open, dense set.

c) The activation-inhibition patterns that generate exact differential forms are the only ones that give rise to systems whose dynamics can be globally described by a set of differential equations. But such patterns are highly nongeneric. Consequently, there are many systems for which we can give an activation-inhibition pattern but for which there does not exist a globally defined dynamical system generating this pattern. Discuss the relevance of this observation for the problem of system complexity.

23. One type of model of cellular differentiation postulates that discrete cellular states could arise out of continuous gradients. The states are specified by the concentration of a gene product g. In this type of model, the

gene itself is activated by a "signal substance" S, with the rate of change of the gene product being given by the differential equation

$$\frac{dg}{dt} = k_1 S + \frac{k_2 g^2}{k_3 + g^2} - k_4 g.$$

Here the quantities k_i are all constants. A bit of fiddling with the various physical units in the problem shows that we can assume that $k_1 = k_2 = k_3 = 1$. Furthermore, it is physically plausible to set $k_4 = 0.4$.

a) The steady-state equation for g is

$$2g^3 - 5g^2(1 + S) + 2g - 5S = 0.$$

Show that when $S = 0$ this equation has three real roots at $g = 0, 0.5, 2$. Prove that only the roots at $g = 0$ and $g = 2$ are stable.

b) Using the above-given values of the constants and the assumption that both g and S are initially zero, give a qualitative argument for how boundaries, i.e., phase transitions, can arise for the gene product g.

c) Interpret the above result as a cusp catastrophe by introducing a second parameter K representing the saturation constant. Show that now the equilibrium equation is

$$2g^3 - 5g^2(1 + S) + 2Kg - 5KS = 0.$$

d) Draw the bifurcation set in (K, S)-space for this problem.

24. Consider a rectangular picture having length $AB = 2a$ and depth $BC = 2b$. Assume the center of mass is at the point G. The picture hangs from a cord of length $2l$ attached at A and B passing over a hook H. The entire situation is shown in the diagram on the following page. Choose axes at a point O within the picture such that the x-axis lies along OB, and let the hook have coordinates (x, y).

a) Show that the hook lies on the ellipse $\mathcal{E}$ given by the equation

$$\frac{x^2}{l^2} + \frac{y^2}{l^2 - a^2} = 1.$$

b) Suppose that the center of gravity G has coordinates $(\alpha, -a^2/(l^2 - a^2)^{\frac{1}{2}} + \beta)$, where α, β are constants proportional to the center of gravity G. Let x be the x-coordinate of the hook H. Show that the square of the potential energy of the system as a function of x is given by

$$V(\alpha, \beta, x) = -GH^2 = -(x - \alpha)^2 - \left(y(x) + a^2(l^2 - a^2)^{-\frac{1}{2}} - \beta\right)^2,$$

where $y(x) = (l^2 - a^2)^{\frac{1}{2}}(1 - x^2/l^2)^{\frac{1}{2}}$.

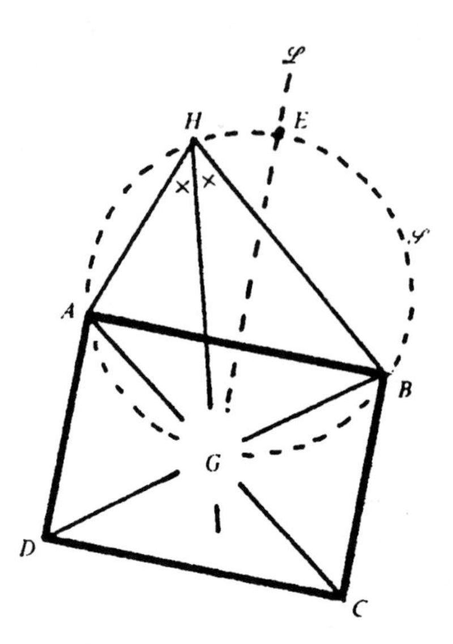

Picture Hanging from a Cord

c) By neglecting terms in x of order greater than $O(x^4)$, show that $y(x)$ can be written as

$$y(x) \simeq (l^2 - a^2)^{\frac{1}{2}} \left[1 - \frac{x^2}{2l^2} - \frac{x^4}{8l^4}\right].$$

d) Substitute the result from part (c) into that of part (b) to show that

$$V(\alpha, \beta, x) \simeq Ax^4 - \beta(l^2 - a^2)^{\frac{1}{2}} x^2/l^2 + 2\alpha x,$$

where $A = (a^2 - \beta(l^2 - a^2)^{\frac{1}{2}})/4l^4$. Hence, conclude that if we neglect the β-dependence of A, this is the potential function associated with a cusp catastrophe. From this, establish the fact that there are at most two stable equilibria for the picture as we vary the coordinates of the center of gravity G.

25. In the text we considered ϵ-stability for the vector field $v(x) = -x$. Generalize this example to arbitrary gradient vector fields, i.e., to fields $v = -\text{grad}\, f$, where $f\colon X \to R$, X being any compact manifold. Show that in this case the steady-state solution of the Fokker-Planck equation is given by the function

$$u = ke^{-f/\epsilon},$$

where the constant k is determined so that $\int u = 1$.

26. Using Fermat's Principle according to which a light ray follows a path of minimal time in going from one point to another, prove that when parallel rays of light are reflected off a circular surface like the inside of a coffee cup, the reflected rays envelope the cusp shown below.

Light Rays, Coffee Cups and Cusps

Notes and References

An excellent overview of all of the topics dealt with in this chapter is available in the lecture notes

Zeeman, E. C., "Bifurcations and Catastrophes," *Contemporary Math.*, 9 (1982), 207–272.

§1. Early in this century, Jacques Hadamard presented the view of a "well-posed" problem in mathematical physics as being one having the following properties: (1) a solution to the problem exists, (2) the solution is unique, and (3) the solution is a continuous function of the initial data specifying the problem. The ideas presented in this chapter cast serious doubt upon the last property, since many of the most important and puzzling questions of modern science like turbulent fluid flow, cellular differentiation and stock market crashes do not seem to adhere to it. Of course, it could be argued that such phenomena are still described by smooth functions of the data in a technical sense, and that the seeming discontinuities are only regions where the derivatives of the function are of very large magnitude. But this seems like a weak argument, at best, and it appears that it is indeed necessary to account in a mathematically explicit fashion for the emergence of discontinuous behavior from smooth descriptions. The ideas and methods of this chapter give an indication of just how this program might be carried out.

A more complete discussion of the Classification Problem is found in

Zeeman, E. C., "A New Concept of Stability," in *Theoretical Biology,* B. Goodwin and P. Saunders, eds., Edinburgh University Press, Edinburgh, 1989, pp. 8–15.

§3. An excellent discussion of Kepler's Laws and their role in the development of the classical deterministic view of the planetary motion is given in the following volume, which is also a fine introduction to the ideas of catastrophe theory and chaos for the layman:

Ekeland, I. *Mathematics and the Unexpected,* University of Chicago Press, Chicago, 1988.

To replace Hadamard's notion of a well-posed problem, an idea that relies upon a concept of stability of the solution to a *fixed* equation, Thom has emphasized the concept of structural stability, which hinges upon stability concepts relative to a *family* of descriptions. Thom's view is that since we can never make exact observations of any physical process, the only mathematical descriptions that can be trusted are those whose behavioral character is preserved when we perturb the description by small amounts. In other words, if a small change in the model results in large changes in the behavior, then the model cannot be accepted as a valid description of a persistent natural process. For an account of Thom's ideas, see

Thom, R., *Structural Stability and Morphogenesis,* W. A. Benjamin Co., Reading, MA, 1975.

Technical discussions of the structural stability and genericity concepts are given in

Golubitsky, M., and V. Guillemin, *Stable Mappings and Their Singularities,* Springer, New York, 1973,

Saunders, P., *An Introduction to Catastrophe Theory,* Cambrdige University Press, Cambridge, 1980,

Arnold, V. I., S. Gusein-Zade and A. Varchenko, *Singularities of Differentiable Maps,* Vol. 1, Birkhäuser, Boston, 1985.

§4. A good account of Morse's Lemma is found in

Poston, T., and I. Stewart, *Catastrophe Theory and Its Applications,* Pitman, London, 1978.

For a development of Morse's Theorem within the more general context of singularity theory, see

Milnor, J., *Morse Theory,* Princeton University Press, Princeton, 1963,

Lu, Y. C., *Singularity Theory and an Introduction to Catastrophe Theory,* Springer, New York, 1976.

For a nice discussion of the concepts of structural stability and genericity within the context of spatial economic pattern formation, see

Puu, T., "Structural Change in Flow-Based Spatial Economic Models: A Survey," *Socio-Spatial Dynamics,* 2 (1991), 1-17.

§5. A proof of the Splitting Lemma can be found in all of the books listed above on singularity theory. The Splitting Lemma is important because it enables us to identify exactly where the essential nonlinearities lie in a given description. For an account of how to use the Splitting Lemma to reduce the dimensionality in nonlinear optimization problems, see the paper

Casti, J., "Singularity Theory for Nonlinear Optimization Problems," *Appl. Math. & Comp.,* 23 (1987), 137–161.

Additional results utilizing singularity theory concepts for optimization problems can be found in

Jongen, H., and G. Zwier, "Structural Analysis in Semi-Infinite Optimization," in *3rd Franco-German Conference on Optimization,* C. Lemarechal, ed., INRIA, Le Chesnay, France, 1985, pp. 56–67,

Fujiwara, O., "Morse Programs: A Topological Approach to Smooth Constrained Optimization," *Math. Oper. Res.,* 7 (1982), 602–616.

§6. A common situation in physics and engineering is to have a finite number of terms in the Taylor series expansion of a function, and then to try to deduce information about the function from this finite amount of information. Often the assumption is made that the function involved is analytic, presumably on the grounds that this will help in determining properties of the function. To see that this is not the case, consider the function $f(x, y) = x^2(e^y - 1)$. The 17-jet of f is given by

$$j^{17} f = x^2 y + \frac{1}{2} x y^2 + \cdots + \frac{1}{15!} x^2 y^{15}.$$

Consequently, the equation $j^{17} f = 0$ has the x- and y-axes as its roots. But

$$j^{18} f = j^{17} f + \frac{1}{18!} y^{18},$$

so the equation $j^{18}f = 0$ has no solution with $y > 0$. Hence, $j^{17}f$ is not sufficient to determine the character of $j^{18}f$, not to mention the character of f itself. Thus, for no finite k does having the information: (1) f is analytic around 0, and (2) $j^k f = j^k(x^2(e^y - 1))$, imply that the roots of f are even approximately those of $j^k f$ near the origin. In fact, analyticity of f has no bearing on the question! This example brings out forcefully the role of determinacy as the relevant notion when we want to know how far out in the Taylor series we need to go in order to capture all the local structure of f. For a fuller discussion of this point, as well as more elaboration on the foregoing example, see Chapter Three of

Poston, T., and I. Stewart, *Taylor Expansions and Catastrophes,* Pitman, London, 1976.

The computational procedure discussed in the text follows that given in the articles

Deakin, M., "An Elementary Approach to Catastrophe Theory," *Bull. Math. Biol.,* 40 (1978), 429–450,

Stewart, I., "Applications of Catastrophe Theory to the Physical Sciences," *Physica D,* 20 (1981), 245–305.

§7. From the standpoint of applications, the idea of a universal unfolding is probably the most important single result from elementary catastrophe theory. The unfolding terms show us the types of perturbations that cannot be neutralized by a smooth coordinate transformation, and give information about the way in which the properties of the function will change when it is perturbed in various ways near a degenerate singularity. Special emphasis is laid on this point of view in the book by Poston and Stewart cited under §4 above. See also many of the papers in the collection

Zeeman, E. C., *Catastrophe Theory: Selected Papers, 1972–1977,* Addison-Wesley, Reading, MA, 1977.

§8. A complete proof of the Classification Theorem may be found in the Zeeman book cited above. It appears that Thom deserves the credit for seeing that such a result must be true, as well as for recognizing the many pieces that would have to be put together in order to actually prove the theorem. He then convinced a variety of mathematicians, including Arnold, Mather, Malgrange and Boardman, to put together the necessary ingredients for the final result. It makes for fascinating reading today to reflect on the dialogue between René Thom and Christopher Zeeman on the future prospects of catastrophe theory for applications in the physical, biological and social sciences. This dialogue is reprinted in the Zeeman book noted above.

Another introductory account of Thom's Theorem for the layman is given in

Woodcock, T. and M. Davis, *Catastrophe Theory,* Dutton, New York, 1978.

§9. For a derivation of the power system dynamics, see

Modern Concepts of Power System Dynamics, IEEE Special Publication #70MG2-PWR, IEEE, New York, 1970.

Other papers harnessing the ideas of singularity theory to the cause of electrical power networks are

Tavora, C. and O. Smith, "Equilibrium Analysis of Power Systems," *IEEE Tran. Power Apparatus and Systems,* PAS-91 (1972), 1131–1137,

Baillieul, J. and C. Byrnes, "A Geometric Problem in Electric Energy Systems," in *Int'l. Symposium on the Mathematical Theory of Networks,* Vol. 4, Western Periodicals, Hollywood, CA, 1981.

§10. A detailed geometrical study of the bifurcation diagrams for all the elementary catastrophes is given in

Woodcock, A. E. R., and T. Poston, *A Geometrical Study of the Elementary Catastrophes,* Springer Lecture Notes in Mathematics, Vol. 373, Springer, Berlin, 1974.

Much more information and speculation about the connections between the spatio-temporal interpretation of catastrophe sets and problems in theoretical biology can be found in the pioneering article

Thom, R., "Topological Models in Biology," *Topology,* 8 (1969), 313–335.

In this same connection, see also the collection of reprints

Thom, R., *Mathematical Models of Morphogenesis,* Ellis Horwood, Chichester, UK, 1983.

§11. The fishery management example follows that in

Jones, D., and C. Walters, "Catastrophe Theory and Fisheries Regulation," *J. Fisheries Res. Board Canada,* 33 (1976), 2829–2833.

Other work along similar lines is reported in

Peterman, R., "A Simple Mechanism that Causes Collapsing Stability Regions in Exploited Salmonid Populations," *J. Fisheries Res. Board Canada,* 34, No. 8, 1977.

The general issue of how to harvest a renewable resource like fish or timber so as to balance out the economic gain with the consumption of the resource is treated in detail in

Clark, C., *Mathematical Bioeconomics,* 2d ed., Wiley, New York, 1990.

§12. The material of this section follows that first presented in

Cobb, L., "Estimation Theory for the Cusp Catastrophe Model," in *Proc. Section on Survey Research Methods,* Amer. Stat. Assn., Washington, D.C., 1980, pp. 772–776.

§13. The dynamics describing the budworm outbreaks for a single site were first put forth in

Ludwig, D., C. Holling and D. Jones, "Qualitative Analysis of Insect Outbreak Systems: The Spruce Budworm and Forest," *J. Animal Ecology,* 47 (1978), 315–332.

The catastrophe theory analysis presented here showing the impossiblity of eliminating budworm outbreaks by manipulation of system parameters is given in detail in

Casti, J., "Catastrophes, Control and the Inevitability of Spruce Budworm Outbreaks," *Ecol. Modelling,* 14 (1982), 293–300.

§14. The initial salvo sparking off the catastrophe theory controversy was fired in

Sussman, H. J., "Catastrophe Theory: A Preliminary Critical Study," *Proc. Biennial Mtg. Phil. Sci. Assn.,* 1976.

A more detailed development of the "case" against Thom & Co. is presented in the critical article

Sussman, H. J. and R. Zahler, "Catastrophe Theory Applied to the Social and Biological Sciences: A Critique," *Synthese,* 37 (1978), 117–216.

For a reasonably balanced account of the whole brouhaha, see Chapter Four of the Woodcock and Davis volume cited under §8 above.

§15. An excellent introductory account of the development of singularity theory as an outgrowth of Morse's work is the Lu book cited under §4 above. For more technical accounts of Whitney's Theorem and its subsequent extension to general smooth maps, see the Golubitsky and Guillemin book referred to in §3 above, as well as

Arnold, V. I., *Singularity Theory,* Cambridge University Press, Cambridge, 1981,

Bröcker, T., *Differentiable Germs and Catastrophes,* L. Lander, trans., Cambridge University Press, Cambridge, 1975,

Gibson, J., *Singular Points of Smooth Mappings,* Pitman, London, 1979,

Martinet, J., *Singularities of Smooth Functions and Maps,* Cambridge University Press, Cambridge, 1982.

The urban spatial structure problem follows the treatment in

Casti, J., "System Similarities and the Existence of Natural Laws," in *Differential Topology, Geometry and Related Fields,* G. Rassias, ed., Teubner, Leipzig, 1985, pp. 51–74.

§16. An undergraduate-level account of dynamical systems from a modern point of view is offered in

Arnold, V. I., *Ordinary Differential Equations,* MIT Press, Cambridge, MA, 1973,

Hirsch, M. and S. Smale, *Differential Equations, Dynamical Systems and Linear Algebra,* Academic Press, New York, 1974.

A somewhat more advanced, but still very readable, account is

Irwin, M. C., *Smooth Dynamical Systems,* Academic Press, New York, 1980.

For an illuminating account of the philosophy, history and methods of dynamical system theory and the use of differential equations to represent nature, it's hard to think of a better source than the survey article

Hirsch, M., "The Dynamical Systems Approach to Differential Equations," *Bull. Amer. Math. Soc.,* 11 (1984), 1–64.

§17. The very major differences between the classification of singularities of smooth functions and the classification of smooth vector fields is brought out in the review of Thom's book by Guckenheimer. See

Guckenheimer, J., "Review of R. Thom, *Stabilité structurelle et morphogénèse,*" *Bull. Amer. Math. Soc.,* 79 (1973), 878–890.

Good accounts of various aspects of the problem of classifying the singularities of vector fields from a variety of perspectives, including excellent discussions of the Hopf and saddle-node cases, are found in

Arnold, V. I., *Geometrical Methods in the Theory of Ordinary Differential Equations,* Springer, New York, 1983,

Guckenheimer, J., and P. Holmes, *Nonlinear Oscillations, Dynamical Systems, and Bifurcations of Vector Fields,* Springer, New York, 1983.

A vitally important tool in the classification of smooth functions is, as we have seen, the Splitting Lemma. The corresponding result for dynamical systems is the Center Manifold Theorem, a good account of which is given in the foregoing volumes. (See also Chapter 4, Problem 5.)

The lake pollution example is taken from

Casti, J., J. Kempf and L. Duckstein, "Modeling Phytoplankton Dynamics Using Catastrophe Theory," *Water Resources Res.,* 15 (1979), 1189–1194.

The more general question of how useful catastrophe-theoretic models are in ecology is taken up in

Loehle, C., "Catastrophe Theory in Ecology: A Critical Review and an Example of the Butterfly Catastrophe," *Ecological Modelling,* 49 (1989), 125–152.

§18. A layman's account of the results of this section is presented in

Stewart, I., "Yin-Yang and the Art of Noise," *Nature,* 335 (September 29, 1988), 394.

The complete results were first presented in

Zeeman, E., "Stability of Dynamical Systems," *Nonlinearity,* 1 (1988), 115–155.

DQ #2. The matter of a "natural law" versus an "empirical relationship" is one fraught with many perplexing epistemological as well as semantic difficulties, some of which we have already considered in Section 3 of Chapter 1. Additional perspectives from a more pragmatic rather than philosophical point of view are found in

Casti, J., "Systemism, System Theory and Social System Modeling," *Regional Sci. and Urban Econ.,* 11 (1981), 405–424,

Kalman, R., "Comments on Scientific Aspects of Modelling," in *Towards a Plan of Action for Mankind,* M. Marois, ed., North-Holland, Amsterdam, 1974, pp. 493–505.

DQ #3. More details on this relativistic view of complexity are given in

Casti, J., "On System Complexity: Identification, Measurement and Management," in in *Complexity, Language and Life: Mathematical Approaches,* J. Casti and A. Karlqvist, eds., Springer, Heidelberg, 1986, pp. 146–173.

DQ #5–6. The issues raised by this cluster of problems have been treated in some detail from the perspective of natural resource management by the ecological systems group at the University of British Columbia. A sampling of their work is given in

Holling, C. S., "Resilience and Stability of Ecological Systems," *Annual Rev. of Ecology and Systematics,* 4 (1973), 1–23,

Walters, C., G. Spangler, W. Christie, P. Manion and J. Kitchell, "A Synthesis of Knowns, Unknowns, and Policy Recommendations from the Sea Lamprey Int'l. Symposium," *Canadian J. Fish. Aquat. Sci.,* Vol. 37, No. 11, (1980), 2202–2208,

DQ #7. For a layman's introduction to the science as opposed to the art of war, see Chapter Five in

Casti, J., *Searching for Certainty: What Scientists Can Know About the Future,* Morrow, New York, 1991.

For more detailed technical information about the use of system-theoretic ideas in modeling international conflict situations, see

McClelland, C., "System Theory and Human Conflict," in *The Nature of Human Conflict,* E. B. McNeil, ed., Prentice-Hall, Englewood Cliffs, NJ, 1965,

Holt, R., B. Job and L. Markus, "Catastrophe Theory and the Study of War," *J. Conflict Resol.,* 22 (1978), 171–208.

DQ #8. More details on this catastrophe-theoretic model of the decline and fall of the Roman Empire, as well as other models of political and military conflict, are given in the Woodcock and Davis book cited under §8 above.

DQ #9. Marchetti has found evidence for logistic growth (and decline) in things as dissimilar as US passenger car registrations, discovery of the chemical elements, mainframe computer development in Japan, and the number of films directed by John Huston. A good summary of this work is given in

Marchetti, C., "Stable Rules in Social Behavior," IBM Conference, Brazilian Academy of Sciences, Brasilia, 1986.

Another excellent source for material on the logistic equation and its application in a wide variety of areas is the survey article

Montroll, E., "On the Dynamics and Evolution of Some Sociotechnical Systems," *Bull. Amer. Math. Soc.*, 16 (1987), 1–46.

DQ #10. For more information about how catastrophes can be used to model embryological development, see the work by Thom cited under §10 above.

DQ #12. The question of what convention to use in applied catastrophe theory situations is a delicate one, ultimately coming down to a consideration of the various time-scales at play in the problem. For a more detailed consideration of these matters, see

Gilmore, R., "Catastrophe Time Scales and Conventions," *Physical Rev. A*, 20 (1979), 2510–2515.

DQ #15. The stock market example is taken from the work reported in

Zeeman, E. C., "On the Unstable Behavior of Stock Exchanges," *J. Math. Economics*, 1 (1974), 39–49.

DQ #18. The question of the degree to which a universal unfolding can account for modes of generation as well as modes of failure is considered in somewhat more detail in

Rosen, R., "How Universal is a Universal Unfolding?," *Appl. Math. Lett.*, 1 (1988), 105–107.

DQ #19. For more of Woodcock's views on discontinuities in societal dynamics, see

Woodcock, A., "Political Landscape Models of Emerging Societal Complexity," *Wash. Area Systems Society Newsltr.*, 1 (1991), 7–13.

PR #11. The various factors involved in the collapse of civilizations are considered further in

Renfrew, C., "Systems Collapse as Social Transformation: Catastrophe and Anastrophe in Early State Societies," in *Transformations: Mathematical Approaches to Cultural Change,* C. Renfrew and K. Cooke, eds., Academic Press, New York, 1979.

A detailed account of the collapse of the Mayan civilization, complete with computer programs simulating the decline, is found in the volume

Lowe, J., *The Dynamics of Collapse: A Systems Simulation of the Classic Maya Collapse,* University of New Mexico Press, Albuquerque, NM, 1985.

PR #12–13. Further discussions of these problems are found in the Zeeman book cited under §7 above.

PR #22. An activation-inhibition network is an important example of a situation for which we can have a dynamical description that cannot be expressed by means of a set of differential equations. In general, an activation-inhibition pattern $\{u_{ij}(x)\}$ does not lead to an exact differential form; hence, there is no differential equation system corresponding to such a pattern. The best that can be done is to *approximate* the pattern in a neighborhood of the state x by means of a system of differential equations.

We have seen that if there is a differential equation description, the activation-inhibition pattern is given by

$$u_{ij}(x) = \frac{\partial}{\partial x_j}\left(\frac{dx_i}{dt}\right).$$

There is no reason to stop here. We can consider the quantities

$$v_{ijk}(x) = \frac{\partial}{\partial x_k} u_{ij}(x),$$

which express the effect of changes in x_k on the activation-inhibition pattern. If $v_{ijk}(x) > 0$, then an increase in x_k tends to accelerate the activation of the rate of production of x_i by x_j, and we call x_k an *agonist* of x_i. If $v_{ijk}(x) < 0$, then x_k is termed an *antagonist.* Note that we can further extend this process by defining the quantities

$$w_{ijk\ell}(x) = \frac{\partial}{\partial x_\ell} v_{ijk}(x),$$

and so forth. If there is a differential equation description of the dynamics, each of these levels of description is obtained from the preceding one by means of a simple differentiation. However, if only the patterns are given and there is no global differential description, then the patterns are independent and cannot be obtained from a single "master" description. This is the situation that Rosen claims distinguishes the "complex" from the "simple." A fuller account of this view of system complexity and its relationship to activation-inhibition patterns can be found in

Rosen, R., "Some Comments on Activation and Inhibition," *Bull. Math. Biophysics,* 41 (1979), 427–445,

Rosen, R., "On Information and Complexity," in *Complexity, Language and Life: Mathematical Approaches,* J. Casti and A. Karlqvist, eds., Springer, Heidelberg, 1986, pp. 174–196.

CHAPTER THREE

PATTERN AND THE EMERGENCE OF LIVING FORMS: CELLULAR AUTOMATA AND DISCRETE DYNAMICS

1. Discrete Dynamical Systems

From the time of Heraclitus and the ancient Greeks, philosophers have contemplated the idea that "all things are in flux." This is our leitmotif as well, and in this book we will encounter the mathematical manifestation of this concept in a variety of forms. All of these forms, however, can be subsumed under the general heading *dynamical system.* So again in this chapter we examine this most important of mathematical "gadgets," exploring in some detail cellular automata, one of the simplest classes of dynamical systems. As we shall see, however, even this seemingly elementary type of dynamical process offers an astonishing variety of behavioral patterns, and serves in some ways as a universal representative for all other dynamical phenomena. We'll return to this point later. But for now let's look at the general features needed to characterize a cellular automaton in mathematical terms.

The key element in the notion of a dynamic is the idea of a change of state. Assume that X is a set of abstract states of the sort discussed in Chapter 1. Then as we saw in the last chapter, a *dynamic* on X is simply a recipe, or a rule, specifying what state to go to next from the state you currently find yourself in. More formally, the dynamic is a family of maps $T_t\colon X \to X$. For example, if we consider our system to be the table lamp on my desk, and if the set $X = \{ON,\ OFF\}$, then the maps T_t might say that the switch should be turned to the ON position every other hour starting at midnight, and to the OFF position on every odd-numbered hour. The dynamic defined on X in this case would be given explicitly by the rule

$$T_t(ON) = \begin{cases} ON, & t = \text{odd-numbered hour}, \\ OFF, & t = \text{even-numbered hour}, \quad t = 0, 1, 2, \ldots, 23, \end{cases}$$

and similarly for the situation when the system is in the state OFF. This simple rule defines a *flow,* or change of state, on the set X.

The first point to note about the definition of the dynamical rule T_t is that both the state set X and the time set can be either continuous or discrete. In the simple example of the light switch they are both taken to be discrete, but this need not necessarily be the case. In fact, we most often encounter the situation when one or both of these sets is continuous. But for the remainder of this chapter we will usually assume that both the time set and the state set are discrete, although not necessarily finite. Thus, a typical state set for us in this chapter will be $\mathbb{Z}^n$, the set of n-tuples of integers. Our standard time set will be the nonnegative integers $t = 0, 1, 2, \ldots$.

It's customary to express the transformation T_t by the dynamical law

$$x_{t+1} = \mathcal{F}(x_t, t), \qquad x_t \in X, \tag{$\dagger$}$$

indicating that the state at time $t + 1$ depends upon the previous state x_t, as well as upon the current time t itself. Sometimes there will also be dependencies upon past states $x_{t-1}, x_{t-2}, \ldots$. As a typical example of this setup, consider the case when $X = \mathbb{Z}_+^5 \mod 2$, i.e., each element of X is a 5-tuple whose entries are 0 or 1. Let x_t^i represent the value of the ith component of x_t at time t, and define the transition rule to be

$$x_{t+1}^i = \left(x_t^{i+1} + x_t^{i-1}\right) \mod 2.$$

In other words, the value at $t + 1$ is just the sum of the values on either side of x^i taken "mod 2." Here to avoid boundary condition difficulties we identify the state components "mod 5," which means that we think of the component x^6 as being the same as x^1, x^7 as the same as x^2, and so forth. We will see later that the behavior of even such a simple-looking dynamic conceals a surprising depth of structure leading to long-term behavior of a bewildering degree of complexity from initial states having as few as a one nonzero entry.

Given the dynamical rule of state-transition $\mathcal{F}$, together with the initial state $x_0 \in X$, there is really just one big question we can ask about the system: What happens to the state x_t as $t \to \infty$? Of course, there are many subquestions lurking below the surface of this overarching issue, queries about what kind of patterns emerge in the long-term limit, how fast these limiting patterns are approached, the types of initial states that lead to different classes of limiting patterns and so on. But all these matters ultimately come down to the question of whether or not certain types of limiting behaviors are possible when the system starts in the initial state x_0 and proceeds according to the rule $\mathcal{F}$. Here are two examples to illustrate the point.

Example 1: Linear Dynamics

Consider the scalar linear system

$$x_{t+1} = ax_t, \qquad x_t,\ a \text{ real}.$$

We consider the long-term behavior of the state x_t as $t \to \infty$. It's evident that at any time t, we have $x_t = a^t x_0$ for any $x_0 \in X$. Thus, there are only three qualitatively different types of limiting behavior:

$$x_\infty = \begin{cases} 0, & \text{if } |a| < 1, \\ \infty, & \text{if } |a| > 1, \\ \pm x_0, & \text{if } |a| = 1. \end{cases}$$

Furthermore, the rate at which these limiting behaviors are approached depends upon the magnitude of a: the closer a is to 1 in absolute value, the slower the corresponding limit is reached. In addition, it's easy to see that the limiting behavior is reached monotonically, i.e., $|x_{t+1} - x_\infty| < |x_t - x_\infty|$ for all $t \geq 0$. Finally, we see that the limiting behavior is independent of the initial state x_0 (except in the special case $|a| = 1$). This decoupling of the final behavior and the initial state is exceptional, being a consequence of the system's linear structure. Usually there will be several different types of limiting behaviors. Which one the system displays usually depends directly upon the particular starting state x_0. Nevertheless, the foregoing example, elementary as it is, displays in a very transparent form the type of information we want to know about all such processes: What do they do as $t \to \infty$? How fast do they do it? In what way does the limiting behavior depend on the starting point and on the structure of the dynamical rule $\mathcal{F}$? These are the issues that we will explore in detail during the remainder of this chapter.

Example 2: Rabbit Breeding and the Fibonacci Sequence

As an example of a situation where the change of state depends not only upon the last state but also upon the state at earlier times, let's consider a problem of rabbit breeding. Assume that we start with a pair of rabbits (one male, one female) that breeds a second pair in the next period. Thereafter, this new pair also produces another pair in each period. Moreover, assume that each pair of rabbits produces another pair in the second period following birth, and thereafter one pair per period. The problem is to find the number of pairs at the end of any given period.

A little calculation shows that if u_i represents the number of pairs at the end of period i, then the dynamical rule governing the development of the rabbit population is

$$u_i = u_{i-1} + u_{i-2}, \qquad i \geq 2,$$

with the initial condition $u_0 = u_1 = 1$. This is a discrete dynamical system in which the state transition rule depends not only upon the immediate past state at period $i-1$, but also upon the state one period earlier at $i-2$. It's also easy to see that the limiting behavior of this system is the well-known population explosion of rabbits, $u_i \to \infty$.

Computing the sequence $\{u_i\}$, we obtain the values $1, 1, 2, 3, 5, 8, 13, \ldots,$ the well-known sequence of Fibonacci numbers. We shall see this sequence turning up in a surprising variety of settings ranging from plant phyllotaxis to cryptographic codes as we make our way through the course of this book.

Exercises

1. Consider the discrete dynamical system $x_{t+1} = ax_t(1 - x_t)$. To ensure that x_t always remains between 0 and 1, we restrict the parameter a to the region $0 \leq a \leq 4$. Using a hand calculator or a computer, explore the long-term behavior of this system for various values of a and various starting points x_0. In particular, consider the case $a = 4$. (Note: We'll see a lot more of the behavior of this system in the next chapter.)

2. The so-called "butterfly" curve is given in polar coordinates as

$$\rho = e^{\cos\theta} - 2\cos 4\theta + \sin^5(\theta/12).$$

(a) Regarding θ as the time variable, plot the "trajectory" of this dynamical process, i.e., plot ρ as a function of increasing θ. Can you see why this curve is called the butterfly curve? (b) How does the behavior of the system

$$\rho = e^{\cos 2\theta} - \frac{3}{2}\cos 4\theta,$$

differ from the standard butterfly curve given above?

3. (a) Suppose a school has an endowment that produces $2n$ dollars per week to pay for seats at a local concert hall. The tickets cost \$2 each for a teacher and \$1 each for a student. In how many ways could a party for the hall be arranged, assuming that neither an interchange of teachers nor an interchange of children, but only an interchange of a teacher with a child constitutes a new arrangement? $\left(\text{Answer: } \sum_{i=0}^{n-1} \binom{n+i}{2i}\right)$. (b) Show that this number is always one of the numbers in the Fibonacci sequence.

2. Cellular Automata: The Basics

Consider an infinite checkerboard in which each square can be colored either black or white at each moment of time t. Assume that we have a rule that specifies what the color of each square should be as a function of its four neighboring squares—above, below, left and right. Now let an initial pattern of black and white squares be given at time $t = 0$. We then turn the system on and let the rule of state transition operate, examining the pattern that emerges as $t \to \infty$. In other words, we watch the squares blink on and off at each time step in accordance with the rule of state transition, looking for the steady-state pattern, if any, that emerges as the time becomes large. This set-up describes the prototype for what is termed a *cellular automaton* or, more accurately in this case, a two-dimensional cellular automaton, since the state space is a planar grid. Let's look at the key ingredients forming such a gadget.

• *Cellular State Space*—The "backcloth" upon which the dynamics of the automaton unfold is a cellular grid of some kind, usually a rectangular partitioning of R^n. Our interest will focus upon the one- and two-dimensional cellular automata for which the grids forming the state spaces will be either the checkerboard structure described above, or its one-dimensional counterpart, which consists of lattice points ("cells") on the infinite real line R^1.

• *Finite States*—Each cell of the state space can assume only a finite number of different values k. So if we have a finite grid of N cells, then the total number of possible states is also finite and equals k^N. Since the cellular grid is assumed to be at most countable, the total number of states is also countable, being equal to $k^{\aleph_0}$, where $\aleph_0$ represents the cardinality of the integers $\mathbb{Z}$.

• *Deterministic*—The rule determining the value of each cell is a deterministic function of the current value of that cell and the values of the cells in some local neighborhood of that cell. On occasion we will modify this feature and allow stochastic transitions. But for the most part we consider only deterministic transition rules.

• *Homogeneity*—Each cell of the system is the same as any other cell, in the sense that the cells can each take on exactly the same set of k possible values at any moment and the same state-transition rule is applied to each cell.

• *Locality*—The state transitions are local in both space and time. The next value of a given cell depends only upon the current value of that cell and those in a local neighborhood at the previous time period. Thus we have no time-lag effects, nor do we have any nonlocal interactions affecting the state transition. For one-dimensional automata, the local neighborhood consists of a finite number of cells on either side of a given cell; for two-dimensional automata, there are traditionally two basic neighborhoods of interest: the *von Neumann neighborhood,* consisting of those cells vertically and horizontally adjacent to the given cell and the *Moore neighborhood,* which also includes those cells that are diagonally adjacent. These neighborhoods are illustrated in Fig. 3.1. Time now for another example.

Example: Racial Integration in Urban Housing

To illustrate the use of cellular automata theory in a social setting, consider the problem of racial integration of housing in an urban neighborhood. Suppose we partition the urban region into a rectangular grid of size 16 by 13 cells, assuming that each cell represents a location that could be occupied by a white or a black family, or is empty. Thus, here we have $k = 3$ and there is a total of $3^{(16\times 13)} = 3^{208} \approx 10^{99}$ possible states of this urban area.

As the rule of state transition, let's postulate that each racial group

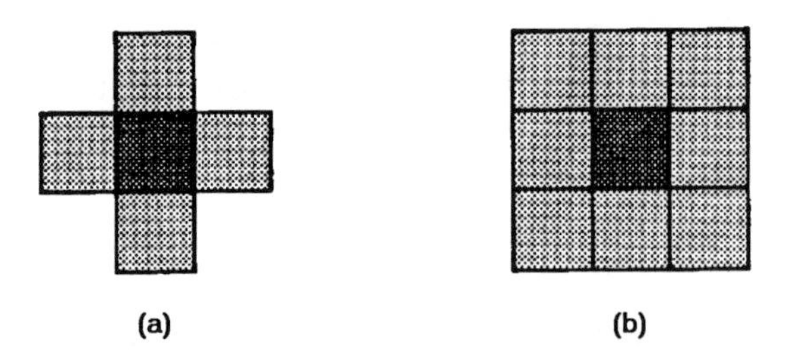

Figure 3.1. (a) Von Neumann Neighborhood, (b) Moore Neighborhood

would prefer to have a certain percentage of its immediate neighbors being of the same group. If this is not the case, we'll assume that each moves to the nearest location where the percentage of like neighbors is acceptable. In order to have a reasonable choice of where to move, it's been empirically observed that around 25%–30% of the housing locations should be vacant. If we use the Moore neighborhood and start with the initial state shown in Fig. 3.2, in which "o" denotes a white family, "#" denotes a black household and a blank space means the housing space is empty, we arrive at the steady-state distribution of Fig. 3.3(a) when we impose the condition that at least half of one's neighbors must be of the same color. Figure 3.3(b) shows the steady-state distribution from the same initial state when the requirement changes to at least one-third of one's neighbors must be of the same group.

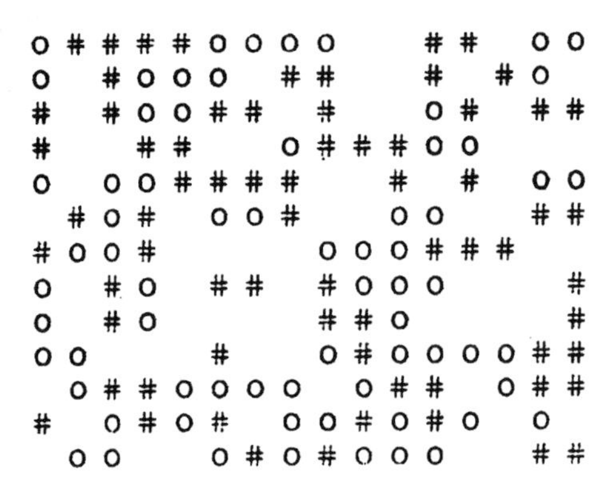

Figure 3.2. Initial Housing Distribution

This vastly overly simplified version of urban housing patterns already suffices to illustrate some important features of cellular automata models. First of all, the rule of state transition is one that has no special analytic structure; it is a linear threshold rule, with the threshold set by the parameter measuring the number of neighbors (in the Moore neighborhood)

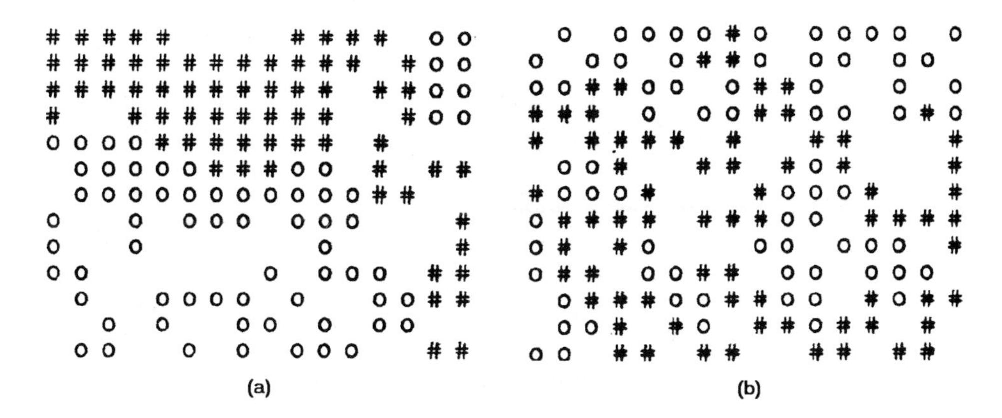

Figure 3.3. (a) $\frac{1}{2}$ of Neighbors the Same, (b) $\frac{1}{3}$ of Neighbors the Same

that are of the same color group. Furthermore, we see from Fig. 3.3 that the structure of the final distribution is critically dependent upon this parameter. Fig. 3.3(b) looks much like the initial distribution of Fig. 3.2, i.e., it's pretty unstructured. But Fig. 3.3(a) shows a considerable amount of structure emerging out of the unstructured initial distribution. Thus, the passage in going from demanding that a third of one's neighbors be of the same color to requiring that half of them be of the same group induces a rather pronounced "phase transition" in the final housing pattern. We shall see more of this type of bifurcation behavior in other cellular automaton models later on in the chapter.

At this juncture it's of interest to devote a couple of paragraphs to some historical facts surrounding the development of cellular automata. The first cellular automaton was dreamt up by John von Neumann around 1950, following up a suggestion due to Stanislaw Ulam. What von Neumann was looking for was a model of computation in which the computer could act on its own "matter," so to speak. Conventional models of computation, like the Turing machine that we'll discuss in Chapter 9, make a distinction between the structural parts of a computer and the data on which the computer operates. These kinds of computers cannot extend or modify themselves, nor can they build other computers. Von Neumann was interested in a model of computing that would more faithfully mirror the behavior of complex, life-like objects. Enter the cellular automaton.

In a cellular automaton, objects that represent passive data and objects that represent computing machinery are both assembled out of the same elements and are subject to the same laws of state transition. As we'll see later in the chapter, von Neumann was able to show that a cellular automaton with a suitable state space and rule of state transition could serve to represent the processes of both construction and computation. Even more remarkably, von Neumann proved that such a device would also be capable of building copies of itself, thereby displaying that most characteristic feature of living things—self-reproduction. And, in fact, the mechanisms von Neumann proposed for achieving self-reproducing structures within a cellular automaton bear a strong resemblance to those that are actually used by wet, squishy real-world biological life. So while most of our discussion in this chapter will focus on the mathematical aspects of cellular automata, it's useful for the reader to keep this very concrete problem of living organisms in mind as motivation for our mathematical development before we return to the biology at the end of the chapter. Now let's look at a couple of very interesting examples that set the stage for the mathematics.

Example 1: Cyclic "Eaters"

Suppose our state space is the infinite discrete set $\mathbb{Z}$, and that each location on this line can be occupied by one of N colors, labeled not very imaginatively $0, 1, 2, \ldots, N-1$. Further, assume that these colors are ordered so that $0 \prec 1 \prec 2 \prec \cdots \prec N-1 \prec 0$. Thus, we can think of these colors as being represented on a "color wheel," so that color 0 and color $N-1$ are adjacent. Now for our rule of state transition.

Consider a fixed, but arbitrary, cell x on the line. We say that the color at a neighboring cell y "eats" the color at x provided that the color at y immediately precedes the color at x on the color wheel. In other words, the rule of transition says that cell x takes on the color of cell y at the next time step provided that

$$\text{color } y - \text{color } x = 1 \mod N.$$

This rule is motivated to some extent by the Lotka-Volterra equations of classical population dynamics, in the sense that the colors can be thought of as species in some food chain. Now let's start the system at time $t = 0$ by randomly assigning the colors to the cells. We then turn on the *cyclic appetite rule* above, sit back in front of our color screen and watch the patterns that appear. Let's consider the possibilities.

Intuitively, what should happen is that if the number of colors N is much larger than the size of the neighborhood we choose for the cells, then each color should have a hard time finding anything to eat and the process should eventually stabilize. So, after a sufficiently large time, every cell should converge to a color characteristic for that cell and retain that color thereafter. In short, the system should converge to a random, fixed configuration. But if N is small, things don't work out this way, at all.

For N sufficiently small, we expect to see the system fluctuate forever. In particular, there should be a critical number of colors N^* at which a phase transition from fluctuation to stabilization takes place. A bit of computer experimentation leads us to speculate that $N^* = 4$, a conjecture that can actually be proved mathematically. Thus, if $N \leq 4$ the system fluctuates, while it stabilizes if $N \geq 5$. We'll return to this one-dimensional cyclic eater in the Exercises. Let's now turn our attention to a two-dimensional automaton whose behavior is also generated by a rule that's motivated by some simple aspects of population dynamics.

Example 2: The Game of "Life"

Undoubtedly, the most famous and well-studied cellular automaton is the two-dimensional system described by John Horton Conway's board game "Life." In this metaphor for birth, growth, evolution and death, the state space is taken to be an infinite checkerboard, each square capable of being either ON or OFF at each time period. Thus, for "Life" there are only $k = 2$ possible values for each cell at each moment. The state-transition rule is set up in order to mimic some basic features of living organisms. Life uses the Moore neighborhood, the state transitions being given by the rule:

A. If the number of ON neighbors is exactly 2, then the cell does not change its current value in the next period.

B. If the number of ON neighbors is exactly 3, then the cell is ON in the next period regardless of its current value.

C. If the number of ON neighbors is any other number, then the cell is OFF in the next period.

There are no other conditions. Basically, the rule says that the cell dies if it has either too few neighbors (isolation) or too many (crowding), but prospers if the number of neighbors is just right (conditions A and B). Note that Life is not a game in the usual sense of the term. The only decision to be made is what the initial configuration of ON and OFF cells will be. In a later section we will return to a detailed study of many of the features of this cellular automaton. For now, it's sufficient just to note that the types of questions of greatest interest in the Life universe center about the possibilities of self-reproducing patterns of various sorts.

Exercises

1. Figure 3.3 shows that in the urban housing problem there is a pronounced clustering of racial types when people demand that at least half their neighbors be of the same color. But when this requirement drops to one-third, the initial, more-or-less random distribution does not settle down to a bipolar pattern. This suggests that there is some crossover value between $\frac{1}{3}$ and $\frac{1}{2}$ at which racial "tipping" occurs. By numerically experimenting with this automaton, see if you can pin down this value.

2. Consider how you might extend the one-dimensional Cyclic Eater example of the text to the plane. Show that if the initial configuration in the plane contains a large square of color 0 surrounded by a random pattern of all colors, then the patterns that emerge will display wave-like clusters.

3. One-Dimensional Cellular Automata

The Game of Life and the urban housing integration examples in the preceding section are illustrations of two-dimensional cellular automata, where the state space is a grid in the plane R^2. But it turns out that the even simpler case of one-dimensional automata already contains many of the characteristic features of the dynamical behavior of these objects, and in a setting in which it is far easier to analyze the patterns that come about. So in this section we will focus our interest upon those cellular automata "living" on the integer points $\cdots - 3, -2, -1, 0, 1, 2, 3, \ldots$ of the real line. Furthermore, we will make the situation even easier by assuming that at each of these grid points the automaton can assume only $k = 2$ possible values, which we shall denote graphically by a 0 (or a blank space) or a 1 (or the symbol $*$). Thus, if we let ${a_t}^i$ be the value at grid point i at time t, we can express the state-transition rule in the form

$$a_t^i = \mathcal{F}({a_{t-1}}^{i \pm r}), \quad i = 0, \pm 1, \pm 2, \pm 3, \ldots; t = 0, 1, 2, \ldots ; r = 0, 1, 2, \ldots, R,$$

where $R \geq 0$ is an integer expressing the size of the neighborhood and $\mathcal{F}$ is the specific transition rule. Usually we will take $R = 1$ or 2. Now let's consider the rule $\mathcal{F}$ in more detail for the case $k = 2$, $R = 1$.

Assume the rule $\mathcal{F}$ is the so-called *mod 2 Rule,* in which the value of a cell at time $t + 1$ is just the sum of its two neighbors taken modulo 2, i.e., we have

$$a_{t+1}^i = (a_t^{i-1} + a_t^{i+1}) \mod 2,$$

for $i = 0, \pm 1, \pm 2, \ldots$. The local rule for the time evolution of this automaton can be represented by the diagram

$$\frac{1\,1\,1}{0} \quad \frac{1\,1\,0}{1} \quad \frac{1\,0\,1}{0} \quad \frac{1\,0\,0}{1} \quad \frac{0\,1\,1}{1} \quad \frac{0\,1\,0}{0} \quad \frac{0\,0\,1}{1} \quad \frac{0\,0\,0}{0}$$

Here the upper part shows one of the eight possible states that a row of three successive cells can be in at time t, whereas the lower half shows the value that the central cell of the trio assumes at time $t + 1$. We can compactly denote this rule by using an idea due to Stephen Wolfram and interpreting the eight binary digits composing the lower-half of the diagram as the binary representation of a decimal number. We then name the rule by this number. Thus for the Mod 2 Rule illustrated in the diagram, we have the binary expression 01011010. This translates into the decimal number 90, i.e., $01011010_2 = 90$, so the name of the mod 2 Rule is Rule 90. It's clear that any eight binary digits defines a particular transition rule. So there are a total of 2^8 possible rules when $k = 2, R = 1$. In general, there are $k^{k^{2R+1}}$ possible rules. The time evolution of the automaton unfolds by simultaneously applying the given rule at each cell for each time step. So, for example, for Rule 90 the following single-step transition leads from the initial state on the upper line to the next state given on the line below:

(time t)	⋯	1	0	1	1	0	1	1	0	1	0	1	0	1	1	0	⋯
(time $t+1$)	*	*	0	0	1	1	0	1	1	0	0	0	0	0	1	*	*

Of the 256 theoretically possible local rules for a one-dimensional cellular automaton with $k = 2$, $R = 1$, we impose some restrictions on the set of rules in order to eliminate uninteresting cases. First, a rule will be considered "illegal" unless the zero-state 000 remains unchanged. This implies that any rule whose binary specification ends in 1 is forbidden. Next, we demand that rules be reflection symmetric. Consequently, 100 and 001, as well as 110 and 011, lead to identical values. Imposition of these two restrictions leaves us with 32 legal rules having the general form $abcdbed0$, where each letter can assume either the value 0 or 1.

Figure 3.4 shows the evolution of some of the 32 legal rules, starting from an initial configuration consisting of a single nonzero cell. The evolution is shown until a particular configuration appears for the second time (a "cycle") or until 20 time steps have taken place. We can see several classes of behavior in these patterns. In one class, the initial 1 is erased (Rules 0 and 32) or is maintained unchanged forever (Rules 4 and 36). Another class of behaviors copies the 1 to generate a uniform structure that expands by one cell in each direction on each time step (Rules 50 and 122). We call these kinds of rules *simple.* A third class of rules, termed *complex,* yields complicated, nontrivial patterns (e.g., Rules 18, 22 and 90).

The patterns of Fig 3.4 were generated from a single nonzero initial cell. In Fig. 3.5 we show the results of the same experiment but now with the initial configuration randomly selected, each cell independently having the value 0 or 1 with equal probability. Again we see a variety of behaviors with complex rules yielding complicated patterns. The most interesting feature of Fig. 3.5 is the fact that the independence of the initial cell values is totally destroyed, as the dynamics generates correlations between values at separated cells. This is the phenomenon of *self-organization,* in which an initially random state evolves to a state containing long-range correlations and nonlocal structure. The behavior of the automata of Fig. 3.5 is strongly reminiscent of the behavior of the dynamical systems that we looked at in the last chapter, where simple rules led to steady-state behaviors consisting of fixed points or limit cycles. The complex rules give rise to behavior

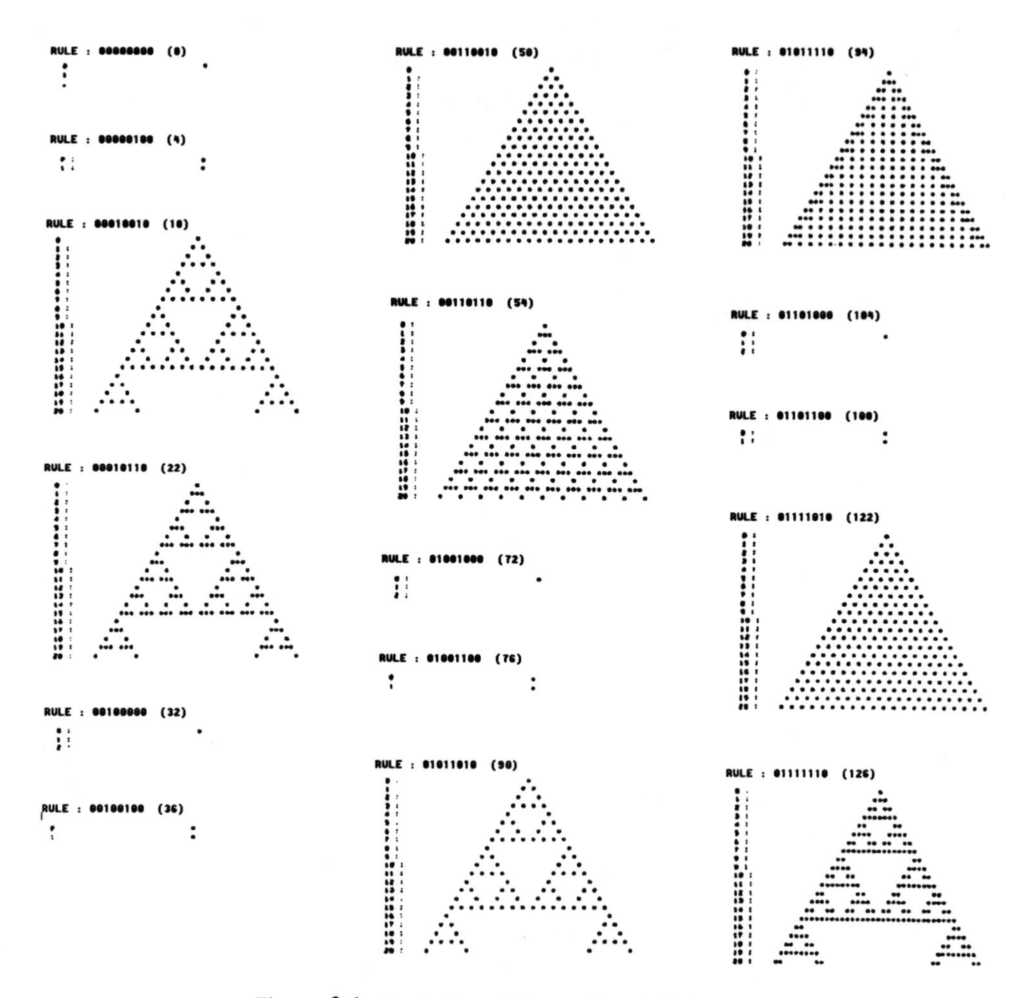

Figure 3.4. Evolution of Some Legal Rules

analogous to more complicated types of "strange attractors," which we'll take up in the next chapter.

In passing, it's important to note that the foregoing experiments were carried out under the assumption of periodic boundary conditions, i.e., instead of treating a genuinely infinite line of cells, the first and last cells are identified as if they were adjacent on a circle of finite radius. Another possibility for "finitizing" the situation would be to impose null boundary conditions, implying that the cells beyond each end always have the value zero rather than evolving in accordance with the local rule.

Example: Rats, Rodents and Cellular Automata

To illustrate the use of one-dimensional cellular automata in an applied setting, let's consider the problem of growth and decline in a population of

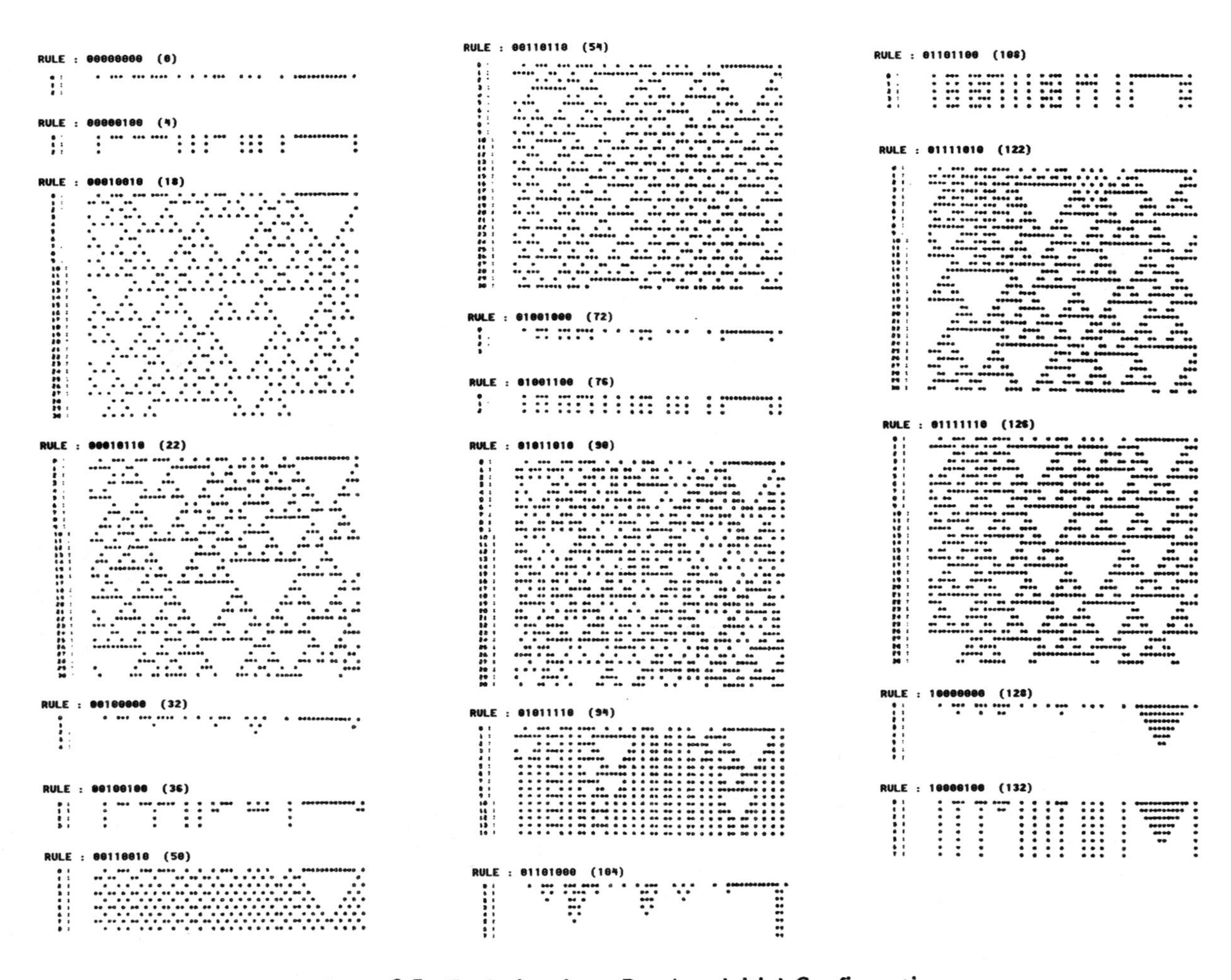

Figure 3.5. Evolution from Random Initial Configurations

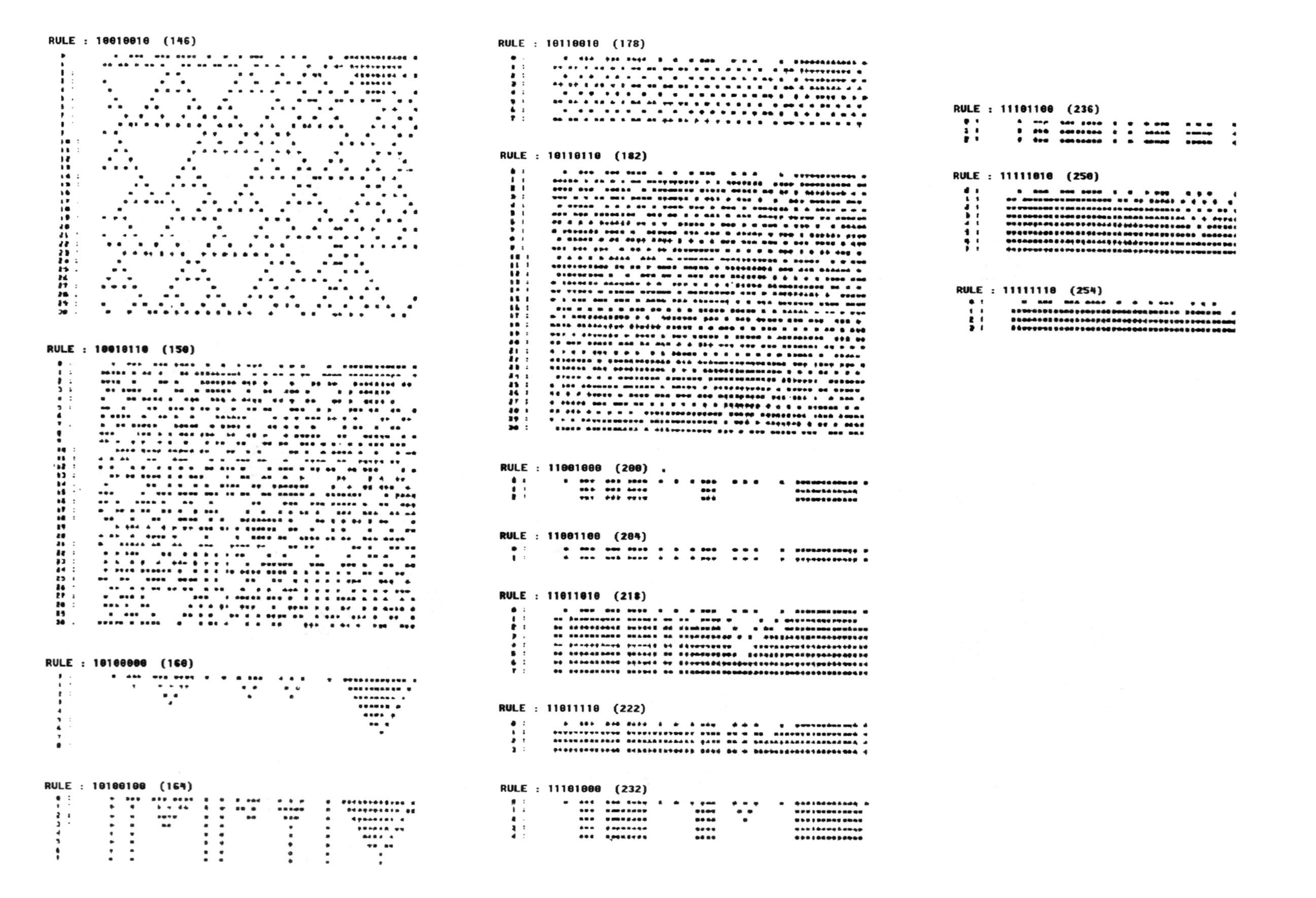

Figure 3.5.(cont'd.) Evolution from Random Initial Configurations

small rodents like lemmings, voles and rats. Ecologists have long been puzzled by the apparent haphazardness of these populations, which seem to defy the neat, cyclic behavior of the standard Lotka-Volterra equations. For instance, some vole populations cycle regularly for a few years and then, for no apparent reason, switch to a phase of random fluctuations. Elsewhere, the same species, under virtually identical circumstances, exhibits regular cycling in one part of its range, and irregular fluctuations in another part. How can we make any sense of this?

To study this question, let's simplify the situation by assuming a closed ecosystem and let's agree to measure population by the local population density at a point in time. We will assume that the rule governing the change of the population density within a given patch of land depends on yesterday's density in both that patch and in neighboring patches. To further simplify matters, assume that the patches are strung out along a one-dimensional grid. Thus, the mathematical situation is exactly the one we have been considering in the abstract throughout this section.

There are basically three physically plausible forms of the population density transition rule, covering situations when the population does best at low densities and medium densities, together with a double-hump rule corresponding to two genetically different types of rodents, the first doing best at low densities, while the second thrives under more crowded conditions. Pictures of these transition rules are shown below in Fig. 3.6, where a given cell's state at time t is shown as a function of the sum of its neighbors at the previous time period. Thus, each of the rules shown in the figure is what we have earlier termed a totalistic rule. We assume that each cell (patch) can be in one of four states, corresponding to having a local population of $0, 1, 2$ or 3 rodents at the site.

Using a line of 29 cells, Helen Couclelis carried out experiments using the above rules with different initial conditions. The results of these experiments are shown in Fig. 3.7, where the density level at each site is indicated by increasingly dark gray tones. Each experiment involved using one of the three growth hypotheses in two variants. In experiments 1–3 a two-cell neighborhood is assumed, while experiments 4–6 extend the neighborhood size to four cells. Furthermore, three kinds of initial conditions are considered: (A) a fairly homogeneous population, (B) an initial population only at one or a few points, and (C) a fairly disordered initial population. Combinations of these categories are indicated by a second letter in Fig. 3.7.

By adding up the population densities at each cell, it's not hard to see that these experiments give rise to population levels that may fluctuate wildly (experiments 2A–2BB and 4B), stabilize (experiments 3AB–3C), go to extinction (experiment 5B), crash (experiments 1C, 5A, 6AB) or even miraculously recover from near-extinction (experiments 2A and 4A). And

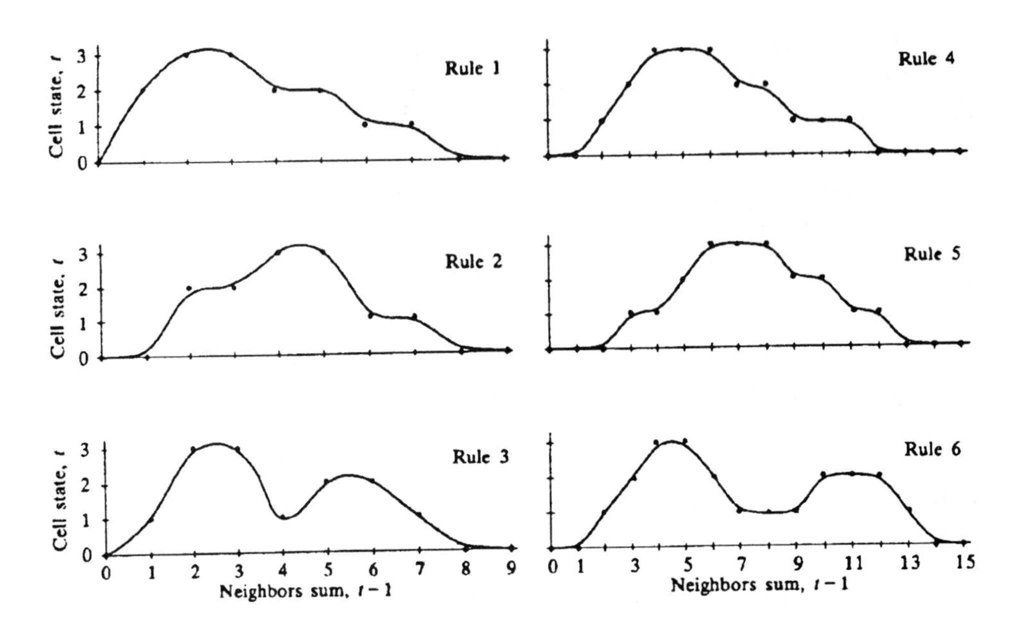

Figure 3.6. Various Population Transition Rules

these effects become even more pronounced when one looks at the fluctuation of population levels at individual sites. With this graphic example in mind, let's now look at some of the finer detail associated with the local behavior of these one-dimensional cellular automata.

Exercises

1. In the case of 1-D cellular automata with $k = 2$, $R = 1$, Wolfram's numbering scheme enables us to attach a decimal number to each transition rule. How would you modify this scheme to do the same thing for the case when $k = 3$, $R = 1$? Can you generalize this idea to arbitrary k and R?

2. In the text we considered only 1-D cellular automata with periodic boundary conditions. Using a computer, experiment with 1-D rules using other kinds of boundary conditions such as absorbing barriers (the boundary cells are fixed at 0) or random barriers, in which the values of the cells at the boundary are selected randomly. Do you see any substantial difference from the periodic case in the patterns that emerge in these cases?

3. Think of some other plausible growth laws for the Rats-and-Rodents situation, and explore them numerically.

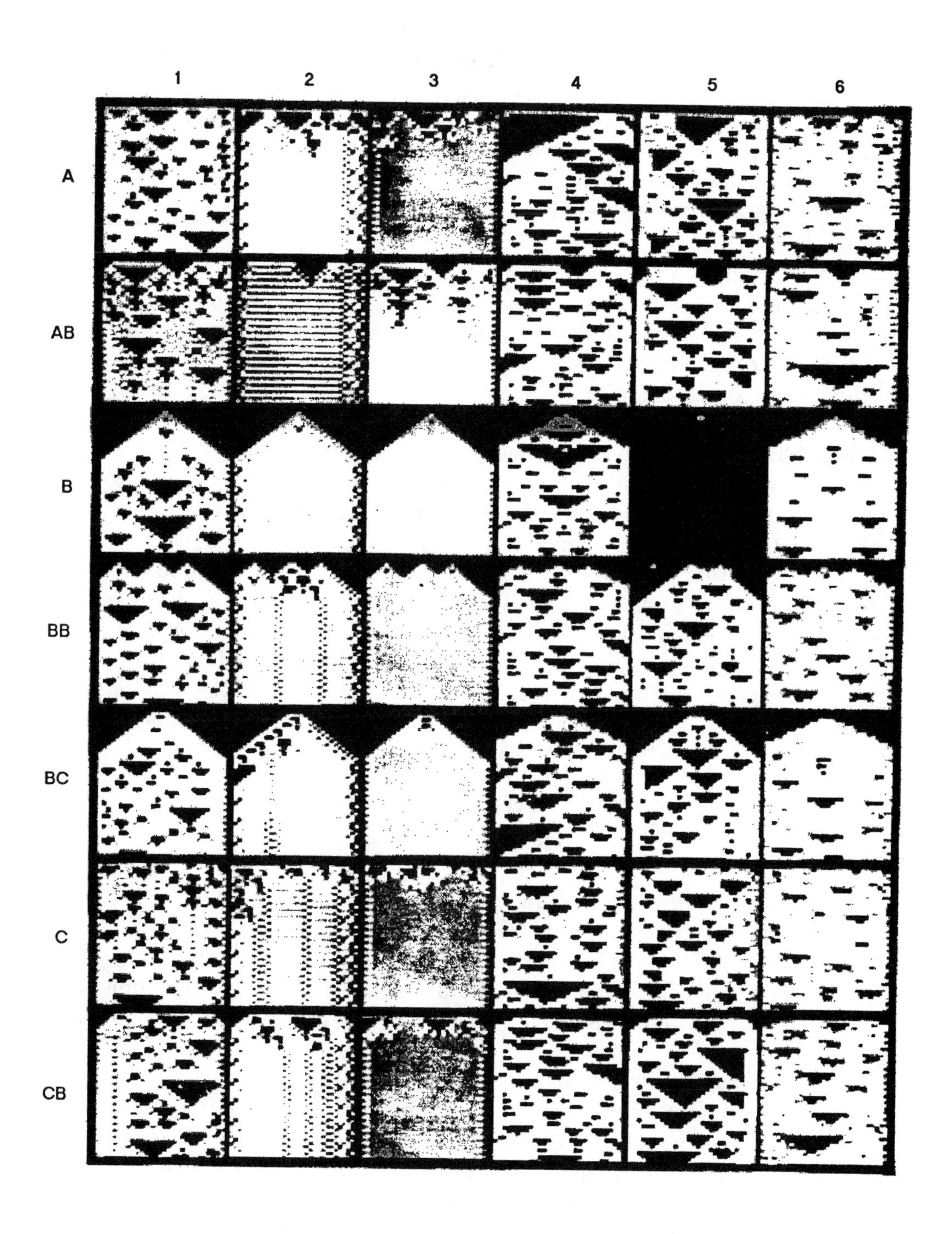

Figure 3.7. Spatiotemporal Structure in Rodent Population Dynamics

4. Local Properties of One-Dimensional Automata

We saw the self-organization property of cellular automata emerge from the application of a variety of complex rules in Fig. 3.5. Now we want to take a more detailed look at means for quantitatively characterizing the self-organization that is so pictorially evident in these computer experiments. In particular, we will examine the statistical properties of the configurations generated during the course of the time evolution of the automata described in the last section.

A configuration (or line of cells) is disordered, or essentially random, if values of different cells are statistically uncorrelated. Deviations of statistical measures of configurations from their values for disordered configurations signify order and indicate the presence of correlations between values at different cells. A disordered configuration is specified by a single parameter p, which represents the probability of each cell taking on the value 1. But we need additional parameters to specify ordered configurations.

The simplest statistical quantity that can be used to characterize a configuration is the average fraction of cells having the value 1. We denote this quantity by ρ. Clearly, for a disordered configuration $\rho = p$.

To begin with, consider the density ρ_1 obtained from a disordered initial configuration after one time step. When $p = \rho = \frac{1}{2}$, a disordered configuration contains all eight possible three-cell neighborhoods with equal probability. Applying a rule specified by a binary sequence $\mathbf{R}$ to this initial state yields, after one time step, a configuration in which the fraction of cells with value 1 is given by the fraction of the eight possible neighborhoods leading to a 1 under the rule $\mathbf{R}$. This fraction is

$$\rho_1 = \frac{\#_1(\mathbf{R})}{(\#_0(\mathbf{R}) + \#_1(\mathbf{R}))} = \frac{\#_1(\mathbf{R})}{8},$$

where $\#_d(S)$ represents the number of occurrences of the digit d in the binary representation of a number S. So, for example, if we take cellular automaton rule $\mathbf{R} = (01111110)$, we have $\#_1(\mathbf{R}) = 6$. Consequently, the density $\rho_1 = 6/8 = 0.75$ (for an infinite line of cells). The generalization of this formula to the case when $p \neq \frac{1}{2}$ is discussed in Problem 3 at the end of the chapter.

Considering the density ρ_τ for time steps $\tau > 1$, we find that the simple argument used for ρ_1 cannot be used any longer because correlations induced by the cellular automaton's temporal evolution interfere with the independence assumption underlying the derivation of ρ. However, exact results for ρ_τ can be obtained with the mod 2 rule (Rule 90), since this rule obeys the property of *additive superposition.* This means that the configurations obtained by evolution from any initial configuration are given by appropriate combinations of those for evolution from an initial configuration having only a single nonzero cell. Cellular automata have the additive superposition property only if their rule is of the form $ab0cbac0$ with $c = (a + b) \mod 2$. Let's calculate ρ_τ for Rule 90.

Let $N_\tau^{(1)}$ be the number of cells with value 1 at time-step τ obtained from an initial state containing a single nonzero cell. It can be shown that

$$N_\tau^{(1)} = 2^{\#_1(\tau)},$$

where $\#_1(\tau)$ is the number of ones in the binary expansion of the integer τ.

The expression for $N_\tau^{(1)}$ shows that the density averaged over the region of nonzero cells (the "light cone") in the evolution according to Rule 90 is given by $\rho_\tau = N_\tau^{(1)}/(2\tau + 1)$, which does not tend to a definite limit as $\tau \to \infty$. However, the time-average density

$$\bar{\rho}_T = \frac{1}{T}\sum_{\tau=0}^{T} \rho_\tau,$$

does tend to the expected limit 0 as T grows large, with the asymptotic behavior being like $T^{\log_2 3-2}$. In general, the value of a cell at time τ is a sum (taken mod 2) of the initial values of $N_\tau^{(1)}$ cells, each of which have value 1 with probability ρ_0. If each of a set of k cells has value 1 with probability p, then the probability that the sum of the values of the cells will be odd (equal to 1 mod 2) is

$$\sum_{i \text{ odd}} \binom{k}{i} p^i (1-p)^{k-i} = \tfrac{1}{2}[1 - (1-2p)^k].$$

Hence, the density of cells with value 1 obtained by evolution for τ time steps from an initial state with density ρ_0 using Rule 90 is given by

$$\rho_\tau = \tfrac{1}{2}[1 - (1-2\rho_0)^{2^{\#_1(\tau)}}].$$

This result is shown in Fig. 3.8 for Rules 90, 18 and 182 as a function of the number of time steps τ when $\rho_0 = 0.2$. For large τ, $\#_1(\tau) = O(\log_2 \tau)$ except at a set of points of measure zero. So $\rho_\tau \to \frac{1}{2}$ as $\tau \to \infty$ for almost all time-steps τ. Note that in Fig. 3.8 the values of cells at time τ depend on the values of $O(\tau)$ initial cells for the nonadditive complex Rules 18 and 182. Moreover, ρ_τ tends smoothly to a definite limit that is independent of the density of the initial disordered configuration.

The very essence of the idea of self-organization is that the evolution of a cellular automaton should generate correlations between cell values at different locations. The average density measure just discussed is much too

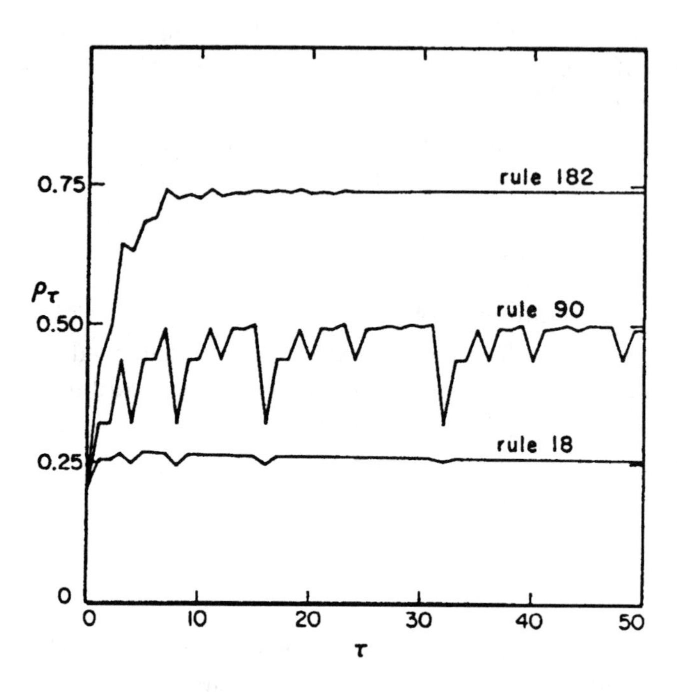

Figure 3.8. Average Density of Nonzero Cells as a Function of Time

coarse to give any information of this sort. So we consider the simplest type of correlation measure, the two-point correlation function

$$C^{(2)}(r) = \langle S(m)S(m+r)\rangle - \langle S(m)\rangle\langle S(m+r)\rangle,$$

where the average (using the $\langle\cdot\rangle$ notation) is taken over all possible locations m in the cellular automaton at a fixed time, with $S(k)$ taking on the values -1 and $+1$ when the cell at location k has values 0 and 1, respectively. As already noted, a disordered configuration gives $C^{(2)}(r) = 0$ for $r > 0$, with $C^{(2)}(0) = 1 - (2\rho_0 - 1)^2$. With the single-cell initial states, the periodicities of the configurations that result from complex rules give rise to peaks in $C^{(2)}(r)$. At time-step t, the largest peaks occur when $r = 2^k$, with the digit corresponding to 2^k appearing in the binary decomposition of t. Smaller peaks occur when $r = 2^{k_1} \pm 2^{k_2}$, and so on. For additive rules, the form of $C^{(2)}(r)$ is obtained by a convolution of the foregoing result with the correlation function for the particular initial configuration. So for these rules we have $C^{(2)}(r) = 0$ for all disordered initial configurations. However, for nonadditive rules there is the possibility of nonzero short-range correlations from disordered initial configurations. A nonzero correlation length is the first sign of self-organization in the evolution of an automaton. Figure 3.9 shows the form of $C^{(2)}(r)$ for the complex Rule 18. Here we see that the correlation appears to fall off exponentially with a correlation length $r \cong 5$.

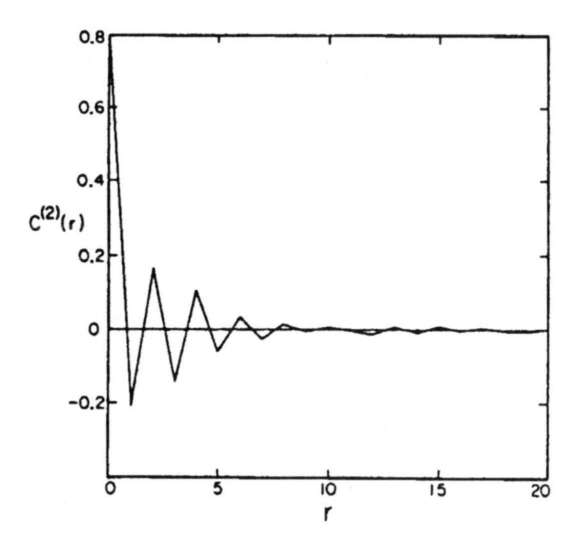

Figure 3.9. The Correlation Function $C^{(2)}(r)$ for Large t Using Rule 18

The preceding measures of structure give some insight into the local patterns that may emerge during the evolution of the automaton's state. However, inspection of Figs. 3.4 and 3.5 shows that considerable additional structure is present that these measures are incapable of identifying. For example, the two-dimensional picture arising from the evolution of all legal rules displays many triangular figures of various sizes. These triangles are formed when a long sequence of cells, which suddenly all assume the same value 1, is progressively reduced in length by local "thermal" fluctuations. If we let $T(n)$ be the density of triangles of base length n, then it can be shown that $T(n) \sim \lambda^{-n}$ for disordered initial states. It turns out that $\lambda \sim 2$ for additive rules, while for nonadditive rules $\lambda \sim \frac{4}{3}$. Thus, the spectrum of triangles generated by complex rules is universal, independent of both the initial state and the particular rule being used. This result should be compared with the estimate $T(n) \sim n^{-\log_2 3} \approx n^{-1.59}$ for Rule 90 (see Problem 4). Other measures like the sequence density can also be used to pinpoint additional local structure in the patterns emerging from the cellular automaton rules. We refer the reader to the Notes and References at the end of the chapter for further details of these tests and now turn our attention to problems of *global* structure.

Exercises

1. Let $N_\tau^{(1)}$ be the number of cells having value 1 at time-step τ. Prove the result stated in the text that for an additive rule, $N_\tau^{(1)} = 2^{\#_1(\tau)}$.

2. Calculate the quantity ρ_τ for Rule 4 and Rule 122, assuming $\rho_0 = 0.2$.

3. Using the definition of the two-point correlation function given in the text, prove that the correlation is zero for all additive rules when the initial configuration is random (i.e., when the initial state itself has correlation zero).

4. Try to generalize the results for the two-point correlation to the case of k points.

5. Global Properties of One-Dimensional Automata

In the preceding section we examined the behavior of cellular automata by considering the statistical properties of the values of cells in individual configurations. Now we want to take an alternate approach and consider the statistical properties of the set of values in all possible final configurations. These are the configurations that the automaton reaches after the "transient" states have died out. This view of the behavior of an automaton allows us to make contact with issues involving the structure of the attractor set of a dynamical process, as well as with certain questions centering upon self-organization that are in some sense complementary to those considered in the last section.

For the most part, in this section we shall consider "finite" automata having $N < \infty$ cells. Thus, there are a total of 2^N possible configurations at each time step for such an automaton, with each configuration being specified by a binary integer of length N whose digits give the values of the corresponding cells. Further, we assume periodic boundary conditions, so that the first and last cells are identified as if the cells lay on a circle of circumference N. This set-up shows that the state-transition rule of the automaton defines a transformation from one sequence of binary digits to another, and provides a mapping from the set of binary digits of length N to itself. Experiments have shown that the final configurations obtained from nearby initial configurations are generally almost completely uncorrelated after just a few time steps.

In order to speak about the ensemble of final configurations and the convergence or divergence of initial configurations, we need to have some measure of distance in the set of possible configurations. It turns out to be convenient to use the *Hamming distance* $H(s_1, s_2)$, which measures the number of places at which the binary sequences s_1 and s_2 differ. Let's first consider the case of two initial configurations that differ only by the value of a single cell, i.e., the Hamming distance between them is 1. After T time steps, this initial difference may affect the values of at most $2T$ cells. However, for simple rules we find that the Hamming distance converges rapidly to a small value. The behavior of the Hamming distance for complex rules differs radically, depending upon whether the rule is additive or not. For additive rules (e.g., Rule 90 or Rule 150), the Hamming distance at time-step t is given by the number of nonzero cells in the configuration obtained from a starting configuration having a single nonzero cell. For Rule 90, this distance takes the form $H_t = 2^{\#_1(t)}$. For nonadditive rules, the difference between patterns obtained over time no longer depends just on the difference between the initial configurations. Generally, we find that in this case $H_t \sim t$, for large t. Thus, a bundle of initially close trajectories diverges over time into an exponentially expanding volume.

We can consider a statistical ensemble of states for a finite automaton by giving the probability for each of the 2^N possible configurations. A collection of states (configurations) such that each configuration appears with equal probability will be termed an "equiprobable ensemble." Such an ensemble may be thought of as representing the maximum degree of disorganization possible. Any transition rule modifies the probabilities for states in an ensemble to occur, thereby generating organization. Figure 3.10 shows the probabilities for the 1,024 possible configurations of a finite automaton with $N = 10$ obtained from an initially equiprobable ensemble after 10 time steps using Rule 126. The figure shows that the automaton evolution modifies the probabilities for the different configurations from the initially equally likely occurrence of all states to a final probability density that has some states' probabilities reduced to zero, while increasing some and decreasing others from their initial levels. Properties of the more probable configurations dominate the statistical averages over the ensemble, giving rise to the average local features of the steady-state configurations that we discussed in the last section.

It's of considerable importance in understanding the long-term behavior of cellular automata to examine the problem of *reversibility*. Most transition rules can transform several different initial configurations into the same final configuration. Thus a particular configuration can have many ancestors, but only a single descendant (e.g., the trivial Rule 0). In a reversible system, each state has a unique ancestor and a unique descendant so that the number of possible configurations must remain constant in time (Liouville's Theorem). But for an irreversible system the number of possible configurations may decrease with time, leading to the observation that final configurations of most cellular automata become concentrated in limited regions and do not fill the available volume of state space uniformly and densely. This kind of behavior is exactly what leads to the possibility of self-organization, as it allows some configurations to occur with larger probabilities than others, even in the steady-state limit. It turns out that as N becomes large, the fraction of possible configurations that can actually be reached by an irreversible rule tends to zero at a rate λ^N, where $\lambda \simeq 0.88$ for nonadditive rules. Such irreversible behavior can also be studied using ideas of entropy and information, but we leave a consideration of these notions to the chapter's Problems section.

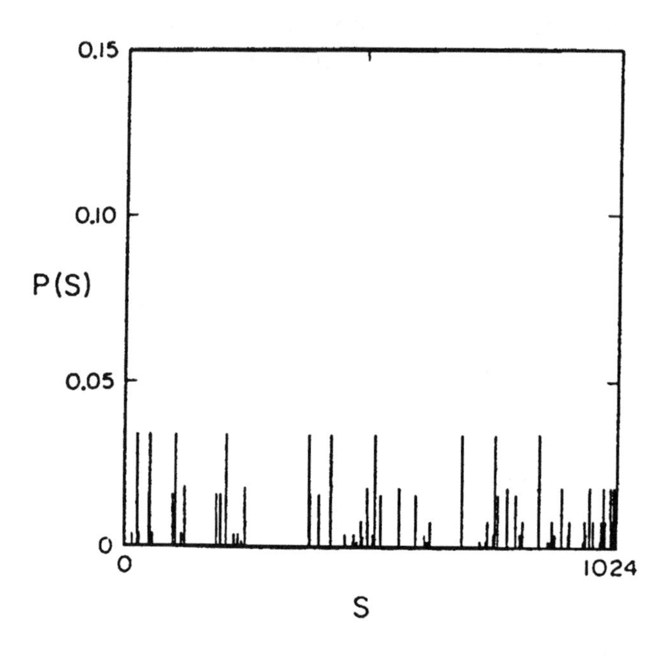

Figure 3.10. Configuration Probabilities with Rule 126

For finite cellular automata with N cells there are a total of only 2^N states. This means that the sequence of configurations reached from any initial state must become periodic after at most 2^N time steps. Thus, after an initial transient, the system must enter a cycle in which a set of configurations is cyclically repeated. Examination of Fig. 3.5 suggests that simple cellular automata give either short cycles involving just a few configurations or evolve after about N time steps to a state consisting of a stationary configuration of cycle length 1. Complex rules yield much longer cycle lengths involving isolated independent regions, each of which follows a short cycle of length at most 2^p, where p is the number of regions. It's also possible for complex rules to give rise to cycles whose length increases without bound as N increases.

Experiments show that of the 2^N possible configurations, only a small fraction are actually visited from any particular initial configuration. For example, using Rule 126 with $N = 8$, it has been found that at most 8 out of the possible 256 configurations are visited from any initial configuration. After a transient of at most two time steps, the system enters a cycle that repeats after no more than six further steps. A total of 29 distinct final configurations appear in these cycles. For $N = 10$, the maximum number of configurations is 38, whereas for $N = 32$ the number is at least 1,547. This type of behavior seems typical for most other complex rules.

Using additive rules, it's possible to give analytic results for transients and cycle lengths, as will be outlined in the next section. Using these techniques for Rule 90, if we let Π_N be the period arising from an initial configuration having a single nonzero cell, then it can be shown that

$$\Pi_N = 1, 1, 3, 2, 7, 1, 7, 6, 31, 4, 63, 14, 15, 1, 15, 14, 511, 12,$$

for $N = 3$ to 20. If we let Φ_N be the length of the transient phase, again starting with a single nonzero initial cell, we find with Rule 90 that $\Phi_N = 1$ for N odd, and $\Phi_N = D(N)/2$ otherwise, where $D(N)$ is the largest power of 2 that divides N. Note that $\Phi_N = 1$ for N odd implies that in these cases exactly half of the 2^N possible configurations appear on cycles when we are using Rule 90.

Finally, and in passing, we should note that, in general, infinite cellular automata do not usually exhibit finite cycles except under exceptional circumstances. Any initial configuration with a finite number of nonzero cells either evolves to the zero state or yields a pattern whose size increases progressively with time. However, it can be shown that if the values of the initial cells form an infinite periodic sequence of period k, then the evolution of the infinite cellular automaton will be identical to that of a finite cellular automaton with $k = N$, and cycles with length much less than 2^k will be found.

We can summarize the foregoing results by observing that cellular automaton configurations can be divided into essentially three classes, according to the circumstances under which they may be generated:

- *Gardens of Eden*—Some configurations can only appear as initial states of an automaton. These states can never be generated in the course of cellular automaton's evolution.

- *Transient States*—These are configurations that can arise only within, say, the first t steps of an automaton's evolution. Such configurations leave no long-term "descendants."

- *Cyclic States*—Such configurations appear in cycles and are visited repeatedly. Cyclic configurations can be generated at any time step and may be considered attractors to which any initial configuration is ultimately drawn.

For the most part, the discussion of the preceding sections has been based upon empirical observations of computer-generated experiments with the legal rules displayed in Figs. 3.4 and 3.5. Before looking at a few applications of these simplest of all cellular automata, let's first look at how *analytic* tools can be brought to bear upon the study of the behavior of cellular automata, at least in the case of additive rules like Rule 90 and Rule 150.

Exercises

1. Compute the quantities Π_N and Φ_N for Rule 150 for several values of N. Do the same calculation for a complex rule like Rule 18.

2. Try to create a reversible 1-D rule. (Hint: Remember that reversibility means the rule is a one-to-one mapping.) Can a reversible rule have cycles? Or fixed points?

3. In the text we stated that for the complex Rule 126 with $N = 8$, at most 8 of the 256 possible configurations are visited from any initial configuration. This means that for this automaton there are 248 Garden-of-Eden configurations. Carry out experiments with additive rules like Rule 90 or Rule 150 to see if for these rules the vast majority of possible configurations are also never reached.

6. Algebraic Properties of Additive Cellular Automata

We consider a one-dimensional cellular automaton consisting of N cells arranged in a circle (to give periodic boundary conditions). Denote the values taken on by the cells at time t by $a_0^{(t)}, a_1^{(t)}, \ldots, a_{N-1}^{(t)}$. As before, these values are either 0 or 1.

The complete configuration of such an automaton can be represented by a *characteristic polynomial*

$$A^{(t)}(x) = \sum_{i=0}^{N-1} a_i^{(t)} x^i,$$

where the value of cell i is the coefficient of the term x^i, each coefficient taking on either the value 0 or 1.

It's convenient to consider generalized polynomials containing both positive and negative powers of x. These polynomials are usually termed *dipolynomials,* and possess the same divisibility and congruence properties as ordinary polynomials. It's clear from the above representation that multiplication of $A(x)$ by $\pm x^k$ yields a dipolynomial that represents a configuration obtained from $A(x)$ by shifting each cell k places to the left $(+x^k)$ or the right $(-x^k)$. Periodic boundary conditions are enforced by reducing the characteristic dipolynomial modulo the fixed polynomial $x^N - 1$ at every stage, i.e.,

$$\Big(\sum_i a_i x^i\Big) \mod (x^N - 1) = \sum_{i=0}^{N-1} \Big(\sum_j a_{i+jN}\Big) x^i.$$

For Rule 90, the dynamics are given by

$$a_i^{(t)} = \left(a_{i-1}^{(t-1)} + a_{i+1}^{(t-1)}\right) \mod 2.$$

We represent the temporal evolution of this automaton by multiplying the characteristic polynomial representing the initial state by the dipolynomial

$$\mathbf{T}(x) = x + x^{-1},$$

i.e., the state at time t emerging from the configuration $A(x)$ at time $t-1$, is given by

$$A^{(t)}(x) = \mathbf{T}(x)A^{(t-1)}(x) \mod (x^N - 1),$$

where all arithmetic is performed mod 2.

The foregoing representation shows that an initial configuration consisting of a single nonzero cell at position $a_0 = 0$ evolves after t time steps to a configuration given by

$$\mathbf{T}^t(x)1 = (x + x^{-1})^t = \sum_{i=0}^{t} \binom{t}{i} x^{2i-t}.$$

For $t < N/2$ the region of nonzero cells grows linearly with time, the position of the nonzero cells being given by $\pm 2^{j_1} \pm 2^{j_2} \pm \dots$, where the j_i are the positions of nonzero digits in the binary decomposition of the integer t. The additive superposition property implies that patterns generated from initial configurations containing more than one nonzero cell may be obtained by addition mod 2 of the patterns generated from initial configurations with a single nonzero cell.

The above formalism gives a neat characterization of "Garden-of-Eden" configurations, as seen by the following result:

LEMMA 3.1. *Configurations containing an odd number of cells with value 1 can never occur during the course of the evolution of the cellular automaton defined by Rule 90. Thus, such configurations can occur only as initial states.*

To show how to work with dipolynomials, let's give the quick proof of this result.

PROOF: Consider any initial configuration given by the polynomial $A^{(0)}(x)$. The next state of the system is given by the polynomial $A^{(1)}(x) = (x + x^{-1})A^{(0)}(x) \mod (x^N - 1)$. Thus,

$$A^{(1)}(x) = (x^2 + 1)B(x) + R(x)(x^N - 1)$$

for some dipolynomials $B(x)$ and $R(x)$. Since $x^2 + 1 = x^N - 1 = 0$ for $x = 1$, $A^{(1)}(1) = 0$. Thus $A^{(1)}(x)$ contains an even number of terms, and corresponds to a configuration with an even number of nonzero cells. Therefore, only such configurations can be reached from an arbitrary initial configuration $A^{(0)}(x)$.

The preceding lemma allows us to establish the following basic result governing the density of reachable configurations for the automaton generated by Rule 90.

THEOREM 3.1. *The fraction of the 2^N possible configurations of a cellular automaton with N cells evolving according to Rule 90 that can occur only as initial states is $\frac{1}{2}$ for N odd, and is $\frac{3}{4}$ for N even.*

Now let's take a look at how the algebraic set-up enables us to speak precisely about the cyclic properties of the Rule 90 automaton.

To characterize the properties and density of cycles, we first need the following technical results:

LEMMA 3.2. *The lengths of all cycles in a cellular automaton of N cells evolving according to Rule 90 must divide the length Π_N of the cycle obtained with an initial configuration containing a single cell with a nonzero value.*

This result follows immediately from the additivity of Rule 90, since any configuration is a superposition of configurations emerging from initial states with a single nonzero cell.

LEMMA 3.3. *For the Rule 90 cellular automaton, $\Pi_N = 1$ if N is a power of 2, while $\Pi_N = 2\Pi_{N/2}$ if N is even but not of the form 2^j.*

Note that the first part of Lemma 3.3 shows that when N has the form $N = 2^j$, then any initial configuration evolves ultimately to a fixed point consisting of all zeros, since

$$(x + x^{-1})^{2^j} 1 \equiv (x^{2^j} + x^{-2^j}) \equiv (x^N + x^{-N}) \equiv 0 \mod (x^N - 1).$$

These two lemmas enable us to characterize cycle lengths for all Rule 90 automata having an even number of cells. The following result establishes the cyclic pattern when N is odd:

THEOREM 3.2. *For Rule 90 when N is odd,*

$$\Pi_N | \Pi_N^* = 2^{\text{sord}_N(2)} - 1,$$

where $\text{sord}_N(2)$ *is the multiplicative* suborder function, *which is defined to be the least integer j such that $2^j = \pm 1 \mod N$.*

Using the above Lemmas 3.2, 3.3 and Theorem 3.2, one can compute directly the maximal cycle lengths that are given in Section 5 for the cases $N = 3$ to 20. Using similar arguments, it's also possible to determine the number and the length of the distinct cycles that emerge from Rule 90. We leave these matters, together with a consideration of extensions to more than two possible states for each cell, to the Problems section. Now we want to examine a few application areas in which one-dimensional cellular automata turn out to be useful.

Exercises

1. Prove Theorems 3.1 and 3.2.

2. (a) Calculate the characteristic polynomial for Rule 150. (b) Use this result to establish an analogue to Theorem 3.1 for Rule 150.

7. Languages: Formal and Natural

Abstractly, human languages can be thought of as a set of rules (grammars) for constructing sequences of symbols in order to convey semantic information (meaning). It makes little difference in this view whether the symbols are actual marks on a piece of paper, sound patterns emitted by the human vocal apparatus, or magnetized "bits" on a computer diskette. It's a small leap from this vision of linguistic structure to the consideration of a cellular automaton as a metaphoric device for studying the abstract properties of languages, in general. The sequence of symbols that represent the state of a one-dimensional automaton at time t corresponds to one of the potentially infinite statements that can be made in a language, and the rule of state transition in the automaton determines how new, grammatically correct statements can be generated from an initial "seed." In this section we will explore this interconnection between automata and languages as a vehicle for illustrating some of the issues involved in human information processing.

To formalize the above considerations, we first need to define more sharply our notion of a finite automaton.

DEFINITION. *A* finite automaton $\mathcal{M}$ *is a composite consisting of a finite input alphabet* $A = \{a_1, \ldots, a_n\}$, *a finite set of states* $X = \{x_1, x_2, \ldots, x_m\}$, *and a state-transition function* $\delta\colon X \times A \to X$, *which gives the next state in terms of the current state and current input symbol. In addition, there is a set* $F \subseteq X$, *termed the set of* final *or* accepting *states of the automaton.*

If $\mathcal{M}$ is an automaton that starts in state x_1 and if $u \in A^*$ is any finite *input string* consisting of a sequence of elements from the alphabet A, then we will write $\delta^*(x_1, u)$ to represent the state that the automaton enters if it begins in state x_1 and then moves across u from left to right, scanning one symbol at a time, until the entire string has been read. We say that $\mathcal{M}$ accepts the word (string) u if $\delta^*(x_1, u) \in F$; otherwise, $\mathcal{M}$ rejects the word u. Finally, the *language* $L(\mathcal{M})$ accepted by $\mathcal{M}$ is the set of all words $u \in A^*$ that are accepted by $\mathcal{M}$. A language L is called *regular* if there exists a finite-state automaton that accepts it. Such formal definitions are always a bit intimidating. So to see what such a gadget looks like in more concrete terms, let's consider a couple of examples.

Example 1: A Regular Language

Consider an automaton $\mathcal{M}$ with input alphabet $A = \{a_1, a_2\}$ and a state space given by $X = \{x_1, \ldots, x_4\}$. Let the rule of state transition be

$$\begin{aligned} &\delta(x_1, a_1) = x_2, \quad \delta(x_1, a_2) = x_4, \quad \delta(x_2, a_1) = x_2, \quad \delta(x_2, a_2) = x_3, \\ &\delta(x_3, a_1) = x_4, \quad \delta(x_3, a_2) = x_3, \quad \delta(x_4, a_1) = x_4, \quad \delta(x_4, a_2) = x_4. \end{aligned}$$

Assume that the set of accepting states is $F = \{x_3\}$, with the initial state being x_1. It's now easy to check that the above automaton $\mathcal{M}$ will accept the input string $u = a_1a_1a_2a_2a_2$ but will reject the string $u = a_2a_1a_2a_1$, since in the first case the termination state is $x_3 \in F$, whereas for the second input string the system terminates in the state $x_4 \notin F$. It's easy to verify that this automaton $\mathcal{M}$ accepts the language

$$L = \{a_1^{[n]} a_2^{[m]} \mid n, m > 0\}.$$

Thus, the above language L is a regular language.

Example 2: A Regular Language (cont'd.)

Here $\mathcal{M}$ has the same input alphabet and the same state set as above, but now we take the accepting states to be the set $F = \{x_1\}$. Let's take the transition rule of $\mathcal{M}$ to be

$$\begin{aligned} &\delta(x_1, a_1) = x_3, \quad \delta(x_1, a_2) = x_2, \quad \delta(x_2, a_1) = x_4, \quad \delta(x_2, a_2) = x_1, \\ &\delta(x_3, a_1) = x_1, \quad \delta(x_3, a_2) = x_4, \quad \delta(x_4, a_1) = x_2, \quad \delta(x_4, a_2) = x_3. \end{aligned}$$

Assume the initial state of $\mathcal{M}$ is x_1.

Let's start from the state x_1 with the input word $u = a_2a_2a_1a_2a_1a_2$. The machine will clearly accept this word, since it will enter the states $x_2, x_1, x_3, x_4, x_2, x_1$ in that order, and the final state x_1 is an accepting state for $\mathcal{M}$. We leave it as an exercise to the reader to prove that the language accepted by this automaton consists of those words having an even number of both a_1's and a_2's.

It's sometimes convenient to represent the state-transition function δ of an automaton by a *directed graph,* in which case the nodes of the graph represent the system states and the arcs represent the possible inputs that can be applied at each state. The direction of an arc determines the state transition when the input on the arc is applied. We show the state-transition graph for Example 1 on the next page.

In the study of formal languages, it turns out to be convenient to extend the idea of a deterministic finite-state automaton to allow for nondeterministic (but not random) state transitions. For this we need the notion of a

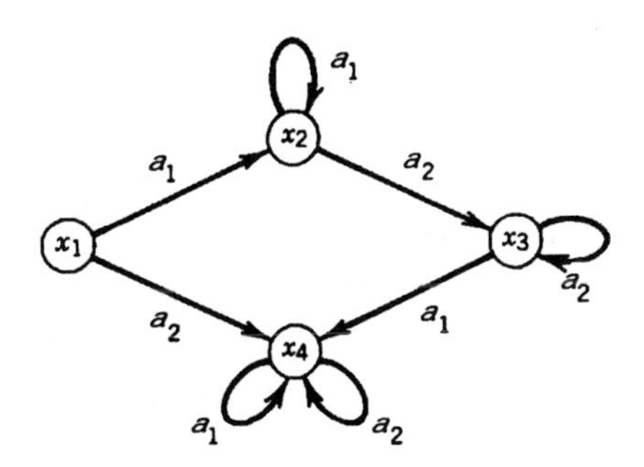

nondeterministic automaton, which comes about by allowing the next state to be a *subset* of X, rather than just a single element. Thus, in calculating $\delta^*(x_1, u)$ we accumulate *all* states that the automaton can enter when it reaches the end of the word u if it starts in the state x_1. We then say that $\mathcal{M}$ accepts u if at least one of these states is in F. It can be proved that a language is regular if and only if it is accepted by a nondeterministic finite automaton. In view of the definition of a regular language, this means that a nondeterministic automaton is equivalent to a deterministic one—at least insofar as accepting languages is concerned. So why do we introduce nondeterministic automata? Basically, the reason is that it's usually easier to design a nondeterministic automaton to accept a particular language. But any actual machine built to accept this language must be a deterministic one. The equivalence theorem then guarantees that the design can be physically implemented.

In the 1950s Noam Chomsky began developing a theory of grammar whose key idea is that of a transformation. According to Chomsky's theory of *transformational grammars,* human beings have an innate language-acquisition mechanism hard-wired into their brains at birth. This mechanism is not completely fixed, however, as various parameters can be set, much like switches in a railroad yard, enabling the individual to acquire whatever language is spoken in the linguistic community in which he or she grows up. However, the overall "template" for language acquisition is fixed for each human being. Chomsky called this template the *universal grammar.*

The mathematical translation of Chomsky's idea leads to a theory of formal languages of the sort outlined above. And in this theory it's possible to identify four qualitatively different types of languages, distinguished by the memory requirement of an automaton capable of accepting that type of language. These language types are:

- *Type 0: Unrestricted languages*—indefinitely large memories.
- *Type 1: Context-sensitive languages*—memory is proportional to the word length.

- *Type 2: Context-free languages*—memory arranged in a stack, with a fixed number of elements available at any moment.
- *Type 3: Regular languages*—no memory.

These languages categories form a hierarchy, the unrestricted languages being the most general. Each of the types is associated with a particular class of automata that accepts it. As it turns out, Type 0 languages are the only class requiring the power of a universal computer (Turing machine) for its implementation. We have already seen that a finite, deterministic automaton suffices to implement a regular language, with context-free and context-sensitive languages requiring what are termed *pushdown automata* and *linear-bounded automata,* respectively, for their implementation.

This classification of languages bears a striking similarity to the classification of one-dimensional cellular automata according to their long-run behaviors. In the preceding sections we have seen that all such automata ultimately settle one of the following sorts into long-term behaviors:

A. The initial pattern disappears.

B. The initial pattern evolves to a fixed finite size.

C. The initial pattern grows indefinitely at a fixed rate.

D. The initial pattern grows and contracts with time.

It's tempting to try to identify the types of automata A–D with the language classes 0–3. To follow up the identification here would require more time and space than we have available in a book of this sort. But it's worthwhile pursuing a few of the simpler aspects of the language $\leftrightarrow$ automata duality just to see how far one can go toward establishing the matchups.

Let $\Omega^{(t)}$ represent the set of configurations generated by a cellular automaton at time t, i.e., the set of configurations that can be reached at time t when any of the admissible initial configurations is used. It's been shown that after any finite number of time steps t, the set $\Omega^{(t)}$ forms a regular language. Thus, the real case of interest is the nature of the set $\lim_{t\to\infty} \Omega^{(t)}$. One approach to getting a handle on this object is to look at the *entropy* of the set $\Omega^{(t)}$.

If we have an automaton with k possible cell values, then the total number of possible sequences of X symbols is k^X. In general, only some number $N(X)$ of these sequences occur in nontrivial regular languages. These sequences correspond to the distinct paths through the transition graph of the automaton. The *topological entropy* of the set $\Omega^{(t)}$ is then defined to be

$$s_t = \lim_{X\to\infty} \frac{1}{X} \log_k N(X).$$

For any regular language, this number is also given in terms of the largest characteristic value λ_{max} of the adjacency matrix of the state-transition graph. This entropy is $s_t = \log_k \lambda_{max}$. By the locality assumption on the state-transition rule of the automaton, it follows that the set $\Omega^{(t)}$ always contracts or remains unchanged in time. Consequently, the entropy of $\Omega^{(t)}$ is nonincreasing with t. Further, it can be shown that Class A automata have $s_t \to 0$ as $t \to \infty$. Class B–D automata have nonzero limiting entropies, although Class B automata yield periodic limiting patterns and, hence, generate zero *temporal* entropy. The entropies of regular languages are always logarithms of algebraic integers, i.e., a integer root of a monic polynomial having integer coefficients. Context-free languages, however, have entropies given by logarithms of general algebraic numbers but can still be computed by finite procedures. The entropies of context-sensitive and unrestricted languages are, in general, uncomputable numbers. So if Class C and D automata do indeed yield limit sets corresponding to context-sensitive or unrestricted languages, then the entropies of these sets are generally not computable. However, we can *estimate* the entropy of any automata from experimental data by fitting parameters in various types of models that reproduce the data.

Another approach to the language $\leftrightarrow$ automata question is through the idea of *complexity*. We define the complexity of the set $\Omega^{(t)}$ to be the number of states in the smallest automata that generates the set $\Omega^{(t)}$. Call this number $\Xi^{(t)}$. (The reader will note, incidentally, that this notion of complexity differs considerably from the "relativistic" view considered in Chapter 1.) It's been experimentally observed that $\Xi^{(t)}$ is nondecreasing in time, and that Class A and B automata appear to give complexities that tend to constants after just one or two time steps, or increase at most quadratically with time. On the other hand, Class C and D automata usually have complexities that increase exponentially with time. In all cases we have the bound

$$1 \leq \Xi^{(t)} \leq 2^{k^{2Rt}} - 1,$$

where R is the size of the neighborhood in the automaton transition rule.

To give some indication of the nature and magnitude of the entropies and complexities, here are the values of these quantities for some typical one-dimensional cellular automata:

Rule	$\Xi^{(0)}$	$\Xi^{(1)}$	$\Xi^{(2)}$	λ_{max}
0	1	1	1	1.0
90	1	1	1	2.0
128	1	4	6	1.62
182	1	15	92	1.89

We can summarize this very brief excursion into formal language theory and cellular automata in the following compact form:

LANGUAGE-AUTOMATA THEOREM. *Class A and B cellular automata have limit sets that form regular languages. For most Class C and D automata, the regular language complexity increases monotonically with time, so that the configurations obtained in the limit usually do not form a regular language. Instead, the limit sets for Class C automata appear to form context-sensitive languages, whereas those for Class D automata seem to correspond to general, unrestricted languages.*

At this juncture an obvious question arises: What about natural languages like Arabic, Russian, English, or Chinese? To what degree do any of the formal languages discussed above adequately mirror the structure of these most human of all communication modes? This question lies at the heart of Noam Chomsky's theory of "generative/transformational" grammars, which, at the very least, has disposed of the view that any normal human language can be modeled as a regular language and, hence, generated by a finite-state automaton. It would take another volume at least the size of this one to even begin to enter into the details of Chomsky's theories, but one of the major conclusions that has emerged is that human languages are also probably not "computable" by any structure less powerful than a full unrestricted language of Type 0. In any case, certainly no automaton with finite memory can account for the myriad meaningful statements that can be uttered in every human tongue. It seems very likely that, at best, only a full-fledged Turing machine (a universal computer) can begin to approach the complexity and structure that we see in even the simplest human languages. The interested reader is invited to consult the volumes cited in the Notes and References for a wealth of detail on the linguistic revolution inspired by Chomsky's work and the deep connections between this work and problems of automata theory.

Exercises

1. Show that in Example 2 of the text, the language accepted by that automaton consists of those words have both an even number of a_1's and and even number of a_2's.

2. Prove that a nonempty language accepted by a deterministic finite automaton must contain some word w that is not longer than the number of states of the automaton. (Hint: Keep in mind that if the input word is longer than the number of states in the automaton, then there must be a state that's entered twice during the course of processing that word.)

3. We mentioned in the text that for every nondeterministic automaton, there is an equivalent deterministic one that accepts exactly the same

languages. Consider the nondeterministic automaton $\mathcal{M}$ having the state set $X = \{x_1, x_2, x_3\}$ and input set $A = \{a_1, a_2\}$. Assume the initial state is x_1, while the set of accepting states is $F = \{x_3\}$. Let the rule of state transition be given by

$$\begin{aligned} \delta(x_1, a_1) &= \{x_1, x_2\}, & \delta(x_1, a_2) &= \{x_2\}, \\ \delta(x_2, a_1) &= \emptyset, & \delta(x_2, a_2) &= \{x_2\}, \\ \delta(x_3, a_1) &= \{x_1, x_2, x_3\}, & \delta(x_3, a_2) &= \{x_1\}. \end{aligned}$$

(a) Show that the state set of the equivalent deterministic automaton is 2^X, the power set of X. (b) What is the state-transition rule for the deterministic automaton? (c) What is its set of accepting states?

4. Draw the state-transition graph for Example 2 of the text.

5. Compute the complexities $\Xi^{(0)}, \Xi^{(1)}$ and $\Xi^{(2)}$ for the cellular automata given by Rule 18 and Rule 150.

8. DNA Sequences and Cellular Automata

While human languages are unimaginably complex, there is another type of language employed by nature that offers many possibilities for analysis using the automata-theoretic ideas we have been discussing here. This is the language of DNA sequences, the so-called Language of Life.

At the simplest possible level, the DNA molecule can be thought of as a one-dimensional lattice with four possible values at each cellular site. Of course such a view is hopelessly incomplete. The DNA molecule is much more than just a linear string of base pairs, having many other mechanisms for information storage. Nonetheless, as a first approximation we can consider the DNA molecule as a one-dimensional cellular automaton having four possible values per cell, which we shall label A, C, G and T, representing the four nucleotide bases Adenine, Cytosine, Guanine and Thymine. Structurally, the confuguration of these symbols on the DNA molecule is a sequential arrangement of A, C, G and T along the strand, which is twisted into an antiparallel double-stranded helix, the alignment of the two strands mediated through hydrogen bonds that hold the base pairs together. Furthermore, the base pairing is quite specific: G always pairs with C, and A always pairs with T. As a result, the base sequence of one strand completely determines the sequence on the other strand. This leads to the celebrated *double helical* structure of the DNA molecule depicted in Fig. 3.11. The left-half of the figure shows a linear representation of the base-pairing of the two strands, together with the sugar and phosphate bonds on the strands themselves. The right-half of Fig. 3.11 depicts the way the strands are twisted to form the characteristic double helix geometry. Let's briefly review the func-

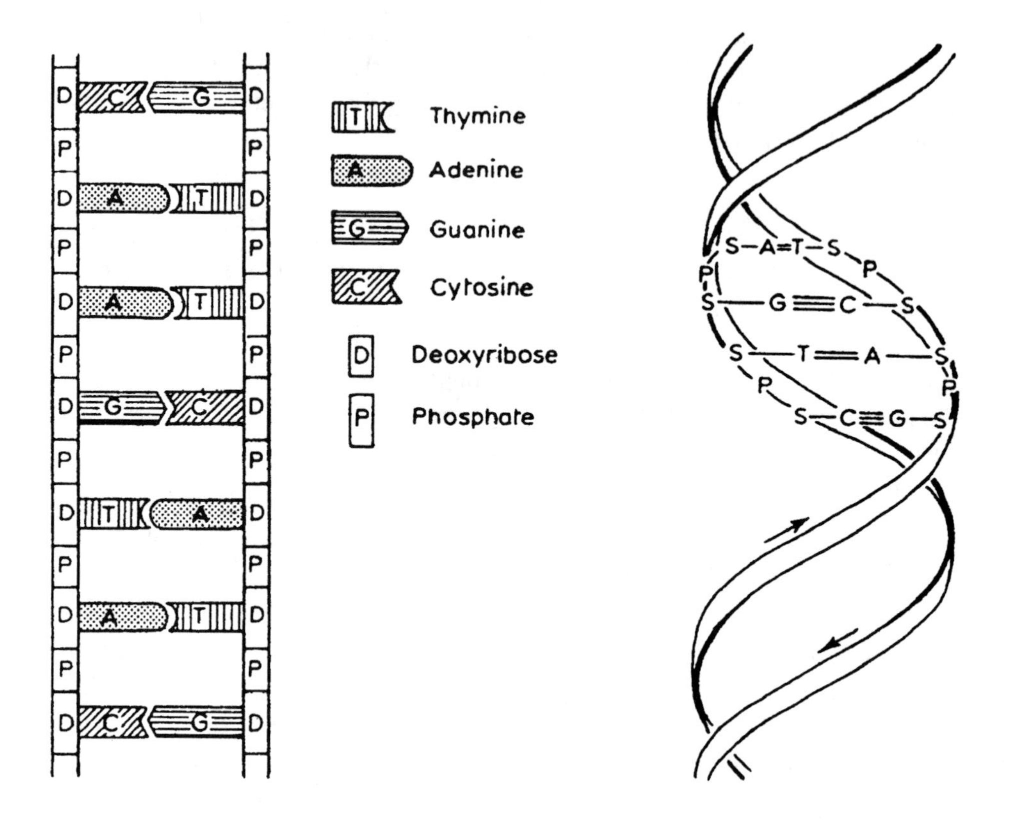

Figure 3.11. The Structure and the Geometry of the Two Helical DNA Strands

tioning of the DNA molecule in the everyday chemical business of cellular metabolism and replication.

While simple organisms have only a single DNA strand, higher organisms contain several separate bundles of DNA called *chromosomes.* These bundles overlap, much like the individual fibers forming a hemp rope. The number of chromosome strands varies from species to species, being 46 for humans, 16 for onions and 60 for cattle. Each of these strands is further subdivided into sections called *genes.* Each such gene is, by definition, the amount of information needed to code for either one protein (a *structural* gene), or to induce or repress certain chemical operations of the cell (*a regulatory* gene). The Central Dogma of Molecular Biology, due to Francis Crick, is a statement about the flow of information in the cell. Compactly, it states that

$$\text{DNA} \longrightarrow \text{RNA} \longrightarrow \text{Protein}.$$

The essential message of the Central Dogma is that there can be no Lamarckian-type of information flow back from the protein (phenotype) to the cellular nucleus (genotype). The Central Dogma is a description of the way in which a DNA sequence specifies both its own copying (replication) and the synthesis of proteins (the processes of transcription and translation). Replication is carried out by means of the base-pairing complementarity of the two DNA strands, with the cellular DNA-synthesizing machinery reading each strand to form its complementary strand. The protein-synthesizing procedure proceeds in two steps: In the *transcription* step, a strand of messenger RNA (mRNA) is formed, again as a complement to one of the DNA strands, but with the base element U (Uracil) taking the place of T (thymine). Usually RNA is a single-stranded molecule. During the second step, the cellular protein-manufacturing machinery (the ribosomes) makes use of the genetic code to *translate* a section of the mRNA strand in order to form a string of amino acids (a protein). The mRNA base sequence is read as a nonoverlapping set of contiguous triplets called *codons.* Proteins are composed of amino acids of which there are 20 different types. Since there are $4^3 = 64$ possible codons, there is some redundancy in the translation of specific codons into amino acids. The overall process is shown in Fig. 3.12.

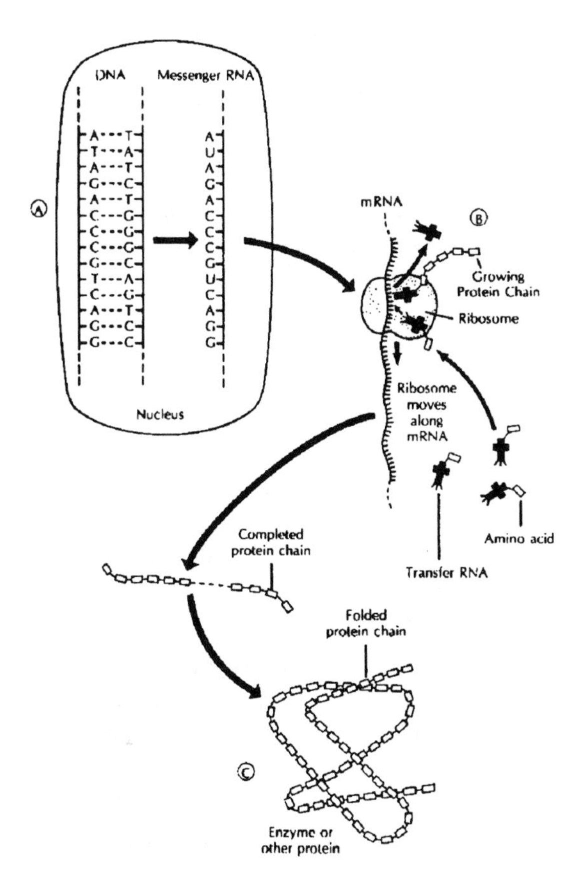

Figure 3.12. The Processes of Genetic Transcription and Translation

The foregoing description of the Central Dogma, oversimplified as it is, already suggests the possibility of modeling the evolution of cellular DNA using one-dimensional cellular automata of the type we have been considering. The sugar and phosphate groups form a one-dimensional lattice of cellular sites, while the value of any given site in the automaton is one of the four elements A, G, C or T from the nucleotide alphabet. Changes in the configuration (mutations) may occur after replication, which we can think of as an application of the particular rule according to which the automaton operates. The nature of the state-transition rule will be determined, of course, by the biological and chemical constraints of the situation.

Before going on, we should caution that the above description of the functional activity of DNA is incomplete in several ways, two of the most important being that not all DNA sequences code for proteins, and that the three-dimensional configuration of the DNA molecule also acts to store dynamic information as to how the DNA sequence should act in the functioning of the cell. Thus, our idea of using the elementary one-dimensional cellular automata set-up to model the action of DNA sequences can be at most only a starting point for much more elaborate structures capable of taking into account these additional activities of the DNA that cannot be encompassed in any simple cellular automaton model. Now let's consider just how the automata framework might be used to address some important questions associated with the evolution of DNA sequences.

At the outset of any DNA modeling exercise, it's necessary to distinguish between the modeling of replication and modeling of protein synthesis (transcription and translation). These quite different goals strongly condition the specific nature of the automata and the rules that ultimately constitute the model.

The most obvious purpose for modeling replication is to study natural selection and/or prebiotic evolution. We could use the initial state of a one-dimensional automaton to represent the germ-line DNA or the pre-DNA macromolecule, with the state-transition rule chosen to mimic base substitution, base insertion/deletion, recombination and the other types of changes that can occur in the DNA over time. The sequence of states that the automaton passes through then represents the genetic changes in the population. It's natural in models of this type to introduce a stochastic element into the transition rule to account for "random" genetic mutations. One way to do this is to randomly shuffle the base sequence according to one of a number of plausible shuffling rules. In more realistic situations, this random element should also take into account chemical constraints as well.

A simplified version of this scheme was carried out by John Holland, who produced an elementary one-dimensional chemistry involving only two base elements, two amino acids, two types of bonds, the above simple form of mutation involving shuffling operators and a primitive notion of catalysis. Starting with a random ensemble of initial states, these experiments showed that catalysis eventually dominates, and the arrangement of amino acids always clusters into groupings that are functionally equivalent to enzymes. Such behavior is strongly reminiscent of the Class D automata discussed in earlier sections. Recall that these were the most complex type of one-dimensional automata, displaying limiting behavior patterns corresponding to the most general type of unrestricted language. Given the evident complexity of living organisms, the result of this experiment with "simplified" DNA is reassuring, lending support to the use of automata-theoretic models in the analysis of DNA evolution.

One of the difficulties in representing DNA replication by cellular automata models is the problem of selection. In the prebiotic stages, selection can be handled simply in terms of competition for raw materials among the competing enzymes. However, at later stages of evolution this is an inadequate criterion, since more highly developed organisms have a phenotype that is quite different in form than the genotype. Moreover, many evolutionary forces tend to act on the phenotype rather than the genotype. Development of an appropriate measure for selection remains one of the major stumbling blocks to the effective use of cellular automata for modeling the process of DNA replication and evolution.

We can also use automata to model the processes of transcription and translation. The purpose of a model of this type is to understand more fully the global mechanisms underlying the functioning of the somatic DNA, as well as to address a variety of problems underlying development. It's essential to incorporate some type of dynamical information storage into such models. The transition rule of the automaton corresponds to the conformational changes that take place in the organism, as well as possible replications. The need to deal with conformational changes introduces an unusual aspect into the cellular automaton construction since, although the automaton is basically a one-dimensional object, it must also have some idea of its own geometry in three dimensions. Changes in this geometry are one of the main processes taking place during the state transition of the automaton. Examples of how this can be done are reported in the papers cited in the chapter's Notes and References.

Our discussion of the uses of cellular automata for studying the dynamics of DNA sequence development, sketchy as it is, shows clearly the value of the theoretical work showing the types of long-term behavior patterns that can emerge from various sorts of one-dimensional automata. Now let's take a look at another way such mathematical "machines" are being used to study the processes of life—plant growth.

Exercise

1. A living cell is a kind of miniature manufacturing plant, transforming raw materials taken from its environment into finished products that the cell needs for its continued existence. Thinking of an automobile plant, for instance, the cellular DNA would correspond to the master blueprint describing all the assemblies needed to put together a finished car, while the cellular ribosomes correspond to the workers in the plant who transform the raw materials into key subassemblies. (a) Try to continue this parallel by finding matchups between all the other cellular components like mRNA, tRNA, amino acids, proteins, the cellular membrane, enzymes and so forth. (b) Within this automobile-manufacturing metaphor, what are the analogues of things like genetic mutations, reproduction, heredity and selection? (c) Can you think of other real-world systems for which this kind of cellular metaphor would be appropriate?

9. L-Systems and the Growth of Plants

To complete our discussion of one-dimensional automata, we briefly consider a model for the development of filamentous plants originally proposed by Aristid Lindenmayer in the late 1960s. This model contains the novel feature that the number of cells in the automaton is allowed to increase with time according to a recipe laid down by the state-transition rule. In this way the model "grows" in a manner mimicking the growth of a filamentous plant like the blue-green algae *Anabaena*.

The simplest version of one of Lindenmayer's *L-systems* involves a one-dimensional lattice that starts at time $t = 0$ with a single active cell having value 1. The state-transition rule is the following: the next value at cell i is given by its current value, together with the value of the cell immediately to the left of cell i, i.e., cell $i - 1$. Thus, the complete state transition at cell i is given by the rule:

$$(i = 0, i - 1 = 0) \Rightarrow i \to 0, \qquad (i = 0, i - 1 = 1) \Rightarrow i \to 1,$$
$$(i = 1, i - 1 = 0) \Rightarrow i \to 11, \qquad (i = 1, i - 1 = 1) \Rightarrow i \to 0.$$

The interesting feature of this rule is the possibility for cell division that occurs when $i = 1, i - 1 = 0$. If we take an initial state such that the first cell at the left has value 0, while the cell next to it has value 1, then the first four state transitions are 01, 011, 0110, 011011. Already we see here the growth of the initial "seed" at cell 2 to several seeds at the cells 2, 3, 5 and 6.

We could change the cell division rule in this example to give a 01 progeny pattern instead of the 11 pattern above. Such a change gives a banded pattern of the following sort emerging from the seed at cell 2: 01,

101, 0101, 101101, 01010101, ... In a different direction, we could think of the cell at site $i-1$ as constituting the "environment" for the cell at site i. Keeping the "banded" progeny rule, consider the case when the environment is constant, i.e., we fix the value of cell $i-1$ at 1. In this situation, site i gives rise to a sequence of values that form an increasing string of 0's that always terminates with a 1.

It must be admitted that the linear string of 0's and 1's above don't look much like what anyone would even charitably call a "plant." So to bring out more clearly the connection between L-systems and the real world of plants, let's soup-up our notation a bit by extending our symbol alphabet to include the new symbols [and]. And while we're at it, let's also jazz-up the state-transition rule so that the four symbols of our system always transform in the following manner:

$$0 \to 1[0]1[0]0$$
$$1 \to 11$$
$$[\ \to\ [$$
$$]\ \to\ ]$$

To see how this rule works, suppose we start with the string consisting of the single symbol 0. With the above transformation rules, the first three steps of this automaton yield:

0
1[0]1[0]0
11[1[0]1[0]0]11[1[0]1[0]0]1[0]1[0]0

This still doesn't look much like a plant. But we can convert this type of string into a tree-like structure by treating the symbols 0 and 1 as line segments, while regarding the two brackets [and] as branch points.

One way to get something that looks faintly plant-like out of this idea is to leave all 1 segments bare, while placing a leaf at the end of each 0 segment. So, for instance, if we have the string 1[0]1[0]0, its stem consists of the three symbols not in brackets. These are a 1 segment beneath another 1 segment, which in turn is topped off by a 0 segment. Two branches, each with a single 0 segment, sprout from this string. The first branch is attached above the first segment, while the second branch occurs after the second segment. As for the direction of the branch, the simplest convention is just to specify that for any given stem the branches shoot off alternately to the left and to the right. Figure 3.13 shows the first three generations of such a "plant" obtained by using the L-system grammar given above, together with the foregoing rule regarding leaves and branches.

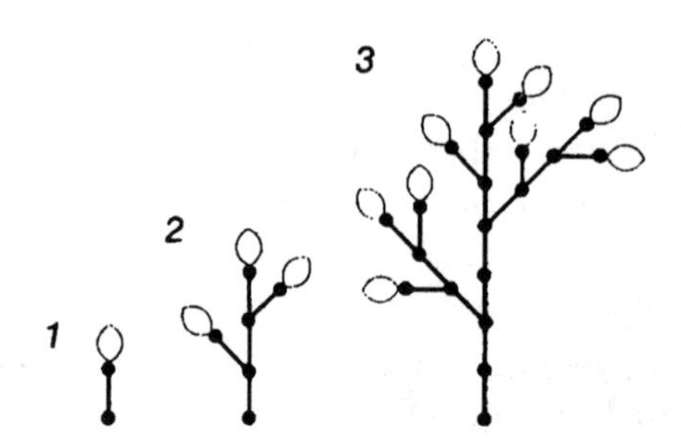

Figure 3.13. The First 3 Generations of an L-System Plant

Earlier we spoke about the Chomsky hierarchy of languages, noting that in this scheme there are four distinct classes of languages characterized by the amount of memory needed for an automaton to be able to accept that type of language. Since the production rule for L-systems gives the appearance of being a systematic procedure for creating new symbol strings from old, it's of interest to ask where the difference is, if any, between Chomsky-style grammars and L-systems.

As it turns out, there is one crucial difference. It resides in the fact that in all the Chomsky grammars, only a single symbol is changed at any particular moment. In other words, productions are applied sequentially. L-systems, on the other hand, involve a simultaneous replacement of all letters in a given word. Thus, the two production schemes mirror perfectly the difference between serial and parallel computing. And this difference shows up in the properties of L-systems. For instance, there are languages that can be generated by context-free L-systems that cannot be generated by context-free Chomsky grammars. So in a very definite sense, the L-systems represent an important extension of the Chomsky scheme.

While the stick-figure tree of Fig. 3.13 is still a pretty feeble excuse for a plant, L-system theorists in collaboration with computer graphic artists have produced some truly stunning "artificial plants," whose fidelity and beauty rival anything that nature has so far whipped up. The reader is strongly urged to consult the volumes cited in the Notes and References for a display of these visual fireworks, which do credit to the maestros of both the keyboard and the blackboard. And with this as our cue, let's now leave the study of one-dimensional automata and double our fun by looking more deeply into the special two-dimensional Life automaton.

Exercises

1. Suppose you have the L-system with the symbols 0 and 1 and the rules: $0 \to 1$, $1 \to 10$. Let the system begin with a single 0. Show that

this L-system generates strings whose lengths coincide with the sequence of Fibonacci numbers $1, 1, 2, 3, 5, \ldots$.

2. Consider the *context-sensitive* L-system with symbols a and b. The rule of state transition is: $b < a \to b$, which says simply that in any word in which a b precedes an a, replace the a by b. (a) Suppose the starting state for this system is *baaaaaaa*. Calculate the next four successive states. (b) Can you see why this system serves as a simple model for one-dimensional diffusion?

3. The blue-green algae *Anabaena catenula* is composed of two types of cells strung out on a filament. Call these types a and b. Under a microscope, a filament of this plant appears as a sequence of cylinders of various lengths, with the a types being longer than those of type b. Moreover, each cellular type has a polarity specifying the positions in which daughter cells of types a and b will be produced. Thus, taking account of polarity, there are a total of four different cellular types. Call these types a_r, a_l, b_r and b_l. Suppose you start with a single cell of type a_r. Determine the transition rule for an L-system that would generate the developmental path of the filament shown in the following diagram:

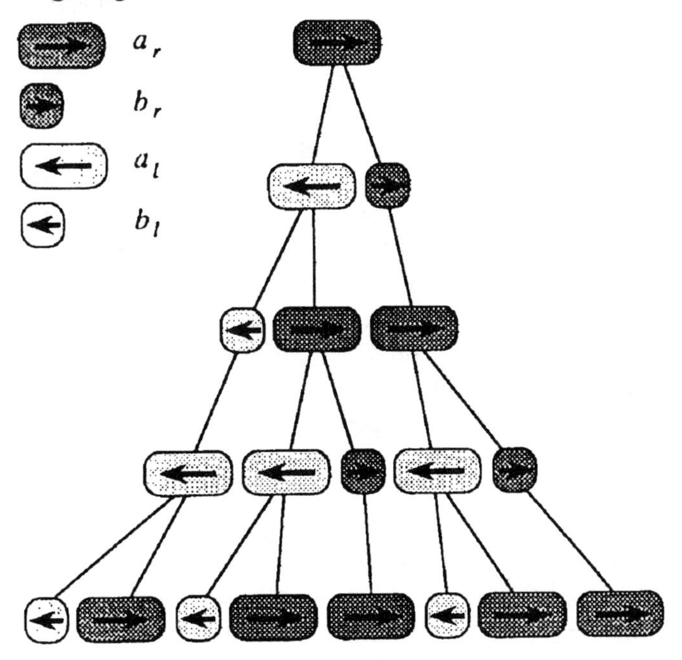

Development of a Filament of *Anabaena*

10. The Game of Life

In Section 2 we introduced a two-dimensional cellular automaton having a state space consisting of an infinite checkerboard, each of whose squares can display one of k symbols at each moment of time. In contrast to the case of one-dimensional automata, where the local neighborhood of a cell is simply a line of cells on either side of the target cell, for two-dimensional automata there are several possible inequivalent neighborhoods, depending upon whether or not we admit cells diagonal to the target cell as being part of the neighborhood. Earlier we spoke about the two most important such neighborhoods, the von Neumann neighborhood, which consists only of the cells adjacent horizontally and vertically to the target cell, and the Moore neighborhood consisting of *all* cells adjacent to the target cell. The state-transition rule at cell (i, j) for a two-dimensional automaton using the von Neumann neighborhood is

$$a_{i,j}^{(t+1)} = \phi[a_{i,j}^{(t)}, a_{i,j+1}^{(t)}, a_{i+1,j}^{(t)}, a_{i,j-1}^{(t)}, a_{i-1,j}^{(t))}].$$

Often we encounter what are termed "totalistic" rules, in which the value of a cell at time $t+1$ depends only on the sum of the values of the cells in the neighborhood at time t. To give some indication of the spectrum of possibilities for two-dimensional rules, we note that for $k = 2$ values per cell, there are a total of $2^{32} \approx 4$ billion rules for the von Neumann neighborhood, although this number drops dramatically to $2^5 = 32$ totalistic rules. For the Moore neighborhood, the corresponding number of rules are $2^{512} \approx 10^{154}$ general rules and $2^9 = 512$ totalistic rules.

In view of the magnitude of these numbers, it's clearly impossible to explore the properties and behaviors of all rules, even with the fastest computers. However, computer experiments have been carried out that sample various rules. These experiments suggest that the long-term behavior of two-dimensional automata tends to fall into the same basic categories that we observed for one-dimensional automata. Thus, in the two-dimensional situation we find the analogues of the Class A–D behaviors discussed earlier for the one-dimensional case. But there are some differences to note.

First of all, two-dimensional processes often generate "dendritic" patterns characterized by noninteger growth dimensions. In addition, for one-dimensional automata we saw that the patterns formed after any finite number of time steps constitute a regular language; in the two-dimensional case this result is no longer true. Consequently, there are questions about configurations generated by two-dimensional automata after a finite number of steps that can be posed but are generally not answerable by any computational process; i.e., they are formally undecidable. Examples of such questions are whether a particular configuration can be generated after a single time step from some initial configuration, or whether there exist configurations that have a particular temporal period and are thus invariant under some number of iterations of the automaton rule. Just to get a feel for a few of the kinds of things that can happen with two-dimensional automata, we now consider the famous "Game of Life."

Recall from Section 2 that Life uses the Moore neighborhood with the following state-transition rule:

$$a_{i,j}^{(t+1)} = \begin{cases} a_{i,j}^{(t)}, & \text{if } \sum_n a_n = 2, \\ 1, & \text{if } \sum_n a_n = 3, \\ 0, & \text{otherwise,} \end{cases}$$

where we have used the notation $\sum_n a_n$ to represent the sum of all the cells in the Moore neighborhood of site (i, j). Using the Life rule, we display the fate of some initial triplets in Fig. 3.14 on the next page.

We see that the first three triplets die out after the second generation, whereas the fourth triplet forms a stable Block, and the fifth, termed a Blinker, oscillates indefinitely. In Figure 3.15, we show the life history of the first three generations of an initially more-or-less randomly populated life universe.

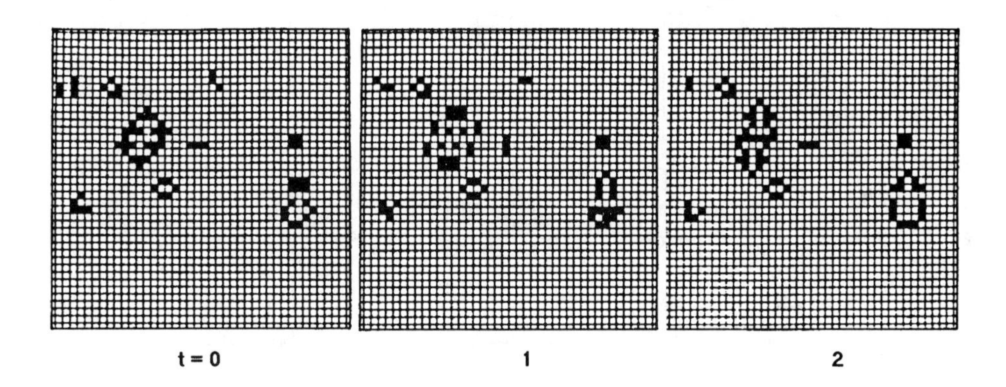

Figure 3.15. A Typical Life History

Of considerable interest in the Life automaton is whether there are initial configurations that eventually reproduce themselves. The first example of such a configuration is the famous "Glider," which is displayed in Fig. 3.16.

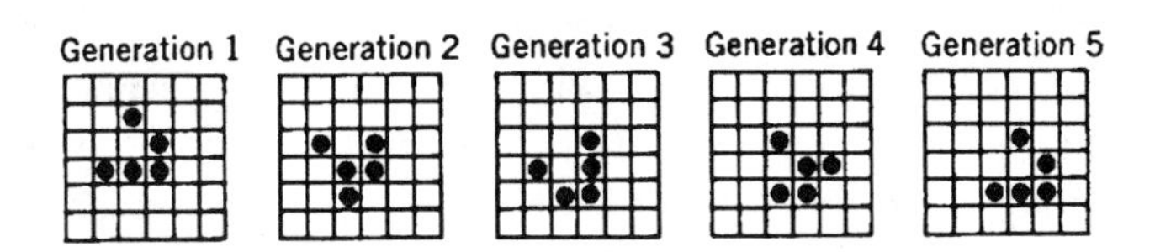

Figure 3.16. The Glider

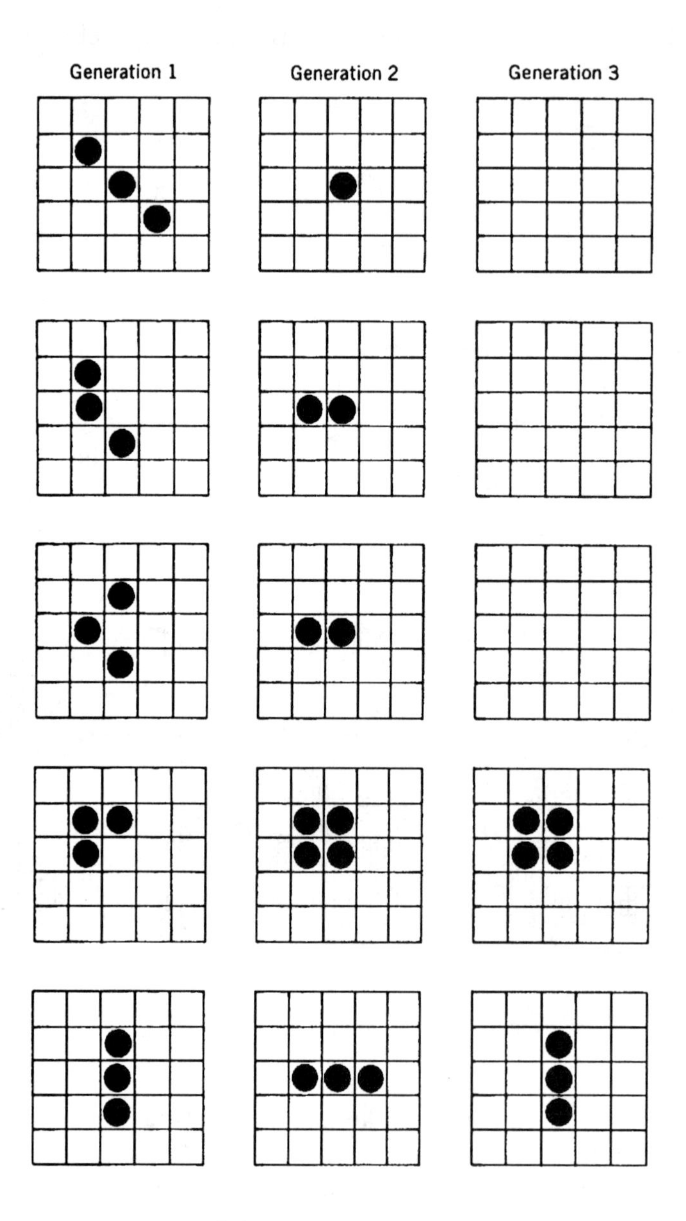

Figure 3.14. Triplet Histories in Life

In the early days of Life, researchers conjectured that because of the overpopulation constraint built into the Life rule, there were no configurations that could grow indefinitely. John Horton Conway, the inventor of the Life rule, offered a $50 reward to anyone who could produce such a configuration. The configuration that won the prize, termed a "Glider Gun," is depicted in Fig. 3.17. Here the "Gun" shown at the lower left of the figure is a spatially fixed oscillator that repeats its original shape after 30 genera-

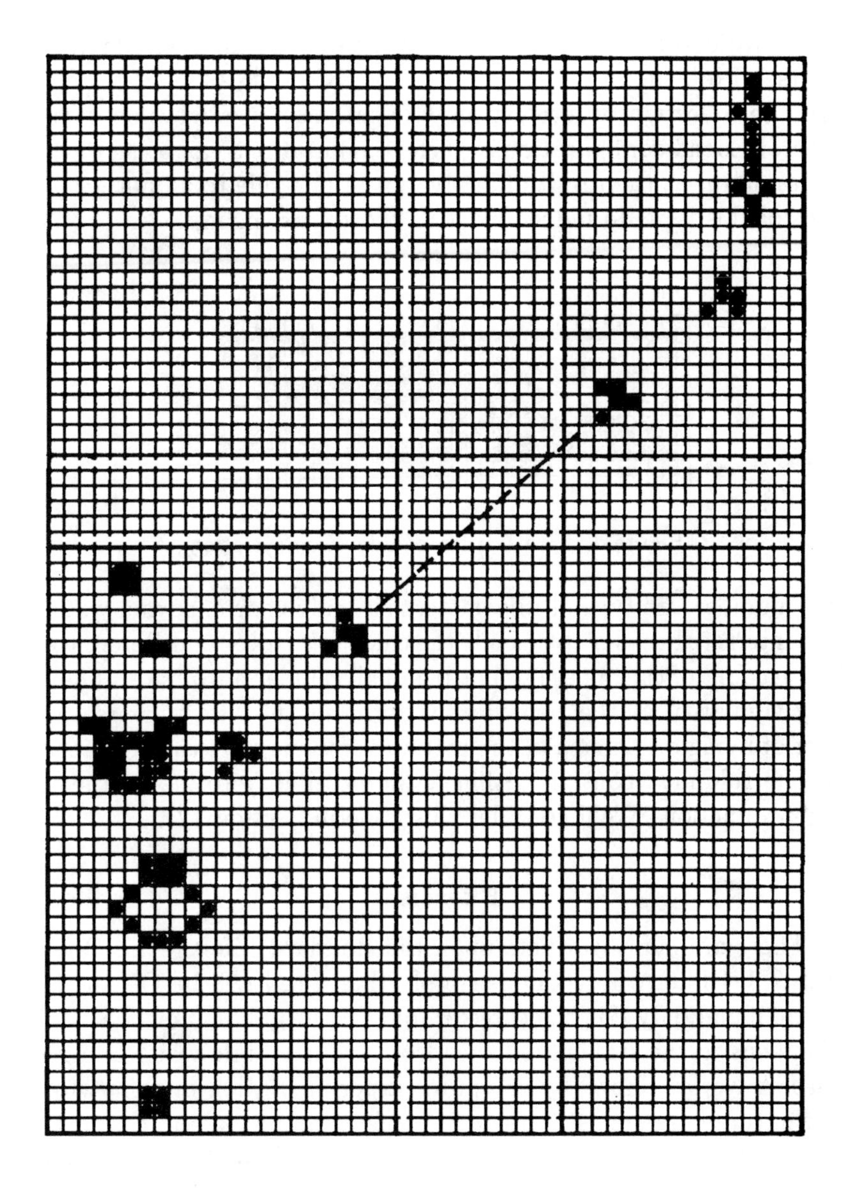

Figure 3.17. The Glider Gun

tions. Within this period, the Gun emits a Glider that wanders across the grid and encounters the "Eater," which is shown at the top right of the figure. The Eater, a 15-generation oscillator, swallows up the Glider without undergoing any irreversible change itself. Since the Gun oscillates indefinitely, it can produce an infinite number of Gliders, implying that there do exist configurations that can grow indefinitely. Another type of unlimited growth pattern emerges from what's called a "Puffer Train." This is a moving configuration like a Glider that leaves behind a trail of stable debris-like objects (see Problem 15).

Another question of interest for Life enthusiasts is whether there exist "Garden-of-Eden" configurations. Such configurations can arise *only* as initial states, never as the result of applying the Life rule to any other pattern. Patterns of this type do indeed exist as shown, for example, in Fig. 3.18.

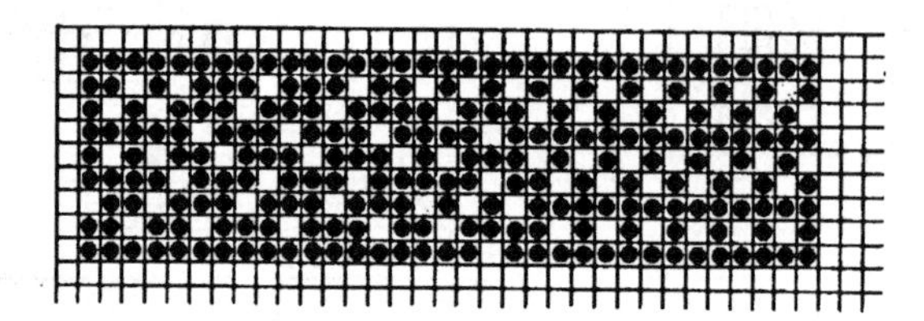

Figure 3.18. A Garden-of-Eden Configuration

The number and type of Life patterns is truly enormous and we strongly urge the interested reader to consult the chapter's Notes and References for a detailed account of this most fascinating of all cellular automata. The almost uncanny similarity of the behavior patterns in the Life universe and the behavior of living organisms in the real universe suggests that the seemingly simple Life rule may contain within it the possibility to model the process of self-reproduction. We'll consider this issue in Section 12.

Exercises

1. There are many variants upon the standard Life theme. One of the simplest is 3–4 Life, whose rule is simply that a cell will be ON in the next generation if and only if it has three or four ON neighbors now. (a) Consider what happens to the stable life forms from ordinary Life when plunged into the 3–4 Life world. Show (by computer experimentation or otherwise) that the only naturally arising still life in 3–4 Life is the Block, and that it is fairly uncommon. (b) On the other hand, another type of Life form is encountered regularly in 3–4 Life. What is it?

2. Fredkin's Game is a primitive relative of Life, which uses the von Neumann neighborhood instead of the Moore neighborhood. In this game, the rule is that a cell will be ON in the next generation if and only if it has an odd number of ON neighbors in the current generation; otherwise, it will be OFF. (a) Show that in Fredkin's Game, no pattern of ON cells can ever fade away. (b) Prove that any pattern whatsoever becomes four copies of itself, with the quadrupling time being a power of 2 (but varying with the size and/or complexity of the pattern). (c) Does this type of replication constitute meaningful self-reproduction?

3. Show that in standard Life, Gliders can collide to form a Glider Gun, thus establishing that a finite number of Gliders can give rise to an infinite number of Gliders. (Hint: Look for a configuration of thirteen Gliders coming together in two streams, one from the northeast and the other from the southeast.)

4. Many Life patterns like the Block and the Blinker are symmetric. Prove that the rules of Life imply that once any symmetric pattern forms, the symmetry can never be lost.

11. A Cellular Automaton Model of Vertebrate Skin Patterns

From an evolutionary standpoint, it's of considerable interest to study how vertebrates acquire skin patterns, such as the zebra's stripes or the leopard's spots, because these patterns are important to the survival of the animal. Most such patterns are spots or stripes formed by specialized pigment cells (melanocytes), and the problem of how skin patterns are formed can be reduced to describing how the colored pigment cells come to be distributed as they are on the embryo's skin.

Current wisdom has it that the patterns are formed by a reaction-diffusion process in which uniformly distributed pigment cells produce two or more species of morphogen molecules. These molecules then react with each other and diffuse in space to produce a concentration pattern having a characteristic wavelength. The morphogen concentrations form a preliminary pattern that then induces a differentiation of the pigment cells resulting in the final skin pattern.

Here we consider only the simplest process of the above type in which there are just two pigment cell types, colored and uncolored, labeled C and U, respectively. We assume that each C cell produces an inhibitor morphogen that stimulates nearby C cells to change to the U state; the C cells also produce an activator morphogen that stimulates nearby U cells to shift to the C state. The two morphogens are diffusable at different rates, with the inhibitor having the greater range. The U cells are passive and produce no active substances. Thus, the fate of each pigment cell will be determined by the sum of the influences on it from all neighboring C cells.

We can describe the production, diffusion and decay of morphogens by the generalized diffusion equation

$$\frac{\partial M}{\partial t} = \nabla \cdot \mathbf{D} \cdot \nabla M - KM + Q,$$

where $M = M(x,t)$ is the morphogen concentration (either activator or inhibitor), with the other terms representing the diffusion, chemical transformation and production of morphogens, respectively. Here x represents spatial position in the plane. To eliminate having to deal with this complicated partial differential equation, assume each C cell produces the two morphogens at a *constant* rate, and that the morphogens diffuse away from their source and are uniformly degraded by the neighboring cells. The two morphogens together constitute a "morphogenetic field" $w(R)$, where R is the distance from the C cell that produces the morphogens. Assume the net activation effect close to the C cell is positive and constant, but that the net inhibition effect farther from the cell has a constant negative value. Thus, the activation-inhibition field has the shape of an annulus, with the C cell at its center and radii $R_1 < R_2$. The morphogenetic field $w(R) = w_1 > 0$ if $0 \leq R < R_1$, while $w(R) = w_2 < 0$ if $R_1 \leq R \leq R_2$.

To model the development of various types of skin patterns using cellular automata, we begin by randomly distributing C cells on a rectangular grid of lattice points in R^2, each lattice point representing a pigment cell. For a grid point at position x, the field values due to all neighboring C cells are then summed. If $\sum_i w(|x - x_i|) > 0$, the cell at x becomes (or remains) a C cell; if the sum is negative, the cell becomes (or remains) a U cell. But if the sum is zero, the cell retains its current state. In this way the continuous problem described by the diffusion equation above is replaced by a discrete two-dimensional cellular automaton operating according to the above simple additive rule. We let the system evolve until the resulting pattern stabilizes, which generally occurs after about five time steps. The only remaining issue is to decide what local neighborhood to use for this cellular automaton.

In experiments carried out by David Young, the radii of the activation-inhibition annulus were taken to be $R_1 = 2.30$, $R_2 = 6.01$, with the activation field value fixed at $w_1 = 1.0$. In these experiments, the inhibition field value w_2 ranged over a variety of negative values. Fig. 3.19 displays the results of some of these calculations. We note that as the inhibition level increases, the spot pattern connects up into a pattern of stripes. In closing, we note that by using an anisotropic neighborhood (ellipses rather than circles), it's possible to obtain patterns exhibiting *both* spots and stripes. Details can be found in the papers cited in the Notes and References.

Exercise

1. The striping patterns shown in Fig. 3.19 are all isotropic with no preferred direction, a consequence of the circular symmetry of the activator-inhibitor model. Consider now an *anisotropic* model, whose activator area is the ellipse $x^2/a^2 + y^2/b^2 = 1$, with $a = 2.30$, $b = 1.38$. Assume the inhibitor area is also an ellipse with $a = 3.61$, $b = 3.01$, and that the activator and inhibitor field values w_1 and w_2 are the same as in the isotropic case. (a) Show that this model leads to striping patterns displaying a preferred direction. (b) How would you extend this model to include multicolored patterns?

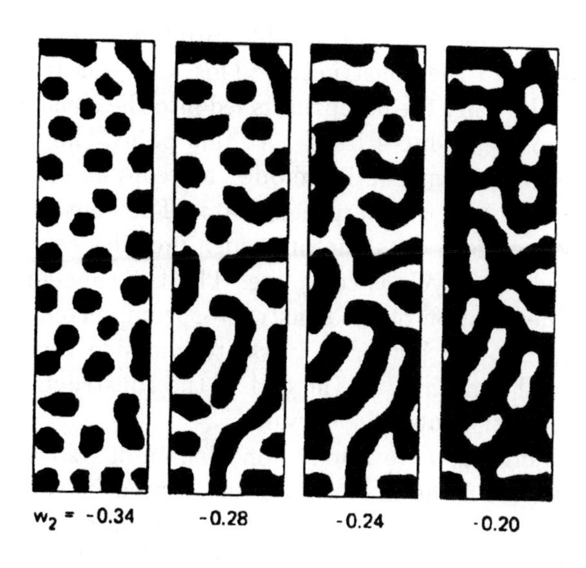

Figure 3.19. Skin Patterns Emerging from Different Inhibitor Values

12. Self-Reproducing Automata

Cellular automata were originally introduced by John von Neumann to serve as an idealized structure for the study of "self-reproducing" machines. The question of interest to von Neumann was to identify the *kind* of logical organization and functional activities that an automaton would have to posses in order to be able to build a copy of itself. Thus, von Neumann wanted to abstract from natural biological self-reproduction the *logical* form of the reproduction process, independent of its material realization in any particular physico-chemical form. He was able to create a universal Turing machine consisting of a two-dimensional automaton (with $k = 29$ states per cell) capable of universal *construction* (using the 5-cell von Neumann neighborhood). Self-reproduction then followed as a special case when the machine described on the constructor's input was the constructor itself.

The key to the self-reproduction problem lies in the way we handle the issue of copying the "blueprint" of the machine. Suppose we've succeeded in building a universal constructor. We then feed the plans for the constructor back into it as input. The constructor will then reproduce itself. But it will not reproduce the instructions describing how to build itself. This is a trivial and nonperpetuating type of reproduction, and not at all what we have in mind when we speak of self-reproducing machines. So how do we arrange it so that the blueprint, as well as the constructor, is reproduced? This was the big difficulty that von Neumann had to surmount and, as we shall soon see, his solution involved using the information on the blueprint in two completely different ways.

Von Neumann's resolution of the blueprint dilemma was to build a "supervisory unit" into the constructor. This unit was to function in the following manner: Initially the blueprint is fed into the constructor as before, and the constructor reproduces itself. At this point, the supervisory unit switches its state from "construction-mode" to "copy-mode," proceeding to copy the blueprint as raw, uninterpreted data. The copy is then appended to the previously produced constructor (which includes a supervisory unit), and the self-reproducing cycle is complete. The key element in this scheme is to prevent the description of the constructor from becoming a part of the constructor itself, i.e., the blueprint is located outside the machine and is then appended to the machine at the end of the construction phase by the copying operation of the supervisory unit.

The crucial point to note about von Neumann's solution is the way in which information on the input blueprint is used in two fundamentally different ways. First, it's treated as a set of instructions to be *interpreted* which, when executed, cause the construction of a machine somewhere else in the automaton array. Thereafter the information is treated as *uninterpreted* data, which must be copied and attached to the new machine. These two different uses of information are also found in natural reproduction, the interpreted instructions corresponding to the process of gene *translation,* while copying the uninterpreted data corresponds to the process of *transcription.* These are exactly the processes we discussed above in connection with the cellular DNA, and it's worth noting that von Neumann came to discover the need for these two different uses of information several years before their discovery by biologists working on the mysteries of DNA. The only difference between the way von Neumann arranged things and the way nature does it is that von Neumann arbitrarily chose to have the copying process carried out after the construction phase, whereas nature copys the DNA early-on in the cellular processes.

Simpler machines than von Neumann's can be shown to be capable of self-reproduction, so the question arises: How simple can a self-reproducing machine be? This is the converse of von Neumann's original question, which involved consideration of what would be *sufficient* for self-reproduction. Now we are concerned with what's *necessary.* As noted above, there are many types of "pseudo–self-reproduction." A simple example is the automaton defined by the mod 2 addition rule using the von Neumann 5-cell neighborhood in the plane. In this case, starting with a single ON cell, a little later we will see 5 isolated cells that are ON. Clearly, this doesn't constitute self-reproduction, since the initial configuration was "reproduced" by the state-transition rule rather than by containing the rules for reproduction within itself. So reproduction here is due solely to the transition rule and in no way resides within the configuration itself.

To rule out this kind of pseudo–self-reproduction, it's now customary to require that any genuinely self-reproducing configuration must have its reproduction actively directed by the configuration itself. Thus, we require that responsibility for reproduction reside *primarily* with the parent structure—but not *totally*. This means that the structure may take advantage of certain features of the transition "physics," but not to the extent that the structure is merely passively copied by mechanisms built into the transition rule. Von Neumann's requirement that the configuration make use of its stored information as both instructions to be interpreted and as data to be copied provides an appropriate criterion for distinguishing genuine from pseudo self-reproduction. It's appropriate to close this discussion of self-reproduction by outlining the manner in which Conway showed that the Life rule admits configurations that are capable of self-reproduction in exactly the sense just described. With this bit of mathematical chicanery, Conway proved that the simple Life rule is actually complicated enough to allow us to compute **any** quantity that can be computed.

Conway's self-reproduction proof is based on the observation that Glider Guns (as well as many other Life objects) can be produced in Glider collisions. He then shows that large constellations of Glider Guns and Eaters can produce and manipulate Gliders to force them to collide in just the right way to form a copy of the original constellation. The proof begins not by considering reproduction *per se*, but by showing how the Life rule allows one to construct a universal computer. Since the Life universe consists of an array of ON–OFF squares on a sufficiently large checkerboard, what this amounts to is showing that one can construct a Life pattern that *acts* like a computer. This means that we start with a pattern representing the computer and a pattern representing its programming. The computer then calculates any desired quantity, which then has to be expressed as a Life pattern itself. For numerical computations this might involve the Life computer emitting the requisite number of figures or, perhaps, arranging to place the required number of figures at some prespecified part of the display area.

The key to the Life computer is the demonstration that any binary number can be represented by a Glider stream, and that other Life patterns can be arranged to function as AND, OR and NOT gates, the building blocks needed for any computer. The biggest problem with these constructions is to show how the various streams representing the "wires" of the Life computer can be made to interpenetrate without losing their original structure. While wires and logic gates are all that's needed for any real-world computer, a universal computer needs something more: a potentially infinite memory! Conway's solution is to use the Life configuration termed a "Block" to serve as an external memory element. The Block consists of a 2×2 array of ON cells with the property that it is a so-called "still life," i.e., it remains invariant under the Life rule (see the fourth row of Fig. 3.11). The idea is to use the Block as a memory element outside the computer pattern, and to use the distance of the Block from the computer to represent the number being stored. To make this scheme work, it's necessary to devise a procedure to move the Block even though it's not in the computer. This can be accomplished by a tricky set of Glider–Block collisions. The end result of all these maneuvers is Conway's proof that the circuitry of *any* possible computer can be translated into an appropriate Life pattern consisting only of Guns, Gliders, Eaters and Blocks. But what about the other part of the self-reproduction process, the universal constructor?

The second part of the Conway proof is to show that any conceivable Life pattern can be obtained by crashing together streams of Gliders in just the right way. The crucial step in this demonstration is to show how it's possible to arrange to have Gliders converge from four directions at once in order to properly represent the circuits of the computer. The ingenious solution to this seemingly insoluble problem, termed "side tracking," is much too complicated to describe here, but it provides the last step needed to complete Conway's translation of von Neumann's self-reproduction proof into the language of Life.

Now what would a self-reproducing Life pattern look like? For one thing, it would be **BIG.** Certainly it would be bigger than any computer or video screen in existence could possibly display. Moreover, it would consist mostly of empty space, since the design considerations require the use of extremely sparse Glider streams. The overall shape of the pattern could vary considerably depending upon design considerations. However, it would have to have an external projection representing the computer memory. This projection would be a set of Blocks residing at various distances outside the pattern's computer. Moreover, at least one of these blocks would be special in that it would represent the blueprint of the self-reproducing pattern. (In actuality, the blueprint is the *number* represented by this block.) For a detailed description of how the reproduction process actually works, we refer to the literature cited in the Notes and References.

Now let's close this discussion of Life by noting the sobering estimate for how big such a self-reproducing Life pattern would have to be. Rough estimates indicate that such a pattern would probably require a grid of around 10^{14} cells. By comparison, a high-resolution graphics terminal for a home computer can display around 10^6 cells (pixels). To get some feel for the magnitude of this difference, to display a 10^{14} cell pattern, assuming that the pixels are 1 mm^2, would require a screen about 3 km ($\approx$ 2 miles) across. This is an area about six times greater than Monaco! Thus, we can safely conclude that it is highly improbable that Conway's vision of living Life patterns will ever be realized on any real-world computer, even those likely to emerge from the rosiest of estimates of future computer technology. From this sad state we now pass to a consideration of how cellular automata can be used to create lifeforms in a machine.

Exercises

1. In 1959 the geneticist L. S. Penrose (father of the physicist Roger Penrose) concocted a system of blocks in a box to illustrate a kind of self-reproduction. The blocks have hooks that can engage other blocks in several different arrangements. When a "seed"—consisting of a pair of blocks hooked together in one of the possible arrangements—is placed into a box full of unhooked blocks and the box is shaken, the seed induces the rest of the blocks to hook up in pairs exhibiting the same configuration as the seed. The figure below shows a reproductive cycle of one of Penrose's devices, in which the seed is the two center columns of blocks in part (a), which are reproduced as the system moves from (a) to (d). Discuss whether or not Penrose's machine is an example of genuine or pseudo self-reproduction.

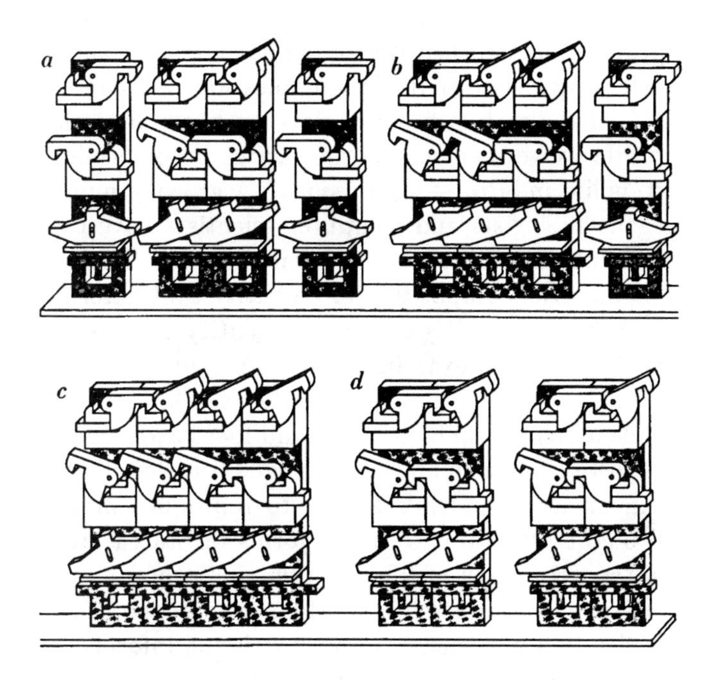

The Workings of a Penrose Machine

2. The capacity for genuine, as distinguished from "pseudo," self-reproduction requires that an entity contain a constructor, a supervisory unit, a blueprint and a copier. Determine whether or not these four components are (or could ever be) present in: (a) a living cell, (b) a computer virus, (c) a national economy, (d) a chemical plant, (e) a robot, (f) a crystal. If they are present, identify each of the components with its real-world manifestation; if some are absent, identify the missing components.

3. In living cells the copying operation takes place before the construction phase, while in von Neumann's scheme of things these two operations are reversed. Can you give any arguments why nature may have chosen the order it did? (Hint: Consider the possible evolutionary advantages to doing the copying first.)

4. Sketch out a scheme for how a machine like a computer could manufacture a copy of itself if placed in an environment containing all the necessary raw materials (e.g., metallic ores for refining into copper and aluminum, sand for making silicon chips, and all the other material elements from which computers are made).

13. Artificial Life

Our discussion about L-systems, plants and self-reproducing machines, together with the examples given earlier showing how cellular automata could be used to capture important features of living things like their DNA structure or skin patterning, raises an intriguing question: Is there any real difference between life of the carbon-based variety we know about here on Earth and the sort of life we've seen scampering about on our cellular automata grids? In short, is life *in vitro* ≡ life *in silico*? A growing number of system theorists, renegade biologists and computer scientists have taken the position that there is no real difference between the two. Their contention is that it's not the material composition of an entity that determines whether it's dead or alive, but rather it's the way the entity processes information. We devote this section to an examination of this audacious claim.

For mainline biochemists, the salient features of life are complex carbon compounds, while a behavioral biologist would focus attention on the ability of an organism to react to stimuli. An ecologist, on the other hand, would probably zero-in on an organism's ability to reproduce in the absence of predators. All of these features of living forms on Earth are irrelevant to an information-theoretic view of life. In this view, the physical structures encoding the information are irrelevant; the printed page, a videotape, and pulses of electricity in a wire mean nothing. It's the information itself that counts. While in earlier sections we've indirectly hinted at what **does** count from an information-theoretic point of view, let's now finally spell it out in detail.

In his work on self-reproducing machines, von Neumann set down the following criteria for life, information-style:

1. A living system contains a complete description of itself.
2. To avoid an internal self-contradiction, a living system does not try to include a description of the description in the description.
3. In order to avoid the contradiction, the description must serve a dual function: (a) it is a coded description of the *rest* of the system,

and (b) it is a working model of the entire system (a model that need not be decoded).

4. Part of the system (the supervisory unit) "knows" about the dual role of the description and makes sure that both roles of the description are used during the process of reproduction.
5. Another part of the system (the universal constructor) can build any of a large class of objects—including the system itself—if it is supplied with the proper materials and is given the correct instructions.
6. Reproduction takes place when the supervisory unit tells the universal constructor to build a new copy of the system, including a copy of the description.

Artificial life enthusiasts take the above criteria as a *definition* of what it means to be alive. So regardless of its material composition, any system satisfying these conditions is "alive."

Part of the attraction of the information-theoretic view of life embodied in the foregoing conditions is that it enables us to settle the life/non-life nature of borderline cases like crystals, computers or extraterrestrials that might have a biochemistry not based on carbon. For instance, we could ask "Could a computer be alive?" By the artificial life criteria above, the answer must be yes, since von Neumann's original self-reproducing model of a machine was actually a robot floating in a sea of parts needed for its own reproduction. He showed how reproduction could be carried out in this kinematic sense, thereby creating an abstract model of a machine satisfying all of the conditions 1–6. On the other hand, a computer could also be arbitrarily powerful as an information processor without being alive. If it just computes without having the capacity for self-reproduction, then it fails to meet up to the criteria for life. So a supercomputer might be intelligent, or even "conscious" in some sense, without being alive by the standards of the artificial lifers. Since the Life game and the plants generated by L-systems are both examples of artificial life forms satisfying the foregoing criteria, let's now look at another corner of the artificial zoo, where evolution rather than development is what's on display.

In September 1987 the Los Alamos National Laboratory was host to the First International Workshop on Artificial Life. One of the more entertaining papers given at the workshop was by the well-known Oxford University zoologist Richard Dawkins. In his presentation, Dawkins gave the audience a short tour of what he calls Biomorph Land, a mythical territory so named in honor of the vaguely lifelike creatures called biomorphs appearing in the surrealistic paintings of the anthropologist Desmond Morris. Let's take a quick tour through Biomorph Land ourselves as an entertaining way making contact with evolutionary processes in the world of information.

The elementary objects of Biomorph Land are simple tree-like figures that we can represent on a piece of paper by a sequence of short lines. A typical object in this universe is shown in Fig. 3.20, which consists of a "seed biomorph" at the top, followed by a sequence of more elaborate relatives. The seed biomorph was generated by having the computer first draw a single vertical line. Next the line branches in two, and then each of the branches splits into two subbranches. Continuing with this same splitting-and-growing process, we obtain the more elaborate biomorphs shown in the rest of the figure, which can be thought of as the "children" of the "parent" seed.

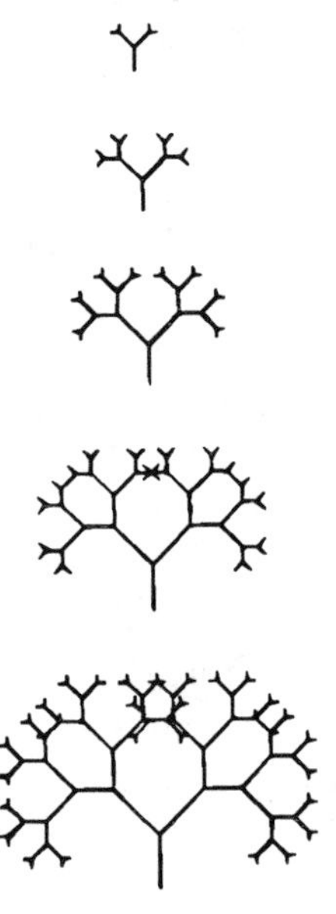

Figure 3.20. A Typical Biomorph Seed and its Progeny

Dawkins introduces genetic influence on the biomorph phenotype by incorporating nine elements—or "genes"—into his program for biomorphic development. These genes influence things like the angle of branching, the length of a branch, the number of subbranches, and so on. For the sake of genetic variability, each gene is assumed to come in both a "plus" and a "minus" form. An idea of how this works can be seen in Fig. 3.21, which shows a central tree (biomorph) in the middle surrounded by eight variants. All biomorphs in the figure are the same as the central one, except that one gene has been changed, i.e., "mutated." For example, the biomorph labeled Gene 1^- shows the effect on the central tree if Gene 1 mutates to its minus form.

Each gene action has its own characteristic "formula" in Dawkins's program, but the formulas themselves are, as Dawkins points out, meaningless. Just as in real life, where genes mean something only when they are trans-

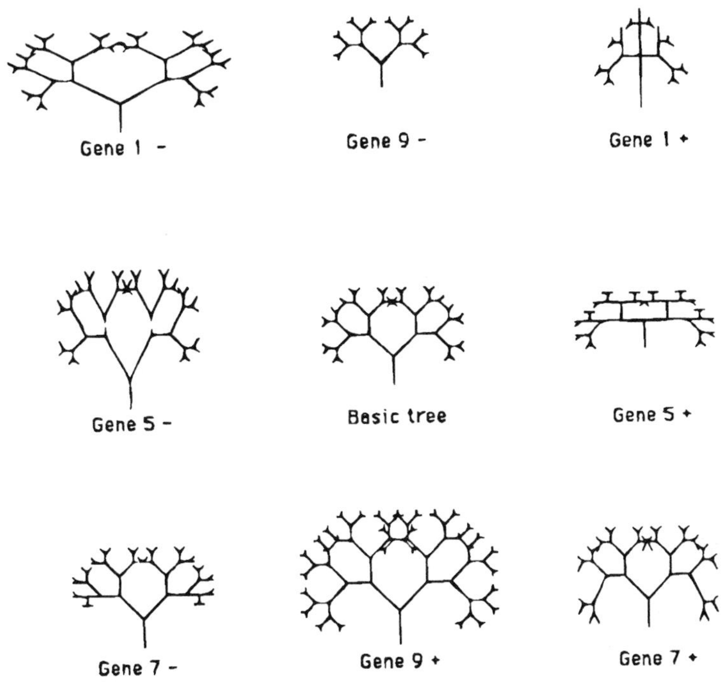

Gene 1 - Gene 9 - Gene 1 +

Gene 5 - Basic tree Gene 5 +

Gene 7 - Gene 9 + Gene 7 +

Figure 3.21. The Effect of Genes in Biomorph Land

lated into proteins and growing rules for developing embryos, in Biomorph Land, too, the genes are meaningful only when they are translated into rules for creating a branching tree pattern.

With these rules for genetic action in hand, Dawkins starts with a parental biomorph and "grows" a litter of children, each differing from the parent by a mutation in one gene. At this stage, evolution steps in to play a hand by selecting just one of the offspring to go on to the next generation. In practice, Dawkins himself plays the role of the "natural selector." Fig. 3.22 shows 29 generations of such a biomorph history, starting from the dot (blob) in the upper left-hand corner. The figure also shows a few of the evolutionary dead ends that Dawkins rejected during the process of selecting those forms that would survive to reproduce for the next generation. Fig. 3.23 displays some of the other inhabitants of Biomorph Land arising out of the combination of random genetic mutations and Dawkins' "natural selection."

But interesting as Biomorph Land is in the variety of its inhabitants, the kind of reproductive mechanism built-in to Dawkins's program is not exactly the kind that would qualify his biomorphs to be termed "alive" by the von Neumann criteria. The problem is that the individual biomorphs don't contain a description of themselves, which is one of the primary requisites for any living organism, von Neumann-style. Rather, the description is superimposed from the outside, so to speak, by Dawkins's rule for genetic

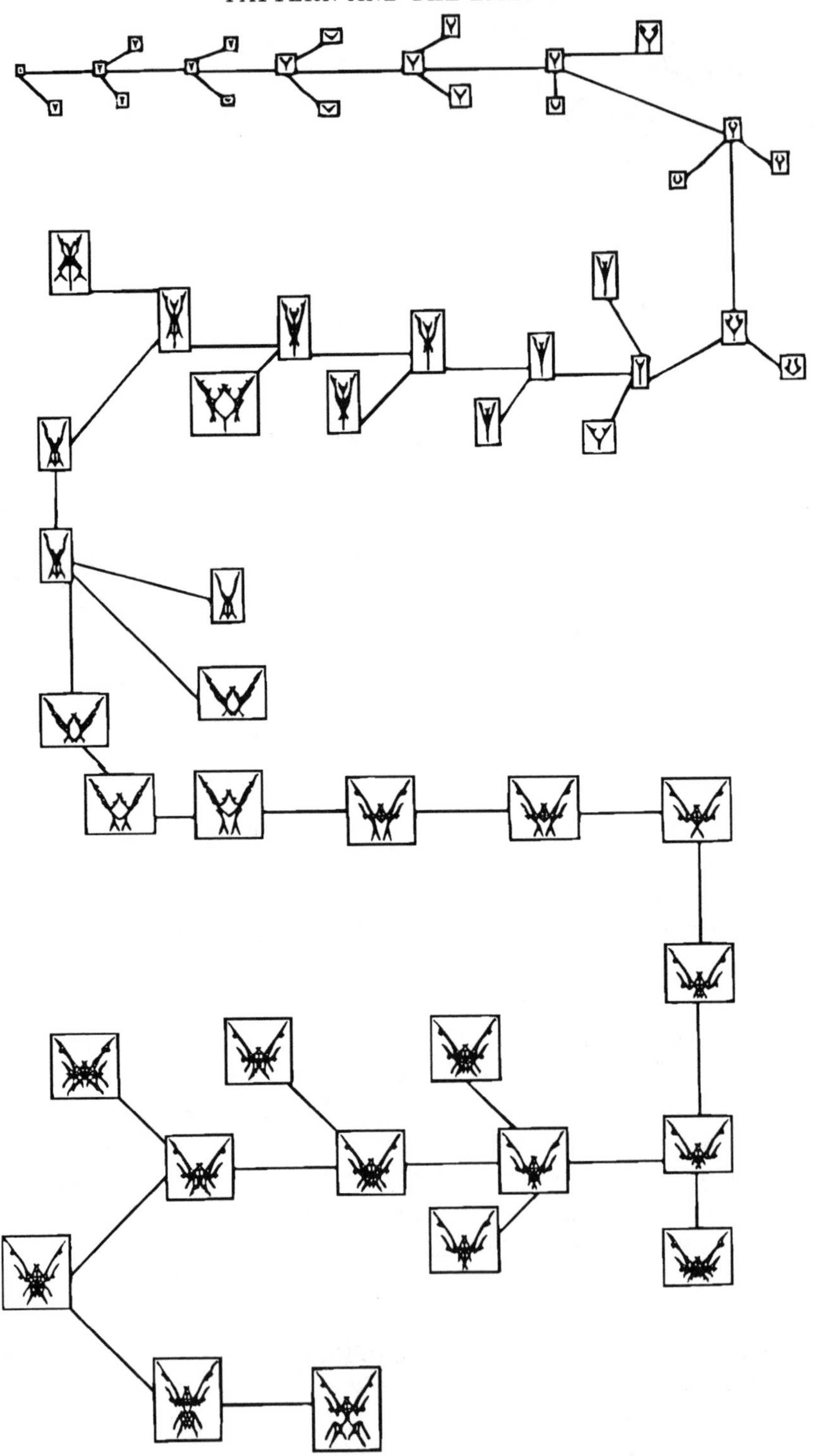

Figure 3.22. A Biomorph Evolutionary Pathway

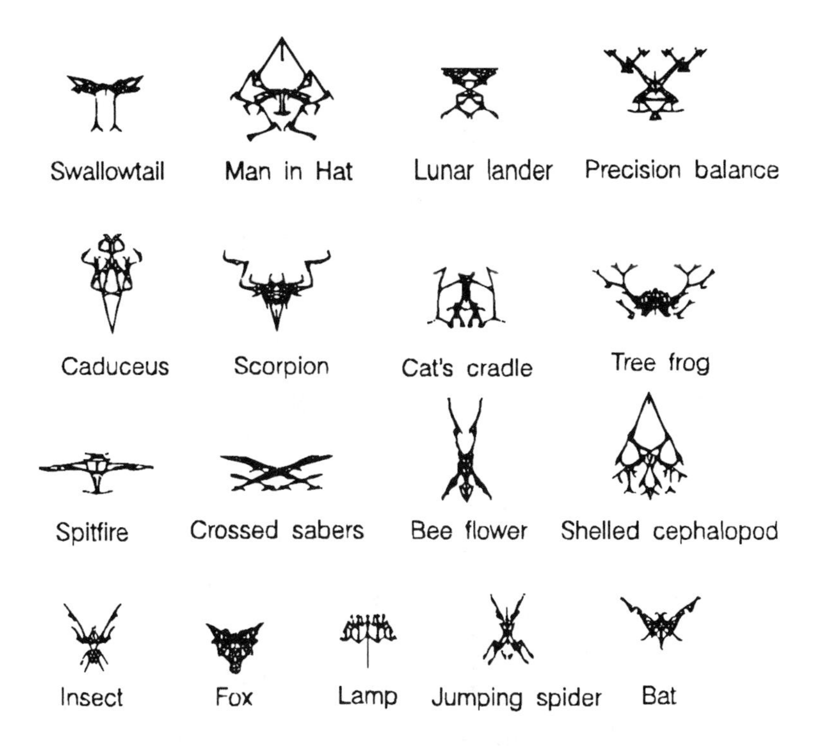

Figure 3.23. Some Residents of Biomorph Land

"translation." Nevertheless, Biomorph Land is an intriguing exercise in the use of information theory to study life-like processes, and I'm sure it could be beefed-up to display the conditions for genuine life discussed earlier.

Exercises

1. At each stage of the developmental process in Biomorph Land, Richard Dawkins played the role of the "natural selector," choosing those candidates to go on to the next generation by personal whim and a sense of what "looked nice." Consider how you might create selection rules that could be built-in to the program directly, and that would make this choice on the basis of things like phenotypical similarities or differences between competing biomorphs, genetic similarities or differences, random selection or whatever.

2. Since Biomorph Land contains many different species, it forms an artificial ecosystem. Think about how population dynamics would unfold in such a system. In particular, introduce the notion of competition for limited resources, and use the selection rules considered in the preceding Exercise to see what path evolution will take in your version of Biomorph Land.

3. Consider how you might extend (literally!) Dawkins's planar life forms into fully-developed 3-D biomorphs.

14. Cellular Automata—Extensions and Generalizations

In the limited space available here we have only begun to scratch the surface of the potential for cellular automata to capture the underlying patterns emerging in processes taking place in the physical, life, social and behavioral sciences. By way of completeness, let's now very briefly indicate some directions in which the basic ideas developed above can be extended to more complicated types of automata.

- *More States*—For the most part, we have confined our attention to the case when each cell can take on only $k = 2$ values. There is no reason why each cell cannot assume any number of different values, an uncountable number even. The problem, of course, is that by allowing a large number of possible values for each cell, it's difficult to interpret whatever structure may be present in the emerging patterns. One way out of this difficulty is to make use of the increasingly sophisticated computer graphics that are becoming more widespread, assigning a different color to each of the possible cellular values. This procedure works well in a number of cases and, in fact, could be the basis for a scheme in which the cells could assume a continuum of values by allowing a spectrum of colors instead of just a discrete distribution. It's tempting to speculate on what kind of cellular "Picassos" or "Chagalls" might emerge from such experiments, as well as the features of the rules that lead to such computational "masterpieces" (cf. Discussion Question 5).

- *Higher Dimensions*—We have shown the type of increased pattern complexity that can arise when passing from one- to two-dimensional cellular automata. But there is no intrinsic reason to stop at planar grids. We can easily imagine defining automata on higher-dimensional grids as well, as was done in the very infancy of cellular automata in studies by Stanislaw Ulam and his co-workers. The problem in dimensions higher than two is again the difficulty in discerning the patterns that may be present in the automaton's behavior, as well as the geometric increase in the number of rules. And this is not to mention the number of different types of neighborhoods that can occur. Here again modern computer graphics may come to the rescue, at least in the sense of being able to display the output of such a higher-dimensional automaton in a form that might enable one to identify interesting structures by looking at lower-dimensional sections of the overall configuration.

- *Complex Rules*—We have imposed severe restrictions on the types of transition rules in order to keep the set of possible automata to a manageable level. However, there are many natural phenomena that seem to call for more elaborate state transitions. In one direction, we could allow the transition rule to be *nonlocal* so that the value of a given cell is determined not just by those cells adjacent to it (in any sense of adjacency), but also by cells that lie in some other part of the array. Rules of this type might be appropriate for certain classes of problems arising in quantum physics and electronics.

In a somewhat different vein, we might want to allow rules that are *time-* or *space-dependent.* Such a rule would actually represent many rules at once, as the change of cell value would then depend upon the spatial location of the cell in the grid as well as upon the particular time step. Clearly, there are an astronomical number of such rules, and the choice of those to explore would have to be narrowed down on physical and/or aesthetic grounds. However, it's quite possible that many phenomena in the life and social sciences might evolve according to rules of this sort that would change as environmental, economic and/or social circumstances shift.

Finally, we may want to consider state-transition rules that are *stochastic* rather than deterministic. With such rules, the actual value of a cell would be determined not just by the values of neighboring cells, but also by the values of random variables obeying some probability distribution. Such a rule might be quite appropriate for a cellular automaton model of, say, genetic mutations or various types of population migration processes. Of course, with such a rule we would have to consider *ensembles* of automaton configurations and study the emergent patterns in some statistical sense.

• *Boundary Conditions*—Cellular automata are assumed to "live" on an infinite grid; computer experiments must be carried out on some finite subset of this infinite space. This introduces the question of how to deal with the behavior of cells that are on the boundary of the finite region. The simplest solution is to adopt the periodic boundary conditions that we have used throughout this chapter. But there are other possibilities, especially when one considers automata in dimensions higher than one. It's of considerable interest to know the degree to which the patterns that arise are artifacts of the particular choice of boundary conditions rather than being genuine properties of the automaton itself. Even in low dimensions, examples can be given showing that the boundary conditions **do** matter. A systematic investigation of this phenomenon would be of considerable value.

• *Control*—The cellular automata we have been considering are discrete versions of the classical dynamical systems of Newton, Laplace, and Lagrange. From the standpoint of an observer, this is a very passive kind of dynamics: You just turn the system on and see what happens (if you live long enough!). A major extension of this Newtonian view is the introduction of *active* control. In cellular automaton terms, this means that the cell's value would depend not only upon the value of its neighbors, but also upon the value of a *control* element chosen by the system controller (manager, decisionmaker or designer).

The idea of active control drastically changes many important features of, and introduces an added level of complexity into, the analysis of the behavior of a system. For instance, now we might want to ask whether it's possible to arrange for the automaton's state to be some prespecified configuration after a predetermined number of time steps by choosing an appropriate sequence of inputs. This is a typical problem of *reachability,* which is now well understood for continuous-state dynamical systems (see Chapter 6). The extension of these results to the cellular automaton world should prove to be interesting, to say the least.

Exercise

1. In this chapter we have focused exclusively on cellular automata defined on *rectangular* grids. Consider what difference it might make if the automata were defined on other types of geometrical regions, e.g., hexagonal or circular lattices.

Discussion Questions

1. We have defined the spatial topological entropy of a sequence of cells as a measure of the likelihood of particular sequences of X cells appearing from random initial configurations. We can also define a spatial *measure* entropy formed from the probabilities of possible sequences, as well as temporal entropies, to count the number of sequences that occur in the time-series of values taken on by each cell. Topological entropies reflect the possible configurations of a system; measure entropies reflect those that are probable. Can you connect up the properties of topological/measure, spatial/temporal entropies with the four classes of one-dimensional cellular automata discussed in the text? That is, can you classify the various types of automata if you know the numerical values of the various entropies?

2. For Class C one-dimensional automata, we see cases of patterns with large triangular regions and low entropies; other rules give highly irregular patterns with no long-range structure. There appears to be no statistical test to distinguish between these different types of Class C structures. How would you go about developing a measure that would identify such subclasses of the four main types of one-dimensional automata?

3. Consider a two-dimensional automaton with a finite state-space consisting of the 64 cells of a checkerboard. Each cell can assume one of $k = 4$ values, which we label red, yellow, blue and green. We begin with an initial configuration consisting of a random distribution of the four colors in equal proportions on the cellular space; i.e., initially there are 16 cells of each color randomly distributed on the board. At each time step, *one* of the cells is selected at random and the following procedure is used to determine the next color of that cell:

A. The color on the cell is noted and the color is then removed.

B. Another cell is selected at random. If one of the cells in the von Neumann neighborhood of the selected cell is occupied by a color *preceding* the color removed from the first cell, then the color from the selected cell is placed in the first cell; otherwise, the first cell remains empty. The cyclical color precedence ordering is:

$$\text{red} > \text{yellow} > \text{green} > \text{blue} > \text{red}.$$

Steps A and B are repeated at each time step until the pattern stabilizes.

Describe what patterns you think are likely to emerge over the course of time. For example, do you think there is an appreciable chance that any color will get "wiped out" as the process unfolds? Or do you think the relative proportions of the colors will oscillate? (Remark: The rules of this automaton are set up to mimic the development of self-reproducing "hypercycles," a popular model for various sorts of autocatalytic chemical reactions that are considered further in Chapter 5.)

4. Andrei N. Kolmogorov and Gregory Chaitin independently introduced the idea of the "complexity" of a sequence of numbers as being the length of the shortest computer program needed to produce the sequence. This idea turns out to be equivalent in many cases to the notion of entropy, which measures the exponential temporal growth rate of the number of distinct orbits of a dynamical system. The set of Lyapunov exponents are a closely related collection of numbers measuring the time-averaged local asymptotic divergence rate of the orbits. The averaged sum of all the positive Lyapunov exponents is always less than or equal to the metric entropy of the system. To what degree do you think either the Kolmogorov-Chaitin or the Lyapunov exponent concept serves as a good characterization of the complexity of a cellular automaton? (Note: The Lyapunov exponents measure only the *local* divergence of trajectories, and are inherently incapable of ascertaining any *global* macroscopic divergence.)

5. The Dutch artist Piet Mondrian was famous for developing a cubist style of painting that emphasized checkerboard patterns in various colors. Displayed on the next page is a black-and-white version of his work *Checkerboard, Bright Colors, 1919,* which involves a seemingly random scattering of eight colors on a rectangular grid of 256 cells, random, that is, to the untutored eye. Imagine a celluar automaton with this grid as its state-space and the $k = 8$ colors as the allowable values of each cell.

a) What kind of state-transition rule leads from an arbitrary initial configuration to that of *Checkerboard, Bright Colors, 1919?*

b) Can you think of any criterion by which to single out the artistically aesthetic patterns from the uninteresting ones among all patterns that might

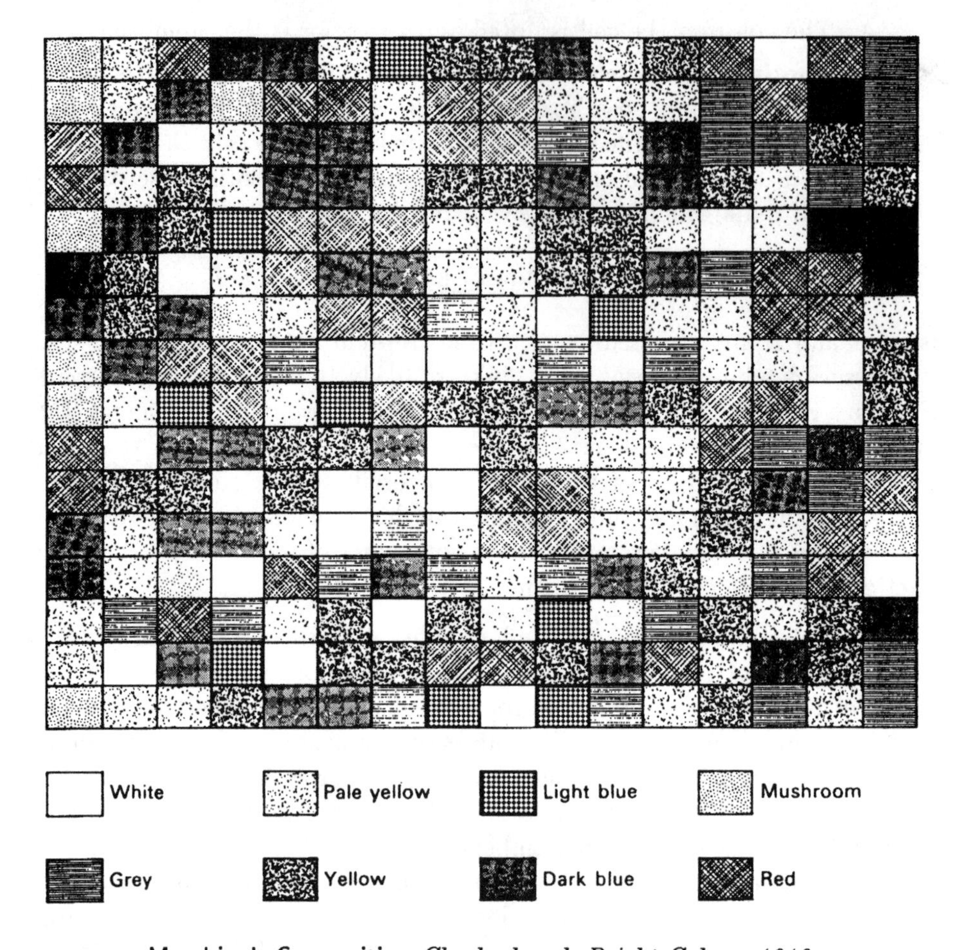

Mondrian's Composition *Checkerboard, Bright Colors, 1919*

emerge from using the particular transition rule of the Mondrian system, starting with different random initial states?

c) Suppose you are given a Mondrian painting and a Picasso, both projected onto the same grid using the same set of basic cell values (colors). Do you think it would be possible to develop a "filter" enabling you to distinguish the Mondrian from the Picasso? That is, is it possible to produce a computable test that would determine the identity of the artist of any particular painting? (Remark: We shall return to a more detailed consideration of these matters in Chapter 8.)

6. Apply now the method of the preceding Discussion Question, which used a two-dimensional state space for paintings, to the the case of musical works. In music we have a linear, sequential structure, so it seems plausible to use a one-dimensional automaton to represent the unfolding of a musical piece, say, a Mozart symphony or a Strauss waltz. Each state of

the automaton would represent the notes that are being played at that moment in time, with the state-transition rule being embodied in the musical score prescribing the notes that are to be sounded at the next moment. Do you think it's possible to develop this idea into a mechanistic procedure for representing and studying musical works? How would you account for things like rhythm and intensity in such a scheme? Presumably, each composer's (and conductor's) style would be represented by a set of invariant features of the state-transition rule for the automata representing this composer's work. How would you go about identifying the specific properties of a transition rule that would stamp it as belonging to a particular composer? Consider the entire issue of computer-generated music within the context of one-dimensional cellular automata, and examine the question of how to represent a musical composition in automata-theoretic terms.

7. The preceding questions on artistic forms raise the issue of how different behaviors are distributed within the space of cellular automata rules. Empirical studies suggest that for symmetric one-dimensional rules, Class A and B automata appear to become progressively less common as the number of state values k and the size of the neighborhood R increase. Class C automata become more common, and Class D slowly less common. For two-dimensional automata, Class C (chaotic) is overwhelmingly the most common, with Class D being very rare.

Consider these observations within the context of the artistic identifications discussed above. Since we may assume that an aesthetically pleasing form does not arise as the result of a random or chaotic rule, it would appear that genuine works of art would be represented by Class A, B or D automata, although Class D also seems rather unlikely as it would involve an infinite time pattern with no periodicity. Discuss the conjecture that artistic "signatures" are represented only by Class A or B rules. How would you establish a metric on the space of rules to measure the "nearness" of one rule to another?

8. We have seen that for one-dimensional automata, the set of configurations reached after a finite time constitutes a regular language, which can be represented by the paths through a finite directed graph. The smallest such graph provides a canonical description of the set of all words in the language, and the number of nodes in the graph represents the *complexity* Ξ of the language.

a) In general, the value of Ξ is bounded above by the quantity $2^{k^{2Rt}} - 1$, and appears to be nondecreasing with time. Discuss the claim that the increase of complexity in time is a principle for self-organizing systems analogous in generality, but complementary in content, to the law of entropy increase in closed thermodynamic systems (the famous Second Law of Thermodynamics).

b) Consider the thesis that the complexity Ξ is a good measure for the artistic merit of a work described by a one-dimensional automaton, in the sense that works of higher complexity are "better" in an artistic sense than those of lower complexity. Examine this contention by comparing the complexities of a rock or C&W tune with a Beethoven or Haydn symphony.

c) How would you go about developing a procedure for calculating the complexity Ξ *directly* from the transition rule and the initial configuration, rather than computing it in "real time" from the unfolding of the automaton's initial state?

d) It's often held that Darwinian evolution involves the passage from a state of lower to increasingly higher complexity. Does the foregoing notion of complexity seem to you to be consistent with this conventional wisdom? Can you tie together this idea of biological evolution and complexity with the evolution and complexity of patterns arising in the two-dimensional Life automaton?

9. The text describes an L-system for modeling one-dimensional plant growth. How could you extend the set-up of the text to two- and three-dimensional L-systems that would account for spatial features of real plant growth, such as sprouting of leaves, reproduction by airborne transmission of seeds, and the budding of flowers?

10. If a system is capable of universal computation, then with appropriate initial conditions its temporal evolution can compute anything that can be computed. But predictions about the behavior of a cellular automaton must be made by performing a computation. So, if the automaton is capable of universal computation, this prediction must reduce, in general, to a direct simulation of the automaton's time history. Consequently, questions about the limiting behavior of such automata may require infinite computations; hence, they are formally undecidable.

a) Universal computation can be used to establish the undecidability of questions about the behavior of a system. What about the converse? That is, can undecidability occur even in systems *not* capable of universal computation? Can you construct an example of such a case? (Hint: Find a system capable of universal computation, and show that a reduction of its capabilities does not affect undecidability.)

b) Rice's Theorem states that almost all questions about arbitrary recursively enumerable sets are undecidable. What do you think about the possibility that for any particular automaton, simple questions that can be stated in a few logical symbols usually *are* decidable? That is, we are not usually interested in *all* propositions that can or cannot be decided about a given automaton, but only in those "natural" questions that can be stated in a relatively compact way.

11. We can think of the evolution of a cellular automaton as a *pattern-recognition* process, in which all initial configurations in the basin of attraction of a particular attractor are thought of as instances of some pattern with the attractor being the "archetype" of this pattern. Thus, the evolution of the different state trajectories toward this attractor constitutes recognition of the pattern. How would you formalize this idea into a practical pattern-recognition device for, say, processing digital images received from satallites?

12. In our discussion of self-reproduction, we saw that any such process must contain four components:

A. A *blueprint* describing the object to be constructed.

B. A *factory* to carry out the construction.

C. A *controller* to make sure the factory follows the plan.

D. A *duplicating machine* to transmit a copy of the blueprint to the offspring.

We also saw how these components entered into biological replication in a living cell. Do you think the same four basic components appear in other "living" organisms such as a society, a manufacturing enterprise or an ecological network?

13. The actual emergence of structure in an automaton's evolution is a consequence of a *level shift* in description. Even though the initial configuration *looks* random, it must fully encode (and hence "know") the final pattern that the computation unveils. But it takes an external observer to appreciate any pattern that may be present in this structure. That is, the emergence of pattern decreases the number of available *microstates* but may increase the number of observed *macrostates.* The ordered states coming about from an irreversible computation are few, but *recognizable.* Discuss how a mathematical procedure could be developed to characterize this recognizability.

14. The propagation of an infectious disease has many striking similarities with the way two-dimensional automata patterns emerge from initial infectives. How would you go about constructing an automaton rule that would mimic the way diseases spread in a population? (Note that here it seems likely that the rule would have to involve *stochastic* state-transitions to account for the fact that not all people exposed to a disease actually contract it.) Do you think a model for disease propagation could be modified to suit a situation involving the migration of human or animal populations from one geographic region to another? How?

15. In Ulam's Coral Reef automaton, the rule of transition is that a cell is ON in the next generation if the following three conditions are met:

A. It borders (in the von Neumann neighborhood) one and only one newly occupied cell, where "newly occupied" means the cell came ON only in the current generation.

B. It must not touch (in the Moore neighborhood) an old occupied cell, where "old" means a cell that came ON prior to the current generation.

C. Among all cells that would qualify to come ON by conditions A and B, eliminate those that would touch each other (in the Moore neighborhood).

a) What kind of patterns do you think would emerge from a single ON seed cell?

b) How would you build in an extension to the above rule to account for the death of cells?

c) What kind of "pseudo–self-reproducing" patterns do you think could come out of the above rule (with death included)?

d) Show by computer experiment that if two or more patterns interpenetrate, they may become involved in a struggle in which the winning pattern destroys the rest. How would you interpret this in terms of competition among living organisms?

16. The Life rule is carefully balanced between allowing too many births (and a consequent population explosion) and too many deaths (and an extinction of the "species"). Consider other Life rules that also balance life and death in order to create nontrivial population patterns over long time periods. For example, a popular variant is *3–4 Life,* in which the state-transition rule is quite simple: a cell is ON (alive) in the next generation if and only if it is surrounded (in the Moore neighborhood) by exactly three or four ON neighbors in the current generation. Discuss the implications of the fact that 3–4 Life is more pro-birth, but at the same time deaths by isolation are more common. Do you think this would lead to more or less the same types of patterns as in Life, or would the two universes be radically different?

17. We have seen that self-reproducing patterns are possible in Life, as well as with other types of automata. This leaves open the question of whether or not a machine can make another machine more complicated than itself. Do you think this is possible? How might it be done? What is the connection between a "complexity-increasing" construction and Darwinian evolution of living organisms?

18. Most of our discussion in the text has been for *irreversible* state-transition rules, i.e., rules for which the state at time t cannot be uniquely determined from the state at time $t+1$.

a) Show that *reversible* rules exist by explicitly displaying one.

b) If the initial configuration for a one-dimensional automaton is selected at random and you apply a reversible rule, what kind of long-term behavior would you expect to see?

c) If you choose an initial configuration and a reversible rule that allow information to propagate, what kind of temporal behavior patterns will emerge?

d) Interpret your answer to part (c) by considering each state of the automaton to be a "message" with the contents of the cells being the characters of the message, assuming only local measures of correlation.

e) Consider the possibility of using cellular automata with reversible rules to model the processes of classical Newtonian physics (space, time, locality, microscopic reversibility, conservation laws, and so forth.)

19. Cellular automata often serve as discrete versions of partial differential equations in one or more spatial variables (as seen, for example, in the animal skin pattern example of the text). What value do you see to the "inverse problem," i.e., development of a partial differential equation whose space-time behavior exactly matches a given cellular automaton at the discrete cellular grid points? Could such an equation be used to predict the long-term behavior of a Class C or D automaton without having to get at it via direct simulation? Why or why not?

20. Comment on the following argument often given to support (or justify) cellular automata research:

A. Novel computational resources are now available that may, on a given task, outperform conventional resources by a great many orders of magnitude.

B. The conceptual developments of mathematical physics must have been strongly influenced by the nature of the computational tools available.

C. Therefore, the new resources suggest new approaches to the modeling and simulation of physical systems, and it should be possible to replace conventional formulations involving real variables, continuity, analytic functions and so forth by more constructive and "physically real" counterparts.

21. In our discussion of self-reproduction, we have assumed that the blueprint (or instructions) for the automaton to be produced was given in advance. Consider the possibility of self-reproduction by means of *self-inspection,* i.e., the machine to be reproduced **creates** the blueprint by means of self-inspection. Do you see any logical contradictions in such a

scheme? If such a reproduction process were possible, what would the implications be for Lamarckian inheritance in biological reproduction, i.e., the inheritance by the offspring of characteristics acquired by the parent(s)?

22. Current DNA research suggests that many subsequences of the DNA strand are there solely "by chance," and serve no intrinsic biological function. Assuming this is indeed the case, how could this "junk DNA" arise in the cellular automata models for DNA evolution outlined in the text?

23. The canonical example of self-organized criticality is a pile of sand. Imagine building the pile by slowly adding sand grains, one at a time (*Nota bene:* This is a thought experiment!). As the pile grows, there will be bigger and bigger avalanches. Eventually, a statistically stationary state will be reached in which avalanches of all sizes occur. In other words, the correlation length is infinite. In analogy with equilibrium statistical mechanics, we call such a state *critical.*

Now suppose we start the Life game with a random distribution of ON cells, and let the system evolve until it comes to 'rest' in a simple periodic state having a distribution of local still life and simple cyclic life. We then perturb this rest state by turning ON a randomly chosen cell. Again we let the system evolve to a rest state and repeat the perturbation process, and so on. As this sequence continues, the system evolves into a statistically stationary state. Experiments on a grid of size 150×150 were carried out to study the properties of this state.

If we measure the total activity s, defined to be the number of births and deaths following a single perturbation, the distribution of clusters of size s averaged over 40,000 perturbations was found to obey the power law $D(s) \propto s^{-1.4}$. Furthermore, the distribution of durations of perturbations (i.e., the number of steps needed before the system returned to the rest state) was also found to satisfy a power law, $D(T) \propto T^{-1.6}$.

a) From the fact that activity does not decay or explode exponentially, what can you conclude about the correlation between births and deaths in both time and space?

b) What do these power law distributions suggest about the criticality of the stationary state?

c) Can you think of anything fishy about this experiment that might tend to cast doubt upon self-organizational properties of the local configurations in the Life universe?

24. A crucial role in showing that the Life rule admits the possibility of universal computation and, hence, self-reproducing patterns is the Glider Gun, which was shown in Fig. 3.17.

A One-Dimensional Glider Gun

a) Construct a Glider Gun for a 1-dimensional automaton. (Just so you'll know what you're looking for, the figure above shows the behavior of such a 1-D Glider Gun.)

b) Can you think of how a 1-D automata might be made capable of acting as a universal computer?

25. In the text we have considered only the classic Life world, which is two-dimensional. In 3-D Life, each cell is a cube instead of a square.

a) What is the analogue of the Moore neighborhood in 3-D Life? How many neighbors does each cell have in the three-dimensional setting?

b) Carter Bays has experimented with a version of 3-D Life called Life 4555. In this notational scheme, the first two numbers dictate the fate of living cells. The first number represents the fewest living neighbors a cell must have to keep from being undernourished and dying; the second indicates the most neighbors the cell can have before dying of overcrowding. The third and fourth numbers govern the fate of dead cells. The third number is the fewest living neighbors a dead cell must have in order to come alive, while the fourth is the most it can have before changing from dead to alive. Thus, in Bays's notation, classic 2-D Life becomes Life 2333.

What is the analogue of a glider in Life 4555? Do there exist stable and cyclic forms in Life 4555 that are analogous to Blocks and Blinkers in ordinary 2-D Life?

c) Now consider Life 5766. Imagine you are looking down on some starting configuration in the planar universe of ordinary 2-D Life. Now place a living cube directly on top and below each living square in the 2-D configuration. Show that the cubes will mimic perfectly the behavior of the sandwiched 2-D cells if and only if the following conditions are satisfied:

i. No living square on the plane ever has five living neighbors.

ii. No dead square cell on the plane ever has six living neighbors.

d) Try to construct a glider gun in either Life 4555 or Life 5766.

26. The Rats-and-Rodents example of the text showed how the use of color (gray-scales, actually) could be used to display the behavior of automata having $k > 2$. Can you think of other ways? (Hint: Consider other types of human sensory inputs besides light.)

Problems

1. Consider a one-dimensional cellular automata with $k = 2$ and a single nonzero initial cell, using the mod 2 rule (Rule 90).

a) Show that the time-history of this automaton leads to patterns in which the value of a site at a given time step is just the value modulo 2 of the corresponding coefficient in Pascal's triangle, with the initial nonzero cell forming the apex of the triangle. That is, the values are the coefficients in the expansion $(1 + x)^n \mod 2$.

b) For Rule 150, in which the value of each cell is the sum of its own value and the values of its nearest neighbors mod 2, show that the sequence of binary digits obtained from a single initial nonzero cell for n time steps is the sequence of coefficients of the polynomial $(1 + x + x^2)^n \mod 2$.

2. Feedback shift-registers consist of a sequence of "sites" carrying the value $\alpha(i)$ at site i. At each time step, the site values evolve by a shift $\alpha(i) = \alpha(i-1)$ and feedback $\alpha(0) = \mathbf{F}[\alpha(j_1), \alpha(j_2), \ldots]$, where the j_i give the positions of "taps" on the shift-register. Show that a one-dimensional cellular automaton of N cells corresponds to a feedback shift-register of N sites with site values 0 and 1 and taps at positions $N-2$, $N-1$, and N, the automaton rule being given by the Boolean function $\mathbf{F}$. Show that for one time step of the automaton, it requires N time steps in the shift-register.

3. The average fraction of cells with value 1 emerging over one time step from a disordered initial configuration has been stated in the text as

$$\rho_1 = \frac{\#_1(\mathbf{R})}{(\#_0(\mathbf{R}) + \#_1(\mathbf{R}))} = \frac{\#_1(\mathbf{R})}{8},$$

where $\#_d(\mathbf{R})$ represents the number of occurrences of the digit d in the binary representation of $\mathbf{R}$. This formula holds only for the case when each cell can take on the initial value 0 or 1 with equal probablility $p = \frac{1}{2}$. Show that the generalization of this formula to the case when $p \neq \frac{1}{2}$ is given by weighting each of the eight possible three-cell neighborhoods σ by the weight $p(\sigma) = p^{\#_1(\sigma)}(1-p)^{\#_0(\sigma)}$, and then adding the probabilities for those neighborhoods σ that yield 1 upon application of the automaton's rule.

4. Self-similar figures in the plane may be characterized in the following fashion: Find the minimum number of squares with side a necessary to cover all parts of the figure (all cells with nonzero values in the cellular automaton case). Call this number $N(a)$. The figure is *self-similar,* or scale invariant, if rescaling a changes $N(a)$ by a constant factor independent of the absolute size of a. If this is the case, $N(a) \sim a^{-D}$, where D is defined to be the Hausdorff-Besicovitch or *fractal* dimension of the figure.

a) Show that a figure filling the plane has $D = 2$, but a line has $D = 1$.

b) In Fig. 3.4 we showed the triangular patterns that emerge from various initial configurations. Show that the pattern for Rule 90 is self-similar with $D = \log_2 3 \cong 1.59$.

c) For Rule 150, the density $T(n)$ of triangles of base length n satisfies the recurrence relation

$$T(n = 2^k) = 2T(2^{k+1}) + 4T(2^{k+2}), \qquad T(1) = 0, \qquad T(2) = 2.$$

Show that for large k this yields

$$T(n) \sim n^{-\log_2(2\varphi)} = n^{-\log_2(1+\sqrt{5})} \sim n^{1.69},$$

and, hence, conclude that the limiting fractal dimension of the pattern generated by Rule 150 is $D = \log_2(2\varphi) \cong 1.69$.

d) Show that when the number of cell values k is prime, the fractal dimension of the emerging triangle pattern of Rule 90 is given by

$$D_k = \log_k \sum_{i=1}^{k} i = 1 + \log_k \left(\frac{k+1}{2}\right).$$

5. Show that irreversibility of one-dimensional cellular automata rules stems from the condition that the next state depends only upon the state at the previous time step, while reversible rules can be obtained by allowing the next configuration to depend upon the *two* previous configurations.

6. A *random mapping* between k elements is defined by the rule that each element is mapped to any one of the k elements with equal probability $p = 1/k$. Thus, all k^k possible mappings are generated with equal probability.

a) Show that under such a random mapping, the probability of a particular sequence of k elements having no predecessor is $(k-1)^k/k^k = (1-1/k)^k$. Thus, this probability approaches $1/e \simeq 0.37$ as $k \to \infty$. That is, a fraction $1/e$ of the theoretically possible states are not reached by iteration of a random mapping.

b) Show that the probability of a cycle of length r appearing in the iteration of such a random mapping is

$$\sum_{i=r}^{k} \frac{(k-1)!}{(k-i)!\, k^i}.$$

c) For complex nonadditive cellular automata rules, almost all configurations become unreachable as the number of cells $N \to \infty$. With this fact in mind, consider the degree to which cellular automata behave like random mappings for large N.

7. If the general rule of transition for a one-dimensional automaton is given by

$$a_i^{t+1} = \mathbf{F}[a_{i-r}^t, a_{i-r+1}^t, \ldots, a_{i+r-1}^t, a_{i+r}^t],$$

show that either one of the following conditions is necessary (but not sufficient) for unbounded growth of the initial configuration:

$$\text{I. } \mathbf{F}[a_{i-r}, a_{i-r+1}, \ldots, 0, 0, \ldots, 0] \neq 0,$$
$$\text{II. } \mathbf{F}[0, \ldots, 0, a_{i+1}, \ldots, a_{i+r}] \neq 0.$$

8. Prove the following "Malthusian" limit regarding the growth of self-reproducing planar configurations: If a configuration is capable of reproducing $f(T)$ offspring by time T, then there exists a constant $k > 0$ such that $f(T) \leq kT^2$. How would you extend this result to n-dimensional configurations?

9. Prove the following assertion: For any two-dimensional automaton with an irreversible rule, there must exist "Garden-of-Eden" configurations, i.e., configurations that have no predecessor.

10. Consider the following one-dimensional version of Life using a neighborhood size of n cells. The state-transition rule is as follows: A cell is ALIVE at time t+1 if it is DEAD at time t and the number of ALIVE neighbors is greater than or equal to $n+1$ and less than or equal to $n+k_0$ (birth), or the cell is ALIVE at time t and the number of neighbors that are ALIVE is greater than or equal to $n+1$ and less than or equal to $n+k_1$ (survival). Here k_0 and k_1 are positive integers. Let's agree to identify any

finite configuration by its *support* (i.e., the cells that are ALIVE) and denote the configuration by

$$c = \bigcup_{1 \le i \le p} [a_i, b_i],$$

where $a_i \le b_i$ and $b_i + 1 < a_{i+1}$. The intervals $\{[a_i, b_i]\}$ are called the *components* of the configuration c. Assume that the above Life game evolves on a finite interval $[a, b]$, i.e., cells x for which $x < a$ or $x > b$ are always DEAD.

a) Show that if $k_0 \ge k_1$, then any finite configuration evolves in a finite number of time steps toward a stable configuration c such that $n \le b_i - a_i \le n + k_1 - 1$ and $n + 2 \le a_{i+1} - b_i$ for all i.

b) Prove that if $n - k_1 + 1 < k_0 < 2k_1 - n$, then one-dimensional Life admits cycles of length 2.

(Remark: These results show that the Life rules exhibiting more complex behavior are those for which birth conditions are more restrictive than conditions for survival.)

11. Consider a three-cell, one-dimensional automaton with the property that each cell can assume a *continuum* of values in the range 0 to 1. Let the rule of the automaton be given by

$$\begin{aligned} x_1(t+1) &= x_2^2 + x_3^2 + 2x_1x_2, \\ x_2(t+1) &= 2x_1x_3 + 2x_2x_3, \\ x_3(t+1) &= x_1^2, \end{aligned}$$

where the values of the three cells x_i, $i = 1, 2, 3$ satisfy the local and global constraints

$$0 \le x_i \le 1, \qquad x_1 + x_2 + x_3 = 1. \qquad (*)$$

a) Show that the state-transition rule maps the region $(*)$ to itself.

b) Refute or verify the conjecture that for *almost every* initial configuration p, the limit set of this automaton is given by a point $p \in R^3$ satisfying the periodicity condition $p_2 = T(p_1)$, $p_3 = T(p_2)$, $p_1 = T(p_3)$, where T is the rule of transition given above. Show by counterexample that this property cannot hold for *all* initial configurations.

c) If the value x_i represents the relative population of species i, interpret in words the state-transition rule, the region $(*)$ and the limiting point p.

12. Consider the generalization of Problem 11 in which the transition rule is

$$x_i(t+1) = \sum_{k,m=1}^{N} \gamma_i^{km} x_k x_m, \qquad 1 \le i \le N,$$

with the coefficients satisfying

$$\gamma_i^{km} = \begin{cases} \gamma_i^{mk} > 0 & \text{if } \min(k,m) \leq i \leq \max(k,m), \\ \gamma_i^{mk} = 0 & \text{otherwise,} \end{cases}$$

$$\sum_{i=m}^{k} i\gamma_i^{km} = \frac{m+k}{2}, \qquad \sum_{i=m}^{k} \gamma_i^{km} = 1.$$

We normalize the set $\{x_i\}$ by requiring

$$0 \leq x_i \leq 1, \qquad \sum_{i=1}^{N} x_i = 1.$$

a) Show that the quantity

$$\sigma \equiv \sum_{i=1}^{N-1} (N-i)\, x_i,$$

is invariant under the above rule.

b) Given an initial vector $\{x_i^0\}$, show that there is exactly one value of the index j such that

$$N - j \geq \sigma \geq N - j - 1,$$

and that in terms of j, every initial vector converges to the vector whose coordinates are

$$x_j = \sigma - (N - j - 1), \qquad x_{j+1} = N - j - \sigma, \qquad x_i = 0, \text{ for all other } i.$$

(Note: This fixed point is *independent* of the coefficients γ_i^{km}.)

13. Show that with the given alphabet A, each of the following languages forms a regular language by constructing a finite automaton that accepts it:

a) $A = \{a, b\}$, $L =$ words whose final four symbols form the string *abab*,

b) $A = \{a, b, c\}$, $L =$ all *palindromes* of length 6 or less. (Recall that a palindrome is a word that reads the same backward or forward.)

c) $A = \{0, 1\}$, $L =$ all binary number strings that are integral multiples of 5.

14. What are the complexities of the regular languages of the previous Problem?

15. In Life we discussed the configuration termed the Glider Gun, which periodically gives off Gliders and, hence, serves as a generator of Life configurations of unbounded growth. There are other configurations having this property called Puffer Trains. These are moving patterns producing debris as they sweep across the plane, with the debris tail growing indefinitely. Construct a specific example of a Puffer Train. (*Warning:* You will need a computer for this Problem!)

16. Consider a 2-dimensional automaton with each cell being capable of taking on one of the eight values $0, 1, \ldots, 7$. Assume that the initial configuration is the loop shown below:

```
  2 2 2 2 2 2 2 2
2 1 7 0 1 4 0 1 4 2
2 0 2 2 2 2 2 2 0 2
2 7 2         2 1 2
2 1 2         2 1 2
2 0 2         2 1 2
2 7 2         2 1 2
2 1 2 2 2 2 2 2 1 2 2 2 2 2
2 0 7 1 0 7 1 0 7 1 1 1 1 1 2
  2 2 2 2 2 2 2 2 2 2 2 2 2
```

Initial State for a 2-D Automaton

Assume that the transition rule for this automaton is given by the table on the next page.

a) Using a computer, show that at time $t = 151$ the resulting configuration is

```
                2
              2 1 2
              2 7 2
              2 0 2
              2 1 2
  2 2 2 2 2 2 2 7 2       2 2 2 2 2 2 2 2
2 1 1 1 7 0 1 7 0 2   2 1 7 0 1 4 0 1 4 2
2 1 2 2 2 2 2 2 1 2   2 0 2 2 2 2 2 2 0 2
2 1 2         2 7 2   2 7 2         2 1 2
2 0 2         2 0 2   2 1 2         2 1 2
2 4 2         2 1 2   2 0 2         2 1 2
2 1 2         2 7 2   2 7 2         2 1 2
2 0 2 2 2 2 2 2 0 2   2 1 2 2 2 2 2 2 1 2 2 2 2 2
2 4 1 0 7 1 0 7 1 2   2 0 7 1 0 7 1 0 7 1 1 1 1 1 2
  2 2 2 2 2 2 2 2         2 2 2 2 2 2 2 2 2 2 2 2 2
```

Generation 151 of 2-D Cellular Automaton

Comparing this configuration with that at time 0, we find that the original loop has reproduced itself.

b) Can you identify the "genetic sequence" this automaton uses to reproduce itself? (Hint: Consider the sequence $410 \to 410 \to 710 \to \ldots$ found by tracing the inner part of the loop in a counterclockwise direction, starting at the top.)

CTRBL->I	CTRBL->I	CTRBL->I	CTRBL->I	CTRBL->I
00000->0	02527->1	11322->1	20242->2	30102->1
00001->2	10001->1	12224->4	20245->2	30122->0
00002->0	10006->1	12227->7	20252->0	30251->1
00003->0	10007->7	12243->4	20255->2	40112->0
00005->0	10011->1	12254->7	20262->2	40122->0
00006->3	10012->1	12324->4	20272->2	40125->0
00007->1	10021->1	12327->7	20312->2	40212->0
00011->2	10024->4	12425->5	20321->6	40222->1
00012->2	10027->7	12426->7	20322->6	40232->6
00013->2	10051->1	12527->5	20342->2	40252->0
00021->2	10101->1	20001->2	20422->2	40322->1
00022->0	10111->1	20002->2	20512->2	50002->2
00023->0	10124->4	20004->2	20521->2	50021->5
00026->2	10127->7	20007->1	20522->2	50022->5
00027->2	10202->6	20012->2	20552->1	50023->2
00032->0	10212->1	20015->2	20572->5	50027->2
00052->5	10221->1	20021->2	20622->2	50052->0
00062->2	10224->4	20022->2	20672->2	50202->2
00072->2	10226->3	20023->2	20712->2	50212->2
00102->2	10227->7	20024->2	20722->2	50215->2
00112->0	10232->7	20025->0	20742->2	50222->0
00202->0	10242->4	20026->2	20772->2	50224->4
00203->0	10262->6	20027->2	21122->2	50272->2
00205->0	10264->4	20032->6	21126->1	51212->2
00212->5	10267->7	20042->3	21222->2	51222->0
00222->0	10271->0	20051->7	21224->2	51242->2
00232->2	10272->7	20052->2	21226->2	51272->2
00522->2	10542->7	20057->5	21227->2	60001->1
01232->1	11112->1	20072->2	21422->2	60002->1
01242->1	11122->1	20102->2	21522->2	60212->0
01252->5	11124->4	20112->2	21622->2	61212->5
01262->1	11125->1	20122->2	21722->2	61213->1
01272->1	11126->1	20142->2	22227->2	61222->5
01275->1	11127->7	20172->2	22244->2	70007->7
01422->1	11152->2	20202->2	22246->2	70112->0
01432->1	11212->1	20203->2	22276->2	70122->0
01442->1	11222->1	20205->2	22277->2	70125->0
01472->1	11224->4	20207->3	30001->3	70212->0
01625->1	11225->1	20212->2	30002->2	70222->1
01722->1	11227->7	20215->2	30004->1	70225->1
01725->5	11232->1	20221->2	30007->6	70232->1
01752->1	11242->4	20222->2	30012->3	70252->5
01762->1	11262->1	20227->2	30042->1	70272->0
01772->1	11272->7	20232->1	30062->2	

Neighborhoods are read as follows (rotations are not listed):

```
  T
L C R      ==>      I
  B
```

Transition Rule for 2-D Automaton

17. Consider 1-dimensional cellular automata of n cells, each of which can take on the value 0 or 1. Assume periodic boundary conditions, i.e., the value at cell i equals that at cell j if $i = j \mod n$. Define the *adjacency matrix* A for this type of automaton as follows: First, index the rows and columns of A by 2-tuples of elements from $\mathbb{Z}/2$ ordered lexicographically. So, for instance, the row indices then become $i = 00, 01, 10, 11$, and similarly for the columns. Then we set the entry $a_{ij} = 0$ if the second component of index i does not match the first component of index j. So, for example, $a_{00,10} = 0$. Otherwise, the indices i and j are combined into a 3-tuple (e.g., $i = 01$ and $j = 11$ combine to produce 011), and the entry a_{ij} is set to 1 if the resultant 3-tuple is such that its middle entry remains invariant under the rule of the automaton.

a) Consider Rule 12, which is defined by the transitions

$$\{000, 001, 100, 101, 110, 111\} \to 0, \qquad \{010, 011\} \to 1.$$

Show that for this rule the adjacency matrix A is

$$A = \begin{pmatrix} 1 & 1 & 0 & 0 \\ 0 & 0 & 1 & 1 \\ 1 & 1 & 0 & 0 \\ 0 & 0 & 0 & 0 \end{pmatrix}.$$

b) Show that the number of fixed points T_n of Rule 12 on a cylinder of size n satisfies the following relations:

$$T_n = \operatorname{trace} A^n = T_{n-1} + T_{n-2}, \qquad T_1 = 1, \qquad T_2 = 3.$$

c) From this, prove that the total number of cycles under Rule 12 increases as a function of the cylinder size n. Can you find any rules for which this is not the case?

18. Let a measure μ be defined on the space of configurations of an n-dimensional automaton by specifying that the symbols that can occur at each cell are independent random variables, the probability of a given symbol occurring being $1/k$. Prove that a cellular automaton preserves the measure μ in the sense that $\mu(S) = \mu[f^{-1}(S)]$ for any measurable configuration S, if and only if the state-transition rule f of the automaton is onto, i.e., if and only if for every configuration $\hat{c}$, there exists a configuration c such that $f(c) = \hat{c}$.

Notes and References

§1. The Fibonacci sequence is one of the most ubiquitous and important in all of mathematics, arising in a bewildering variety of settings ranging from the spiral pattern of pinecones and sunflower seeds to branching patterns on trees and on to the helical pattern of the DNA molecule. Much of this universality ultimately derives from the relationship between the sequence of Fibonacci numbers and the "golden ratio" ϕ of the ancient Greek geometers given by

$$\phi = \lim_{n \to \infty} \frac{u_n}{u_{n-1}},$$

where u_i = ith Fibonacci number. For more information on this sequence and its connections to the morphology of living things, see

Huntley, H., *The Divine Proportion,* Dover, New York, 1970,

Cook, T., *The Curves of Life,* Dover, New York, 1979,

Stevens, P., *Patterns in Nature,* Penguin, London, 1976.

§2. A good introduction to cellular automata from a computational point of view is the volume

Toffoli, T. and N. Margolus, *Cellular Automata Machines,* MIT Press, Cambridge, MA, 1987.

The example of urban housing patterns in a racially-mixed neighborhood is adapted from

Schelling, T., "Dynamic Models of Segregation," *J. Math. Socio.,* 1 (1971), 143–186.

An extended, "spreadsheet version" of this segregation problem, termed the Game of Tolerance, is presented in the paper

Cartwright, T., "Simulating Tolerance: Integration and Disintegration in a Chaotic World," presented at the Int'l. Conf. on Oper. Res., Vienna, Austria, August 1990.

Conway's Life game was brought to the attention of the general public in a series of articles by Martin Gardner in *Scientific American.* The complete set of Life articles, as well as the treatment of a number of related topics, can be found in

Gardner, M., *Wheels, Life and Other Mathematical Amusements,* Freeman, San Francisco, 1983.

An extensive popular account of the Life game, together with computer programs for playing it, is given in

Poundstone, W., *The Recursive Universe,* Morrow, New York, 1985.

For additional information, see also Section 4 of

Dewdney, A. K., *The Armchair Universe,* Freeman, New York, 1988.

A more mathematically detailed account of the cyclic life example, as well as many further references, is available in

Griffeath, D., "Cyclic Random Competition: A Case History in Experimental Mathematics," *Notices Amer. Math. Soc.,* 35 (December 1988), 1472–1480.

§3–5. The results of this and the next few sections are due primarily to Stephen Wolfram, who almost single-handedly revived the interest of the of the mathematical and computational communities in cellular automata. The material of these sections has been adapted from his review paper

Wolfram, S., "Statistical Mechanics of Cellular Automata," *Rev. Mod. Physics,* 55 (1983), 601–644.

This paper, along with many others, is reprinted in the following work which is essential reading for all cellular automata *aficionados:*

Wolfram, S., ed., *Theory and Applications of Cellular Automata,* World Scientific, Singapore, 1986.

The example on rats and rodents populations is adapted from the article

Couclelis, H., "Of Mice and Men: What Rodent Populations Can Teach Us About Complex Spatial Dynamics," *Env. and Plan. A,* 20 (1988), 99-109.

More recent work has substantiated the results of the above paper, showing that chaotic behavior in space is just as common as it is in time. For a cellular-automaton view of this question, see

Hassell, M., H. Comins and R. May, "Spatial Structure and Chaos in Insect Population Dynamics," *Nature,* 353 (19 September 1991), 255–258,

Ives, A., "Chaos in Space and Time," *Nature,* 353 (19 September 1991), 214–215.

§6. For these results, as well as many more, see

Martin, O., A. Odlyzko and S. Wolfram, "Algebraic Properties of Cellular Automata," *Comm. Math. Phys.,* 93 (1984), 219–258.

§7. Good introductory treatments of the relationship between formal languages and automata are

Davis, M., and E. Weyuker, *Computability, Complexity and Languages,* Academic Press, New York, 1983,

Révész, G., *Introduction to Formal Languages,* McGraw-Hill, New York, 1983,

Moll, R., M. Arbib, and A. Kfoury, *An Introduction to Formal Language Theory,* Springer, New York, 1988.

A somewhat more mathematically sophisticated account is provided in

Eilenberg, S. *Automata, Languages and Machines,* Vols. A and B, Academic Press, New York, 1974 and 1976.

The treatment of cellular automata, languages and complexity follows

Wolfram, S., "Computation Theory of Cellular Automata," *Comm. Math. Phys.,* 96 (1984), 15–57.

The pioneering work responsible for taking the study of natural languages from the realm of descriptive analysis and taxonomy to that of a formal scientific discipline is due to Noam Chomsky. Introductory accounts of his life and work are found in

Leiber, J. *Noam Chomsky: A Philosophic Overview,* St. Martin's Press, New York, 1975,

Lyons, J., *Noam Chomsky,* rev. ed., Penguin, London, 1977,

Salkie, R., *The Chomsky Update,* Unwin and Hyman, London, 1990.

For the ideas of the master himself, see

Chomsky, N., *Language and Mind,* Harcourt, Brace, Jovanovich, New York, 1972,

Chomsky, N., *Reflections on Language,* Pantheon, New York, 1975.

Chomsky's ideas have by no means met with universal love and admiration. An assessment of some of the shortcomings of his views, as well as some of the strengths, is given in the collection

Harman, G., ed., *On Noam Chomsky: Critical Essays,* University of Massachusetts Press, Amherst, MA, 1982.

For a layman's account of Chomsky's views on the specific question of language acquisition, together with the views of his opponents, see Chapter Four of

Casti, J., *Paradigms Lost: Images of Man in the Mirror of Science,* Morrow, New York, 1989 (paperback edition: Avon, New York, 1990).

§8. The classic popular account of the discovery of the double helix structure of the DNA molecule is

Watson, J., *The Double Helix,* Athenaeum, New York, 1968.

The complete text of Watson's book, as well as commentaries on the relevance of the discovery for the sociology of science and reprints of the original papers on the double helix structure, may be found in

G. S. Stent, ed., *The Double Helix: Text, Commentary, Reviews, Original Papers,* Norton, New York, 1980.

Good references for understanding the operation of the DNA in both protein production and replication are

Hofstadter, D., "The Genetic Code: Arbitrary?" *Scientific American,* March 1982 (reprinted in Hofstadter, D., *Metamagical Themas,* Basic Books, New York, 1985),

Rose, S., *The Chemistry of Life,* 2d ed., Penguin, London, 1979,

Rosenfield, I., E. Ziff and B. van Loon, *DNA for Beginners,* Norton, New York, 1982.

For an introductory account of the origin of life and the genetic code, see Chapter Two of the Casti book cited above, as well as the volumes

Shapiro, R., *Origins: A Skeptic's Guide to the Creation of Life on Earth,* Summit Books, New York, 1986,

Scott, A., *The Creation of Life: Past, Future, Alien,* Blackwell, Oxford, 1986.

Our discussion of DNA modeling via cellular automata follows that given in

Burks, C., and D. Farmer, "Towards Modeling DNA Sequences as Automata," *Physica D,* 10D (1984), 157–167.

§9. Lindenmayer's original work on modeling plant growth using cellular automata is

Lindenmayer, A., "Mathematical Models for Cellular Interactions in Development, Parts I and II," *J. Theor. Biol.,* 30 (1967), 455–484.

Detailed, up-to-date accounts, complete with stunning computer graphics, are

Prusinkiewicz, P. and J. Hanan, *Lindenmayer Systems, Fractals, and Plants,* Springer, New York, 1989,

Prusinkiewicz, P. and A. Lindenmayer, *The Algorithmic Beauty of Plants,* Springer, New York, 1990.

§10. In addition to the popular accounts of Life cited under §2 above, a more technical account is given in Chapter 25 of

Berlekamp, E., J. H. Conway and R. Guy, *Winning Ways for Your Mathematical Plays,* Vol.2, Academic Press, London, 1982.

§11. For a much more detailed discussion of skin pigmentation modeling problem using cellular automata, as well as further computational results, see

Young, D., "A Local Activator-Inhibitor Model of Vertebrate Skin Patterns," *Math. Biosciences,* 72 (1984), 51–58.

Other work along the same lines is reported in

Bard, J., "A Model for Generating Aspects of Zebra and Other Mammalian Coat Patterns," *J. Theor. Biol.*, 93 (1981), 363–385,

Meinhardt, H., *Models of Biological Pattern Formation,* Academic Press, London, 1982,

Murray, J., "A Pre-Pattern Formation Mechanism for Animal Coat Markings," *J. Theor. Biol.*, 88 (1981), 161–199,

Swindale, N. V., "A Model for the Formation of Ocular Dominance Stripes," *Proc. Roy. Soc. London, Ser. B,* 208 (1980), 243–264.

§12. Von Neumann's proof of the possibility of self-reproducing automata is given in

von Neumann, J., *Theory of Self-Reproducing Automata,* University of Illinois Press, Urbana, IL, 1966.

Von Neumann's original work is rather difficult to follow. A simpler account of his ideas given from several perspectives is available in the collection

Essays on Cellular Automata, A. Burks, ed., University of Illinois Press, Urbana, IL, 1970.

The idea of a living organism as a machine has proven irresistably attractive to scientists and philosophers since the time of Aristotle. For some more recent perspectives on this eternal question, see

Laing, R., "Machines as Organisms: An Exploration of the Relevance of Recent Results," *Biosystems,* 11 (1979), 201–215,

Laing, R., "Anomalies of Self-Description," *Synthese,* 38 (1978), 373–387.

§13. The artificial lifer's "bible" is the volume containing the proceedings of the historical 1987 Los Alamos conference, which brought the various strands of the AL community together for the first time. This work contains reports on theoretical aspects of self-reproduction and what it means to be "alive," as well as many accounts of artificial life forms that are currently cavorting about inside the memory banks of computers across America and around the world. For anyone who wants to know about AL, this is the place to start. The precise citation is

Artificial Life, C. Langton, ed., Addison-Wesley, Redwood City, CA, 1989.

On the principle that one good conference deserves another, a second workshop on AL was organized by the Santa Fe Institute in February 1990. The published output is

Artificial Life–II, C. Langton, et al, eds., Addison-Wesley, Redwood City, CA, 1992.

For an introductory account of Biomorph Land, see Richard Dawkins's book on evolution:

Dawkins, R., *The Blind Watchmaker,* Longman, London, 1986.

A more detailed summary of the rules of Biomorph generation and selection is available in the 1989 Langton volume noted above.

§14. For a fuller account of the various intricacies and subtleties of cellular automata, see the Wolfram volume cited under §3 above as well as

Farmer, D., T. Toffoli and S. Wolfram, eds., *Cellular Automata,* North-Holland, Amsterdam, 1984,

Demongeot, J., E. Golès and M. Tchuente, eds., *Dynamical Systems and Cellular Automata,* Academic Press, London, 1985.

DQ #3. This Hypercycle game, as well as many other simple cellular automata illustrating a wide variety of prototypical situations in biology, language and life, can be found in

Eigen, M., and R. Winkler, *The Laws of the Game,* Knopf, New York, 1981.

DQ #4. The idea of describing the randomness of a sequence of numbers by the length of the shortest program needed to reproduce the sequence seems to have been hit upon independently by the great Russian mathematician Andrei Kolmogorov and by Gregory Chaitin (while he was still a high-school student!). The original papers are

Kolmogorov, A. N., "Three Approaches to the Quantitative Definition of Information," *Prob. Info. Transmission,* 1 (1965), 1–7,

Chaitin, G., "Information-Theoretic Limitations of Formal Systems," *J. Assn. Comp. Mach.,* 21 (1974), 403–424,

Chaitin, G., "A Theory of Program Size Formally Identical to Information Theory," *J. Assn. Comp. Mach.,* 22 (1975), 329–340.

For a complete summary of Chaitin's work on the question of complexity, as well as its relations to the development of life, see the Notes and References for Chapter 9, which includes the collection

Chaitin, G., *Information, Randomness and Incompleteness,* 2nd edition, World Scientific, Singapore, 1990.

DQ #15. The Coral Reef automaton is discussed in a paper by Ulam given in the Burks book cited under §12 above, as well as in the Poundstone volume cited under §2.

DQ #18. The idea of a reversible rule leads immediately to the possibility of an information-lossless computer, i.e., one that would dissipate no heat during the course of its computations. For a discussion of this possibility, see the papers

Bennett, C., and R. Landauer, "The Fundamental Physical Limits of Computation," *Scientific American,* July 1985,

Bennett, C., "The Thermodynamics of Computation," *Int. J. Theor. Physics,* 21 (1982), 905–940,

Landauer, R., "Irreversibility and Heat Generation in the Computing Process," *IBM J. Res. Dev.,* 5 (1961), 183–191.

DQ #23. For an account of the experiments discussed in this question, see

Bak, P., K. Chen and M. Creutz, "Self-Organized Criticality in the 'Game of Life,' " *Nature,* 342 (14 December 1989), 780-782.

Others have argued that these results are merely artifacts of the relatively small grid size chosen for the experiment, and that there is in fact no such criticality in the Life rule. For these dissenting views, see

Bennett, C. and M. Bourzutschky, " 'Life' Not Critical?", *Nature,* 350 (11 April 1991), 468.

DQ #24-25. For more information on these matters, see the Dewdney book cited under §2 above.

PR #4. An extensive treatment of the theory of fractals is given in the well-known book

Mandelbrot, B., *The Fractal Geometry of Nature,* Freeman, New York, 1983.

Other accounts of a somewhat more technical nature are found in

Falconer, K., *Geometry of Fractal Sets,* Cambridge University Press, Cambridge, 1985,

Fischer, P., and W. Smith, eds., *Chaos, Fractals, and Dynamics,* Marcel Dekker, New York, 1985.

PR #11. For a more extensive discussion of this problem, as well as many other problems of a similar nature, see

Stein, P. R., and S. M. Ulam, "Nonlinear Transformation Studies on Electronic Computers," *Rozprawy Matematyczne,* 39 (1964), 1–66 (This paper is reprinted in *Stanislaw Ulam: Sets, Numbers, and Universes,* W. A. Beyer, J. Mycielski and G. C. Rota, eds., MIT Press, Cambridge, MA, 1974.)

PR #16. Details of the self-reproducing loop are found in the paper

Langton, C., "Self-Reproduction in Cellular Automata," in *Cellular Automata,* D. Farmer, et al, eds., North-Holland, Amsterdam, 1984, pp. 135-144.

PR #17. Many more results about the existence of cycles in 1-D automata with periodic boundary conditions, as well as their connections to linear shifts, are given in

Jen, E., "Limit Cycles in One-Dimensional Cellular Automata," in *Lectures in the Science of Complexity,* D. Stein, ed., Addison-Wesley, Redwood City, CA, 1989, pp. 743-758.

CHAPTER FOUR

Order in Chaos: Variety and Pattern in the Flow of Fluids, Populations and Money

1. The Circle-10 System

Consider a dynamical process involving a particle moving on a cirle of circumference 1. The state manifold M of this system is simply the circle S^1, so we can describe the position of the particle at any moment by a number between 0 and 1. Suppose that the vector field determining the particle's movement on the circle is given by the rule: "If you're currently at the point x, move to the point $10x$ at the next moment." With this rule, the circle is stretched to ten times its length, and then wrapped ten times around itself. To keep things as simple as possible, let's assume that time changes in discrete units. For future reference, we'll call this process the *Circle-10 system.* As we now show, this simple-looking dynamical system has some very nontrivial properties.

First of all, we divide the circumference of the circle into ten sectors, labeled $0, 1, \ldots, 9$. In terms of numbers between 0 and 1, sector 0 corresponds to all numbers between 0 and $0.099999\ldots$, while sector 1 runs from 0.1 to $0.19999\ldots$, and so on to sector 9, which runs from 0.9 to $0.9999\ldots$. To get things started, suppose the particle begins its itinerary around the circle at the point $x_0 = 0.379762341$. This is a point living in sector 3, nearly 80 percent of the way to sector 4. Let's now follow the path of a given point as this Circle-10 rule is iterated, i.e., applied over and over again.

When we apply the rule of the vector field to x_0 and wrap the circle ten times around itself, the circle's length expands by a factor of 10. Thus, the point x_0 moves to the point 3.79762341. But note that one time around the circle just gets you back to where you began, and so do two tours, and so do three. So the result of applying the vector field is just the same as coming to the point $x_1 = 0.79762341$. This is a point in sector 7. Therefore, on the first iteration the starting point x_0 moves from sector 3 to sector 7. As we continue to iterate the vector field in this manner, we generate the following results:

Time	Number From Rule	Point on Circle	Sector
0	0.379762341	$x_0 = 0.379762341$	3
1	3.79762341	$x_1 = 0.79762341$	7
2	7.9762341	$x_2 = 0.9762341$	9
3	9.762341	$x_3 = 0.762341$	7
4	7.62341	$x_4 = 0.62341$	6
5	6.2341	$x_5 = 0.2341$	2
6	2.341	$x_6 = 0.341$	3
7	3.41	$x_7 = 0.41$	4
8	4.1	$x_8 = 0.1$	1
9	1	$x_9 = 0$	0

This listing of the trajectory of x_0 under the vector field of the Circle-10 system shows that the action of this rule is the ultimate in simplicity: Just multiply by 10 at each stage and delete the digit to the left of the decimal point. In more compact mathematical terms, we can express the effect of this rule by "integrating" the dynamics, which yields the state at time n as being the quantity

$$x_n = 10^n x_0 \mod 1.$$

Examining the sequence of sectors that the system trajectory visits, we find the starting point's itinerary is sectors $3, 7, 9, 7, 6, 2, 3, 4, 1, 0, 0, \ldots$. If these numbers look familiar, they ought to—they're just the decimal digits of the starting point x_0! And this is no accident, either. For *any* initial point x_0, the itinerary of sectors visited will match exactly the decimal digits of the starting point for the simple reason that the rule, "Multiply by 10 and chop off the first digit," corresponds to nothing more than shifting the decimal point one position to the right. It's hard to imagine a more straightforward, easy-to-calculate, deterministic dynamical system than this.

But simple to describe and calculate doesn't necessarily mean simple in behavior. As tt turns out, the Circle-10 system captures just about all the interesting behavioral features of the sort of systems that have given rise to the theory of chaotic processes. Let's look at these characterizing "fingerprints" of chaos in a bit more detail.

- *Divergence:* Just for fun, consider Champernowne's number

$$C = 12345678910111213141516\ldots,$$

which is obtained by writing down the positive integers in order. Now suppose the decimal digits of the starting point in our Circle-10 system agree with Champernowne's number in the first 100 million places, but thereafter continue with ... 33333... forever. Call this initial state c. Thus the itinerary of the system starting from C agrees with the trajectory starting from c for the first 100 million steps. But thereafter the itinerary from c stays put in sector 3 forever, while that from C goes its merry way, whatever that might be, but which certainly is not complete confinement to sector 3.

This example shows that two starting points C and c, closer together than we could ever hope to measure, end up following completely independent paths. More formally, we say that the itineraries *diverge,* and that dynamical systems with this kind of divergent behavior are displaying *sensitivity to initial conditions.*

• *Randomness:* Let's wander down to the local casino and watch a few rounds of play at the roulette wheel. Suppose we record the numbers as they come up, obtaining the sequence 5, 23, 12, 30, 2, 18, 4, 17, 31, 24, 1, 00, 11, 32, 25, 17, 27, and 33. Intuition suggests that this sequence is about as random as a sequence can be. Just as we formed Champernowne's number by writing down the nonnegative integers in order, we can form a random number by writing down this sequence in the same way. This procedure yields the random number $r = 52312302184173124100113225172733$. There is a point x on the circle whose decimal expansion mimics this sequence, namely, the point in sector 5 given by $0.52312302184173\ldots$. So if we iterate the dynamical system using this value r as the starting point, we generate the sequence from the roulette wheel. But remember that this is a random sequence of numbers. Yet we have succeeded in generating it from what to all appearances is a totally deterministic dynamical system. What this experiment shows is that a deterministic mapping applied to the point r can lead to a sequence that's every bit as random as the outcome of the spin of a roulette wheel.

As we'll spell out in some detail in Chapter 9, almost *every* real number has a decimal expansion that's random. Therefore, our purely *deterministic* dynamical system behaves in a random manner not just for a few special initial states, but for almost every starting point.

• *Instability of Itineraries:* If almost all itineraries are random, it's of considerable interest to ask which ones are not. That is, which starting points are periodic, leading to itineraries that repeat themselves over and over again? The answer is clearly those points whose decimal expansion is either finite, like our test sequence, or repeats with a finite period (technically speaking, they're actually the same thing). And it is a well-known mathematical fact that a number will have such a repeating expansion if and only if it is rational, i.e., if it is a number like 1/2, 7/16, or 207/399, each of which is the ratio of two whole numbers.

In the interval between 0 and 1 there are an infinite number of rational numbers, as well as a much larger infinity of irrational numbers like $\sqrt{3}, \pi$

and Champernowne's number. It can be shown that between any two irrational numbers there is a rational one, although they do not alternate (almost all the numbers are irrational). Thus, the starting points leading to periodic orbits are totally mixed up with those *aperiodic points* that do not lead to such cyclic trajectories. This fact also shows that the periodic points are unstable, since if we move just a little bit to a nearby irrational starting point, we're led to an aperiodic trajectory. As it turns out, it's the case that *all* possible motions are unstable. That is, regardless of whether we start at a periodic or an aperiodic point, the itinerary is unstable (technically, in the sense of Lyapunov, as discussed in Chapter 2).

The set of itineraries followed by the starting points on our Circle-10 system is an example of what's called a strange attractor. Such an attractor is characterized by the properties just discussed: instability of all motions, deterministic randomness, and sensitivity to initial conditions. Furthermore, it turns out that most dynamical systems have a strange attractor for some region of the parameter values describing the system. In the more general setting of a continuous-time dynamical system, a typical strange attractor looks geometrically something like the object shown in Figure 4.1.

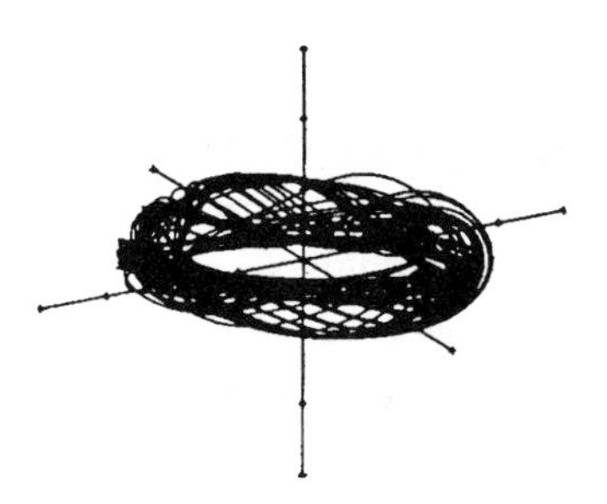

Figure 4.1. A Strange Attractor

If asked to describe this figure, the closest we can come is to say it looks a lot like a bowlful of spaghetti. Each strand, or orbit, is as close as we like to any other, yet separate from them. This picture makes it clear that as the dynamics unfold on such an attractor region, even a small nudge of the system can push the current point from one strand to another from which the system will take off on a totally different course. This kind of pathological sensitivity to perturbations is one of the warning signals by which chaotic dynamics are often discovered. Sensitivity of this sort already shows up in the Circle-10 system above, where a mistake in, say, the 30th digit of the initial point (one part in a million trillion trillion) will cause the system trajectory to be in the "wrong" sector after only 30 steps. Thus, unless the starting point is known *very* accurately, we can't even predict what sector of the circle the system will be in after just 30 steps. But in most real problems we're lucky if we can measure things to an accuracy of even one part in ten, let alone a million trillion trillion. Thus, the Circle-10 example shows us just how bad things can get, even in simple systems, if we want to use such a mathematical model to make long-range forecasts. Since we'll see plenty of real-life examples of strange attractors and deterministic randomness as we proceed, let's finish our introduction to such behavior by considering just what it is *exactly* about the dynamics of the Circle-10 system that gives rise to the pathological sensitivity in its behavior.

Basically, the sensitivity in the Circle-10 system stems from a combination of two competing factors in the system dynamics: stretching and folding. The transformation $x \to 10x$ stretches distances locally near the point x by a factor of 10. In this way nearby points are moved far apart. But the circle is a bounded region, so we can't stretch it everywhere. Thus, to fit it all in after distances have been expanded by a factor of ten, the circle's circumference has to be folded around itself many times. This means that some points that were previously far apart move closer together. Thus the stretching moves nearby points apart, while the folding moves distant points closer together. The stretching aspect shows how points that begin close together can lose sight of each other as the stretching continues, eventually failing to "keep in touch." Of course, the folding means that some points move closer together again. But it's impossible to know beforehand which points these will be. This is as good a definition as any I know of for what constitutes the essence of "chaos."

Keeping these general ideas in mind, let's now look a bit more deeply into the analytic underpinnings of these kinds of deterministic mechanisms that give rise to such wild, unpredictable behavior.

Exercises

1. In 1941 the late Argentine poet and writer Jorge Luis Borges published "The Library of Babel," a short story that some critics regard as his greatest work. This haunting tale gives an account of a library composed of an infinite array of hexagonally shaped levels, or floors, containing every book that has or ever could be written. In one passage, Borges writes that the discovery of the 25 orthographic symbols in which the books are written

> made it possible, three hundred years ago, to formulate a general theory of the Library and solve satisfactorily the problem which no conjecture had deciphered: the formless and chaotic nature of almost all the books. One which my father saw in a hexagon on circuit fifteen ninety-four was made up of the letters MCV, perversely repeated from the first

> line to the last. Another (very much consulted in this area) is a mere labyrinth of letters, but the next-to-last page says *Oh time thy pyramids.* This much is already known: for every sensible line of straightforward statement, there are leagues of senseless cacophonies, verbal jumbles and incoherences.

This passage is a precursor to the kind of problem that the discovery of chaotic dynamical processes poses for prediction schemes in science.

Chaotic dynamical systems are, in effect, mathematical models that "read" the initial conditions or starting points. The Circle-10 system illustrates this behavior. (a) If such systems are like the librarians in Borges's library, who read every word and character of every book in their care, what kind of readers do regular or nonchaotic systems correspond to? (b) In this book metaphor, what does sensitivity of chaotic systems to the starting point correspond to?

2. Another aspect of the inherent unpredictability of chaotic processes is that their time evolution is what is sometimes termed *computationally irreducible.* In other words, there is no faster way of finding out what such a process is going to do than just to turn it on and watch it unfold. In short, the system itself is its own fastest computer. If we place this idea within the context of Borges's library, describe what it means in connection with the problem of finding a particular book by consulting the card catalogue in the Library of Babel.

3. Consider the system $x_{t+1} = (x_t + a) \mod 1$, where a is an irrational number. (a) Show that the solution to this equation is $x_t = (at + x_0) \mod 1$. (b) If there is a small error Δx_0 in the initial condition for this system, does that error grow exponentially as t increases?

2. Deterministic Mechanisms and Random Behavior

Suppose we consider the growth over time of an insect population. Let's denote the insect population at time t by y_t, and assume the insects reproduce at a rate proportional to the size of their current population. Moreover, we further assume that the insects die by mutual competition for scarce resources. If the birth rate proportionality constant is r, while the carrying capacity of the environment is given by the constant K, then the population dynamics can be described by the quadratic difference equation

$$y_{t+1} = ry_t - \frac{y_t^2}{K}, \qquad y_0 = y^0.$$

If we assume that the environment can support only a population of finite size, then we can scale the population relative to its maximum level using a new variable x, which represents the *relative size* of the population with respect to its theoretical maximum. Thus, x is constrained to lie in the interval $0 \leq x \leq 1$. Incorporating the constant K into the scaling, we can now rewrite the dynamics as

$$x_{t+1} = \alpha x_t(1 - x_t) \doteq F(x_t, \alpha), \qquad x_0 = x^0, \qquad 0 \leq x_t \leq 1.$$

Now suppose we fix α in the range $0 < \alpha < 1$. In this range, every initial population eventually dies out ($x_t \to 0$). On the other hand, if $\alpha > 4$ it's easy to see that all initial populations (except the fixed points $x = 0$ and $x = 1$) diverge to $-\infty$. Thus, nontrivial dynamical behavior is possible only for $1 \leq \alpha \leq 4$.

Let's consider what happens in the neighborhood of a fixed point, i.e., a point x^* for which $x^* = F(x^*, \alpha)$. Such a fixed point is given as a function of α by the relation $x^* = (1 - \frac{1}{\alpha})$. We want to know under what circumstances x^* is stable, i.e., when do all points sufficiently close to x^* approach it as $t \to \infty$? The answer depends upon the slope of $F(x, \alpha)$ at $x = x^*$. If the magnitude of this slope is less than one, then x^* is attracting; otherwise, x^* is repelling. An easy calculation shows that the slope is given by

$$\lambda_{(1)}(x^*) \doteq \left.\frac{dF}{dx}\right|_{x=x^*} = 2 - \alpha,$$

so that the point x^* is stable when $1 < \alpha < 3$. But what happens when $\alpha > 3$? Now x^* is no longer attracting, but is repelling. So what is the ultimate fate of a point near x^*?

To address this question, we first note that the previously stable fixed point x^* can be regarded as a period-1 cycle that becomes unstable at $\alpha = 3$. So it's natural to consider whether there might be a cycle of period-2 that *is* stable for $\alpha \geq 3$. To examine this possibility, we look at the population at successive intervals that are two generations apart, i.e., we examine the function relating x_{t+2} to x_t. This function is just the second iterate of the map $F(x, \alpha)$. Thus,

$$x_{t+2} = F(F(x_t, \alpha), \alpha) \doteq F^{(2)}(x_t, \alpha).$$

Population levels that repeat every *second* generation are now fixed points of the map $F^{(2)}(x, \alpha)$, and can be found by solving the equation

$$F^{(2)}\left(x^*_{(2)}, \alpha\right) = x^*_{(2)},$$

where we have used an obvious notation for the period-2 fixed point(s). By simple geometric arguments or by direct algebraic calulations, it can be seen that there are two solutions to the above equation, both representing stable fixed points if

$$\lambda_{(2)} \doteq \left| \frac{dF^{(2)}(x^*_{(2)}, \alpha)}{dx} \right| < 1.$$

But a direct calculation shows that $\lambda_{(2)} = (\lambda_{(1)})^2$ when $x^* = x^*_{(2)} = F(x^*_{(2)}, \alpha)$, so that two new stable fixed points of period-2 are born out of the old fixed point x^* when x^* becomes unstable. This process is perfectly general. As we continue to increase α, the period-2 fixed points become unstable and give rise to new stable fixed points of period-4. A further increase of α gives stable fixed points of period-8, and so on. This sequence of births and deaths of stable fixed points of periods 2^k continues for all $k = 1, 2, \ldots$. However, the "window" of parameter values α for which any one cycle is stable gets progressively smaller, so that the entire process grinds to a halt at a critical parameter value $\alpha^* \approx 3.57\ldots$.

As we go beyond α^*, there are an infinite number of fixed points with different periodicities, as well as an uncountable number of initial populations x^0 that give rise to totally aperiodic, but bounded, trajectories. Beyond the value $\alpha \approx 3.8284\ldots$, there are unstable cycles of every integer period, as well as an uncountable number of aperiodic trajectories. This is the so-called *chaotic* regime for the system. This terminology suggests that the dynamical trajectories in this regime are observationally indistinguishable from a stochastic process, and many analytical and computational results confirm this intuition. We shall deal extensively with this point below, so for now let's only note that even though values of α beyond α^* give rise to aperiodic trajectories and "deterministic randomness," it is still the case that almost all initial points x^0 are attracted to a unique cycle. Thus, the set of initial populations that belong to any of the other infinite number of cycles, or to an aperiodic trajectory, though uncountable in number, still form a set of measure zero on the unit interval.

The justification for describing the motion of our system as stochastic comes from the fact that any given stable cycle is likely to occupy a vanishingly small window of parameter values. Also, the long time required for the transients associated with the initial conditions to damp out means that in practice numerical and/or observational errors are likely to cause the trajectory to "wander away" from this cycle and "randomly" attach itself to one of the other orbits. Thus, a stochastic description seems appropriate despite the completely deterministic nature of the underlying dynamical equations. In passing, it's interesting to note that the above iteration of the function $F(x, \alpha)$ was often employed in the early days of computing as a random-number generator, using values of α near 4. And in fact, this scheme is still useful today for pocket calculator computations.

Since pictures always speak louder than words, Fig. 4.2 shows the entire period-doubling route to chaos as the parameter α is allowed to vary between 0 and 4. Note that in the figure where the long-term evolution of the logistic map converges to a periodic cycle of period k, the diagram shows k discrete values for x. At $\alpha = 3.5$, for example, there are four discrete values for x, with the system eventually settling down to a periodic oscillation among those four values. In the chaotic regime, the values of x cover continuous intervals, the darkness of the shading representing the relative amount of time that the system spends in that particular region. It's also important to note the "windows" of periodic behavior embedded within the chaotic regime. The most prominent of these windows corresponds to a period-3 cycle for $\alpha \approx 3.83$.

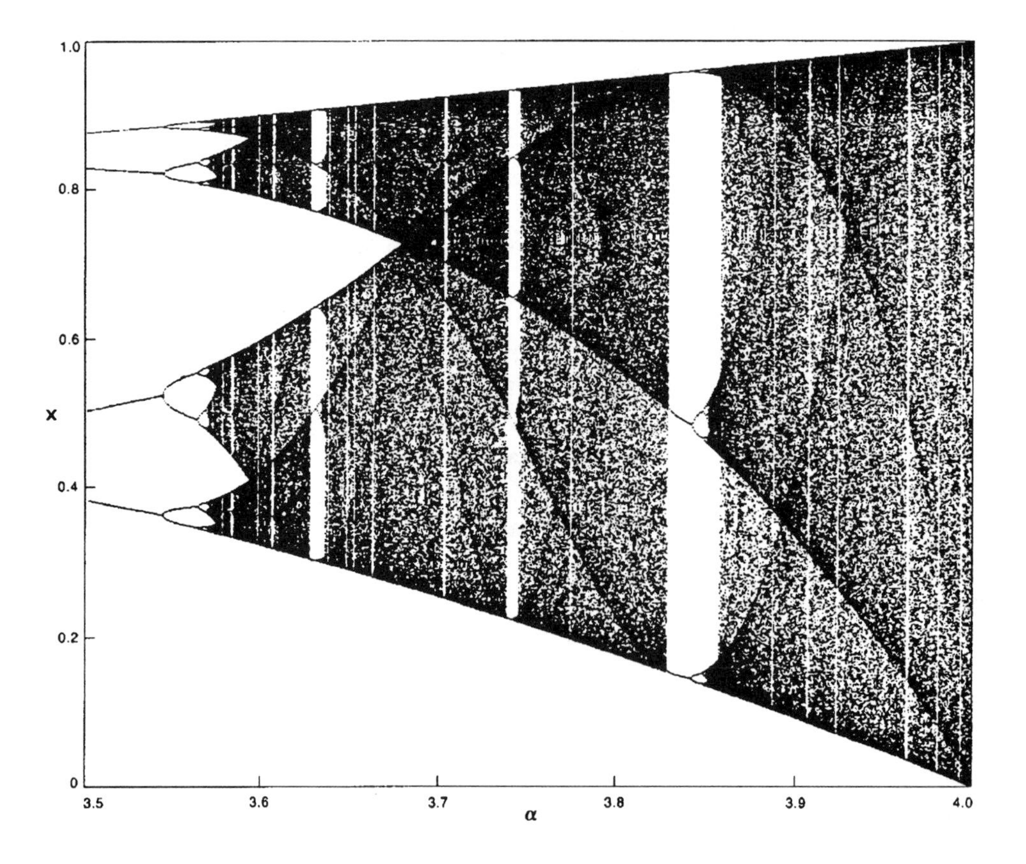

Figure 4.2. Bifurcation Diagram for the Logistic Map

The foregoing example, simple as it is, exhibits all of the types of long-term behavior that can be displayed by a dynamical system: stable and unstable fixed points, periodic orbits and aperiodic trajectories. Our goal in this chapter is to characterize the circumstances under which each of these qualitatively different types of behavior can emerge, as well as to give some indication as to how such dynamical systems can be used to model a spectrum of situations arising in the natural sciences, economics, biology and ecology.

Exercises

1. Consider the logistic map with $\alpha = 4$, and introduce the new variable y_t via the relation $y_t = \sin^2 \pi x_t$. (a) Show that in the y variable the logistic equation becomes $y_t = 2y_{t-1} \mod 1$. (b) What does this transformation have to do with the Circle-10 system discussed in Section 1?

2. As noted in the text, chaotic behavior involves a combination of local stretching and global folding. (a) Identify these two competing processes in the Circle-10 system. (b) Why is it that no *linear* map can display chaotic behavior?

3. Consider the system $x_{t+1} = ae^{x_t}$, where $a < 0$. (a) Show that when $a > -e$ the fixed point of the system is attracting, while for $a < -e$ it is repelling. (b) Thus, show that as a decreases below $-e$, the system undergoes a period-doubling bifurcation.

4. Let $I = [a, b]$ be an interval on the real line and let $f\colon I \to I$ be a differentiable map. Assume that $p \in I$ is a periodic point for f of period n. Prove that for each $i = 0, 1, \ldots n-1$, we have the relation $(df^{(n)}/dx)(x_i) = (df^{(n)}/dx)(p)$, where $x_i = f^{(i)}(p)$ is the $(i+1)$st point on the orbit of p.

3. Quasi-Periodic Trajectories and the KAM Theorem

Consider a classical mechanical system consisting of n particles whose position and momentum are described in local coordinates by the variables $p = (p_1, p_2, \ldots, p_n)$ and $q = (q_1, q_2, \ldots, q_n)$, respectively. We call such a system *Hamiltonian* if there exists a function $H(p, q)$ such that the dynamical behavior of the system can be written as

$$\dot{q} = \frac{\partial H}{\partial p}, \qquad -\dot{p} = \frac{\partial H}{\partial q}.$$

Heading the long list of reasons why such systems are important are the following:

- *Liouville's Theorem*—Hamiltonian systems preserve the phase volume; i.e., an initial volume element $\{dp_1 \times \cdots \times dp_n \times dq_1 \times \cdots \times dq_n\}$ in the phase space R^{2n} is preserved by the flow of a Hamiltonian system.

- *Conservation of Energy*—The function H is a first integral for the flow of the dynamics, i.e., H is constant along trajectories of the system.

The second property leads to the idea of an *integrable* Hamiltonian system as one for which there exist n independent first integrals, i.e., functions $I_1(p, q), \ldots, I_n(p, q)$, such that $dI_i/dt = 0$, $i = 1, 2, \ldots, n$. For such systems we can obtain an explicit characterization of their dynamical behavior, as will be shown in a moment.

Example: The Harmonic Oscillator

The prototypical example of a Hamiltonian system is the simple harmonic oscillator, which describes the small oscillations of a pendulum. The Hamiltonian for this system is

$$H(p, q) = \frac{\omega}{2}(p^2 + q^2),$$

where ω is the frequency of the pendulum. A first integral for this system is $I = p^2 + q^2$, leading to the well-known result that the trajectories of the system lie on the circles

$$p^2 + q^2 = k^2, \qquad k \in R.$$

The constant k^2 represents the energy of the pendulum, and is fixed by the initial displacement $p(0)$ and initial momentum $q(0)$.

The circular motion of this linear system in its phase space (or periodic motion in the state space) is typical of integrable *nonlinear* Hamiltonian systems, the circular motion corresponding in higher dimensions to motion on a torus instead of a circle. Since this generalization is important for what follows, let's take a moment to show how it goes.

Assume we are given a Hamiltonian system as described above. By the integrability assumption, there exists a coordinate transformation from the (p, q)-variables to a special set of generalized position and momentum variables x and y, respectively. In these new variables, the Hamiltonian function H depends only upon the position variables x, i.e., $H = H(x_1, x_2, \ldots, x_n)$. Thus, in the new variables the dynamical equations become

$$\dot{y} = \frac{\partial H}{\partial x} \doteq \Omega(x), \qquad -\dot{x} = \frac{\partial H}{\partial y} = 0.$$

This system can be easily integrated, yielding

$$\begin{aligned} y(t) &= \Omega(x)t + y(0), \\ x(t) &= x(0). \end{aligned}$$

Returning to the old variables, we obtain $2n$ combinations of $\{p_i, q_i\}$ and t:

$$\begin{aligned} y(0) &= y(p(t), q(t), t), \\ x(0) &= x(p(t), q(t), t). \end{aligned}$$

Since they are constant, these variables cannot, in fact, depend upon t. This means that we have solved the equations of motion, in the process obtaining $2n$ constants (integrals, invariants) of the motion. This result has been obtained under the assumption that there exists a coordinate change that will make the new Hamiltonian independent of the generalized momenta variables. It can be shown that the existence of n independent integrals of motion is all that's needed in order for this assumption to be satisfied.

So what does the foregoing result tell us about the dynamical behavior of an integrable Hamiltonian system? Basically, it says that the motion is similar to that of n coupled oscillators, as can be seen from the above equations. In order to keep the motion confined to a finite segment of R^{2n}, it must be the case that the linearly growing terms in y appear as arguments of periodic functions. We have seen in the preceding example that when $n = 1$ the motion is circular with constant angular velocity, hence periodic. When $n = 2$ there are two radii determined by the initial values $x_1(0)$ and $x_2(0)$, with angular velocities Ω_1 and Ω_2. Thus, the motion is confined to a two-dimensional torus. However, in addition to periodic motion, we now have the possibility for a new kind of behavior. When the ratio Ω_1/Ω_2 is irrational, the motion cannot be periodic and the trajectory winds about the surface of the torus endlessly and covers it densely. Such motion is termed *quasi-periodic.* In general, the motion of an integrable Hamiltonian system with n degrees of freedom is quasi-periodic and takes place on an n-torus.

The two most immediate questions surrounding integrable Hamiltonian systems are the following:

- How "typical" is it for a Hamiltonian system to be integrable?
- If an integrable Hamiltonian system is perturbed "slightly," does the qualitative nature of the motion remain the same? That is, if we use the new Hamiltonian $H = H_0 + P$, where H_0 is Hamiltonian and $P(x, y, t)$ is an analytic perturbation that is of period 2π in the x variables and contains a small parameter, does the system trajectory still remain quasi-periodic?

Both of these questions can be given rather clear-cut answers: If $n \geq 3$ it is extremely rare for a Hamiltonian system to be integrable. In fact, it's so rare that a nonintegrable system cannot usually even be *approximated* by a sequence of integrable systems. The second question is answered by the celebrated Kolmogorov-Arnold-Moser (KAM) Theorem, a rough statement of which is as follows:

KOLMOGOROV-ARNOLD-MOSER THEOREM. *Take the perturbation P to be small, and assume that the frequencies $\Omega(x)$ of the unperturbed system satisfy the nonresonance condition*

$$\det \frac{\partial(\Omega_1(x), \ldots, \Omega_n(x))}{\partial(x_1, \ldots, x_n)} \neq 0.$$

Then the motion of the system is still confined to an n-torus, except for a set of initial conditions of small measure.

There are several comments in order regarding this important result:

1) The n-tori, termed *KAM surfaces,* if seen in plane section appear as slightly distorted versions of those for the $P = 0$ case. Nevertheless, the qualitative nature of the motion remains basically the same as for the integrable situation.

2) The set of special initial conditions that lead to trajectories not on an n-torus may wander about freely on the energy surface for the system, i.e., the surface for which H is constant.

3) The boundary of the constant energy surface must be of dimension $2n - 1$; hence, when $n \geq 3$ the n-dimensional KAM surfaces cannot serve as boundaries dividing the energy surface into regions that confine the wandering trajectories (as shown in Fig. 4.3 for the case $n = 2$). Consequently, these trajectories may wander about on the whole energy surface, giving rise to a type of randomness termed *Arnold diffusion.*

Figure 4.3. Invariant Tori for a Three-Dimensional Energy Manifold

Example: Convex Billiard Tables

As an amusing illustration of the application of the KAM Theorem, consider the closed trajectories of a billiard ball moving on a frictionless table of any convex shape. An example of a closed trajectory that is stable in the linear approximation is the minor axis of an ellipse when the table has an elliptical shape. The KAM Theorem enables us to conclude that a closed trajectory that is near the minor axis of an ellipse is also stable on any billiard table that is a small perturbation of the original ellipse.

There are a number of additional applications of the KAM Theorem to problems in classical mechanics involving the rotation of heavy rigid bodies, planetary motion, and the like. Since these matters are somewhat outside our interests in this book, we leave it to the interested reader to consult the literature cited in the Notes and References for further details of these applications of KAM theory. We now move on to a consideration of the type of random motion suggested by the process of Arnold diffusion, namely, randomness of the sort that's generated by a completely deterministic mechanism.

Exercises

1. The dynamical equations for the simple harmonic oscillator are the same as those for a pendulum when the initial displacement x_0 of the pendulum bob is small (so that the approximation $\sin x_0 \approx x_0$ is valid). Suppose instead we use the *real* equation for the pendulum, which is

$$\ddot{x} + \lambda^2 \sin t = 0, \qquad -\pi < x < \pi.$$

(a) Determine the Hamiltonian for this system. (b) Show that the system is integrable in terms of Jacobian elliptic functions.

2. The Hamiltonian for the so-called *Toda lattice* is given by

$$H = \frac{1}{2}(p_x^2 + p_y^2) + \frac{1}{24}(e^{2x-2y\sqrt{3}} + e^{2x+2y\sqrt{3}} + e^{-4x}) - \frac{1}{8}.$$

(a) Show that H is an integral of the motion, i.e., $dH/dt = 0$. (b) Using H, write down Hamilton's equations of motion for this system. (c) Show that there is a second integral of the motion given by

$$I = 8\dot{y}(\dot{y}^2 - 3\dot{x}^2) + (\dot{y} + \dot{x}\sqrt{3})e^{2x-2y\sqrt{3}} + (\dot{y} - \dot{x}\sqrt{3})e^{2x+2y\sqrt{3}} - 2\dot{y}e^{-4x}.$$

4. A Philosophical Digression: Randomness and Determinism in Dynamical System Modeling

The type of aperiodic behavior displayed by the Circle-10 system of Section 1 or by the simple quadratic "logistic type" map discussed in Section 2, as well as the Arnold diffusion type of instability shown by the trajectories of nonintegrable Hamiltonian systems of higher dimension, makes it clear that very simple, totally deterministic mechanisms may give rise to very complicated behaviors. In fact, from an observational standpoint such behaviors are totally indistinguishable from what one would see if the underlying generating mechanism were what probabilists call a *stochastic process.* Before discussing some of the epistemological aspects of this observation, let's look at two very specific examples of this phenomenon as vehicles upon which to focus our subsequent remarks on the stochastic vs. deterministic dichotomy.

Example 1: Piecewise-Linear Maps and Coin-Tossing

Consider the piecewise-linear map

$$x_{t+1} = \begin{cases} 2x_t, & 0 \leq x_t \leq \frac{1}{2}, \\ 2 - 2x_t, & \frac{1}{2} \leq x_t \leq 1. \end{cases}$$

It should be noted that this map is topologically (but not diffeomorphically!) equivalent to the quadratic map $x_{t+1} = 4x_t(1 - x_t)$ considered in Section 2 by making the continuous (but not differentiable) change of coordinates $x \to (2 \sin^{-1} \sqrt{x})/\pi$.

Suppose we divide the unit interval into two halves and label the segment $0 \leq x \leq \frac{1}{2}$ by "H" and the segment $\frac{1}{2} \leq x \leq 1$ by "T." Now pick an arbitrary starting point x_0 in the unit interval and carry out the iteration specified above, labeling each successive iterate by the half of the interval into which it falls. Thus, each such experiment will result in a sequence like HTTHHHTTHTHTH.... It has been shown that after a large number of such iterations, the density function for the variable x representing the fraction of the time that there are more Heads than Tails is given by

$$p(x) = \frac{1}{\pi \sqrt{x(1 - x)}} \, .$$

This is exactly the probability density function that we would have obtained if each iteration of the above map was thought of as being the flip of a fair coin, and we kept track of the resulting sequence of Heads and Tails. In short, the iterates of the deterministic, piecewise-linear map given above are mathematically and *observationally* indistinguishable from the sample path of a sequence of Bernoulli trials with probability $p = \frac{1}{2}$.

Example 2: Quadratic Optimization and Stochastic Filtering

Suppose we wish to minimize the quadratic form

$$\int_0^T (x^2 + u^2)\, dt,$$

over all scalar functions $u(t)$ defined on $[0, T]$. Here we assume that x and u are related by means of the linear differential equation

$$\dot{x} = x + u, \qquad x(0) = c.$$

By means of arguments to be given in Chapter 7, it's easy to see that the solution of this problem is given by

$$u^*(t) = -p(t)x^*(t),$$

where p is the solution of the Riccati equation

$$\dot{p} = 1 - p^2, \qquad p(T) = 0.$$

The minimizing trajectory $x^*(t)$ is the solution of the differential equation

$$\dot{x}^* = (1 - p(t))x^*, \qquad x^*(0) = c.$$

This is a purely deterministic optimization problem.

On the other hand, in the theory of linear systems it is known that the solution of the above optimization problem is the precise mathematical dual of the problem of determining the optimal estimate of a signal observed in the presence of Gaussian noise, i.e., a problem the essence of which involves an *intrinsic* stochastic component. The solution of this estimation problem leads to the well-known *Kalman filter,* one of the most important tools in the arsenal of the modern system theorist. Thus what appears at first glance to be an inherently stochastic problem turns out to be solvable by means that don't involve any statistical considerations, at all! For more details on this duality, we refer the reader to the material cited in the Notes and References, as well as to Chapter 6.

These examples (and there are many others) call into question several axioms of faith in applied mathematical modeling. Two of the most important such axioms involve the degree to which nature is *truly* stochastic, and the contention that the uncertainty always present in our knowledge of the "true" system is best handled by incorporating correction factors involving random terms into our models. Since both of these assertions seem to be firmly entrenched in the modeling *Weltanschauung* of the generic applied mathematician and system modeler, it's worth taking a longer look at the epistemological foundations upon which these prejudices are based, examining the degree to which such claims hold water in view of the kind of "deterministic randomness" we are discussing here.

As to the inherent randomness of nature, this appears to be as much a question of subjective psychology as it is a matter of physics and mathematics. Since the advent of the Copenhagen interpretation of quantum mechanics, most physicists have adhered to the view that nature is indeed intrinsically stochastic—at least in the quantum realm. Einstein, and more recently David Bohm, objected to this interpretation, claiming, in effect, that the necessity of a probabilistic interpretation in the Copenhagen view was a consequence not of nature but of the limitations on our ability to carry out measurements. This led to the claim that the supposed inherent randomness could be accounted for by the introduction of hidden variables, i.e., the seeming intrinsic uncertainty of nature would be eliminated if we had knowledge of these unobservables. In short, what looks like randomness in nature is due to our methods of observation, not to any inherent stochastic mechanisms in nature itself.

This entire circle of thought was given added spice in 1964 with the announcement of Bell's Theorem, asserting that there could be no *local* hidden-variable theories. The initial experiments of John Clauser at Berkeley, and more recently Alain Aspect in Paris, confirmed the predictions of Bell's result, leaving us in the uncomfortable position of saying that nature can be intrinsically deterministic only if the hidden variables are nonlocal in character, i.e., involve some kind of faster-than-light connections. Such superluminal theories are currently under development by Bohm, Basil Hiley, and others, utilizing the idea of a global quantum field encompassing, in principle, the entire universe.

For our purposes here, the main conclusion to be drawn from this body of work is that nature might just be deterministic as Einstein believed. But if so, that kind of determinism is far different from that with which we are familiar from everyday life. So contrary to popular belief and opinion, the way is still open for a totally deterministic reality—but it must be a nonlocal type of deterministic mechanism.

Matters of quantum reality are, for the most part, far removed from the concerns of everyday life for engineers, economists, ecologists, and others of a more macroscopic professional orientation. Such practitioners generally tend to a quite different position on the issue of the inherent randomness of nature. They are usually quite willing to concede that nature may indeed be totally deterministic—even at the micro-level—but that no models or measurements can ever be free of error. Consequently, we must build into our models some additional terms to account for these modeling "hidden variables." The argument then continues by asserting that since we don't *know* what has been left out of the model, the best way to represent this ignorance is to throw a few random variables into the modeling soup. Such an approach is based, of course, on the entirely unjustified thesis that adding a random "fudge factor" will somehow bring the model closer to reality. When stated in such bald fashion, the nonsense underlying such a position is patently obvious; however, there is the germ of a real issue here, i.e., to what degree is it *necessary* to use the machinery of probability theory and stochastic processes to account for our uncertainty in the underlying mechanisms and/or observed behavior?

The examples given above, as well as what we will see in the remainder of this chapter, show that even with *perfect* measuring instruments it's quite possible for a system to display behavior that looks as if it were the output of a random device or, perhaps, the behavior of a deterministic device whose output is corrupted by noisy measurements. Neither of these situations need be the case. A deterministic mechanism can give rise to random-looking behavior, even when the measurements are exact! Further, as the Quadratic

Optimization example showed, it's perfectly possible for a process formulated in terms of stochastic processes to be mathematically identical to a completely deterministic process. These are facts, not opinions, and they cast serious doubt upon the "stochastics are necessary" school of thought in system modeling. In fact, adopting Occam's Razor, another venerable principle of sound scientific practice, we might be tempted to conclude that the unnecessary introduction of random terms into a deterministic framework constitutes an inadmissible multiplication of causes beyond what the facts warrant. While we hold no brief against the *utility* of such procedures in carefully controlled and well-understood situations, our main point here is that they are far from being either necesary or sufficient as a principle of good modeling practice. Even further, Occam's Razor demands that we consider deterministic mechanisms of the above sort before resorting to the *ad hoc* addition of random "correction" terms into a model. So we take the position that the introduction of stochastic components into a model is more a matter of taste and convenience than it is an axiom of necessity, and it will be our goal in the remainder of the chapter to provide examples, in addition to precepts, in support of this contention.

End of digression. Now let's return to mathematical modeling and problems of deterministic chaos.

Exercises

1. Show that the coin-tossing example is completely equivalent mathematically to the following "Circle-2" process on the unit interval: $x_{t+1} = 2x_t \mod 1$. (This shows that the tent map, the logistic map, and the Circle-10 and Circle-2 systems are for all intents and purposes the same system.)

2. We have seen that the difficulty in computing the exact trajectory of each of the chaotic processes considered thus far comes down to our inability to describe completely the initial state x_0. Consider the kind of number system that we would have to use to overcome this obstacle.

5. Scalar Maps and the "Period-3" Theorem

The period-doubling and transition to chaotic motion that we saw in Section 2 for the quadratic map $x \to \alpha x(1-x)$ is not a curiosity confined to this map alone, but is characteristic of *almost all* maps of the unit interval possessing a single "hump." Here we want to examine just what kind of one-dimensional maps lead to this period-doubling and chaotic type of behavior.

We consider a mapping $f\colon I \to I$, where I is some closed interval containing the origin. Let α be a real parameter, and assume that we generate a sequence of elements from I according to the rule

$$x_{t+1} = f(x_t, \alpha), \qquad x_0 \in I.$$

As in Section 2, our interest is in identifying the long-term behavior of the sequence $\{x_t\}$ for different values of α and different initial states x_0.

Let's agree to confine our attention to functions f that are *unimodal*, i.e., have only a single local maximum on I. Without loss of generality, we can rescale f and x so that:

1) the maximum is located at $x = 0$, $f(0, \alpha) = 1$ and $f'(0, \alpha) = 0$;

2) $f'(x, \alpha) > 0$ for $x < 0$, while $f'(x, \alpha) < 0$ for $x > 0$.

Thus any such f is a function with a single "hump." In addition, we assume that the the Taylor series of f near the origin looks like

$$f(x, \alpha) = 1 - ax^k + \dots,$$

with k being a positive *even* integer.

After a sufficiently large number of iterations of the map f, the behavior of the sequence $\{x_t\}$ usually settles into either a periodic or aperiodic stationary pattern, where a fixed point is thought of as a periodic orbit of period 1. By definition, all points on an orbit of period r must be fixed points of the rth iterate of the map f. We denote this rth iterate of f as $f^{(r)}(x, \alpha) \equiv F(r, x, \alpha)$.

The first question to address concerning any fixed point is whether or not it is stable. Let x^* be a fixed point so that $f(x^*, \alpha) = x^*$. Then x^* will be linearly stable if $|f'(x^*, \alpha)| < 1$. Similarly, a periodic orbit consisting of the points $x_1, x_2, \dots, x_r$ is stable if

$$|F'(r, x_i, \alpha)| = \prod_{j=1}^{r} |f'(x_j, \alpha)| < 1, \qquad i = 1, 2, \dots, r.$$

These stability conditions will hold for a finite interval on the α-axis, termed the *periodic window*.

Arguing as in Section 2, for $r = 1$ there will be some periodic window $0 < \alpha < \alpha_1$ for which x^* will be a stable periodic orbit of period r (i.e., a fixed point of F). At $\alpha = \alpha_1$, the stability of x^* will break down and a stable cycle of period 2 will emerge, having a periodic window $\alpha_1 \leq \alpha < \alpha_2$. This cycle, of course, is a fixed point for the map $F(2, x, \alpha)$. So, at $\alpha = \alpha_2$ the stability of this 2-cycle will break down and a new stable cycle of period 4 will emerge, with its own periodic window $\alpha_2 \leq \alpha < \alpha_3$. This 4-cycle will be a fixed point of the iterate of $F(2, x, \alpha)$, accounting for the reason why the period always doubles in length from the previous cycle. This process continues indefinitely, a period of order $r = 2^k$ emerging at the parameter value $\alpha = \alpha_k$. It can be shown that the sequence $\{\alpha_k\}$ converges rapidly for *any* function f satisfying the "one-hump" conditions stated above, and that for quadratic maps we have

$$\alpha_k \approx \alpha_\infty - \frac{A}{\delta^k},$$

where A is a constant depending on f, while $\delta = 4.6692016\ldots$ is a universal constant, termed the *Feigenbaum constant,* which is characteristic of all quadratic maps of the type considered above. Another way of interpreting this universal feature of period-doubling maps is to note that

$$\delta = \lim_{n\to\infty} \frac{\alpha_{n-1} - \alpha_n}{\alpha_n - \alpha_{n+1}}.$$

Thus, even though the numbers $\{\alpha_k\}$ themselves *do* depend upon the particular function f, the relative size of the gaps in the sequence between one periodic window and the next is a universal constant for all one-dimensional quadratic maps. It's clear from this fact, as well as from the earlier estimate for α_k, that the sequence of bifurcation values converges very rapidly to a value α_∞ at which entirely new phenomena emerge.

Note that if the map is not quadratic, e.g., if $f = 1 - ax^k + \ldots$, $k = 4, 6, \ldots$, there is a different universal constant for the appropriate class of maps. For instance, if $k = 4$, δ is $7.248\ldots$, while for $k = 6$ the value is $\delta = 9.296\ldots$. What is important is that the number is universal for all maps of the corresponding class.

For $\alpha > \alpha_\infty$, there are an infinite number of periodic windows immersed in the background of an aperiodic regime. A careful examination of the iterates $\{x_t\}$ shows them jumping among 2^k subintervals of I, where k decreases from ∞ to 0 as α increases from α_∞ to $\alpha_{\max}$ (the maximum value of the parameter that will keep all iterates within the interval I). Before entering into a more detailed discussion of this aperiodic regime, let's have a look at a few of the more important properties of these one-dimensional mappings.

- *Stable Periods*—A unimodal mapping can have at most one stable period for each value of α; in fact, it may have *no* stable period for many values of α. A necessary condition for f to have a stable period is that the *Schwarzian derivative* of f be negative on the interval I, i.e.,

$$\frac{f'''(x)}{f'(x)} - \frac{3}{2}\left(\frac{f''(x)}{f'(x)}\right)^2 < 0,$$

for all $x \in I$. This condition is not sufficient for the stable orbit to exist. So even when the Schwarzian derivative is negative, it's possible to get different aperiodic orbits starting from different initial states x_0, with no initial points leading to a stable periodic orbit.

• *Scaling Properties*—Consider the function $F(2^{k-1}, x, \hat{\alpha}_k)$, where $\hat{\alpha}_k$ is the value of the parameter α at which

$$\frac{dF(2^k, x, \alpha)}{dx} = 0.$$

Such a value of α_k is called a *superstable* value of α. There is an invariance relation between the function above and the next iterate of f at the superstable value of α for $f^{(2^k)}$, i.e., the function $F(2^k, x, \hat{\alpha}_{k+1})$. To see this relation, introduce the operator

$$\begin{aligned} Tf(x, \hat{\alpha}_k) &= -\beta f\left(f\left(-x/\beta, \hat{\alpha}_{k+1}\right), \hat{\alpha}_{k+1}\right), \\ &= -\beta F\left(2, -x/\beta, \hat{\alpha}_{k+1}\right). \end{aligned}$$

Here the parameter β is just a scaling factor in the domain and range of the mapping f. It's also easy to see that

$$T^m f(x, \hat{\alpha}_k) = (-\beta)^m F\left(2^m, x/(-\beta)^m, \hat{\alpha}_{m+k}\right), \qquad m \geq 1.$$

It can be shown that there exists a limiting function

$$g(x) = \lim_{m \to \infty} T^m f(x, \hat{\alpha}_k),$$

and that $g(x)$ is *independent* of the initial function f. Thus, g provides a universal scaling function for the iterates of any unimodal one-dimensional map, and the properties of g enable us to see that the region around any periodic orbit (for a given k) is self-similar to the same region around the periodic orbit corresponding to *any* value of k. The local character of each iterate in the vicinity of its fixed point is universal, being determined by the properties of the function g. So it's only the scaling factor β that depends on the particular map f.

Employing the recursive nature of $F(2^k, x/(-\beta)^k, \hat{\alpha}_{k+m})$, a little algebra yields the functional equation for g as

$$g(x) = -\beta g\left(g(-x/\beta)\right).$$

The normalization $f(0, \alpha) = 1$, together with the superstable condition that the derivative of F vanish at the superstable value of α, leads to the boundary conditions for g as

$$g(0) = 1, \qquad g'(0) = 0.$$

It should be noted that these conditons are *not* sufficient to determine the function g uniquely. However, a series expansion of g near the origin can be used to get a handle on the local behavior, at least numerically.

• *Sarkovskii's Theorem*—In the study of periodic orbits of scalar maps, it's of considerable interest to determine when an orbit of a given period implies cycles of other periods. A powerful result in this direction was given in the mid-1960s by Sarkovskii and later greatly extended by others. This result provides an actual test for chaotic behavior.

SARKOVSKII'S THEOREM. *Order the positive integers according to the following scheme:*

$$3 \to 5 \to 7 \to 9 \cdots \to 3 \cdot 2 \to 5 \cdot 2 \to 7 \cdot 2 \to 9 \cdot 2 \to \cdots \to 3 \cdot 2^2 \to 5 \cdot 2^2$$
$$7 \cdot 2^2 \to 9 \cdot 2^2 \to \cdots \to 3 \cdot 2^n \to 5 \cdot 2^n \to 7 \cdot 2^n \to 9 \cdot 2^n \to \ldots\ldots 2^m$$
$$\to \cdots \to 32 \to 16 \to 8 \to 4 \to 2 \to 1,$$

where the symbol "$\to$" means "precedes." If f is a unimodal map with an initial point x_0 leading to a cycle of period p, then for every q that follows p, i.e., for every q such that $p \to q$, there is an initial point x_0^* leading to a cycle of period q.

(Note: Sarkovskii's Theorem says nothing about the stability of the various periodic orbits, and is only a statement about the cycles emanating from various initial values of x for a fixed value of the parameter α.)

An extremely important corollary of Sarkovskii's result is that if a map has a cycle of period 3, then it has cycles of *all* integer periods. This observation forms the basis for one of the most celebrated results in the chaos literature, the famous "Period-3 Theorem" of Li and Yorke.

PERIOD-3 THEOREM. *Let $f\colon I \to I$ be continuous. Consider a point $\hat{x} \in I$ such that the first three iterates of $\hat{x}$ are given by $f(\hat{x}) = b$, $f^{(2)}(\hat{x}) = c$ and $f^{(3)}(\hat{x}) = d$. Assume that these iterates are ordered as*

$$d \leq \hat{x} < b < c \text{ (or } d \geq \hat{x} > b > c).$$

Then for every positive integer k, there is an initial point in I leading to a cycle of period k. Furthermore, there is an uncountable set $U \subset I$ containing no periodic points. In addition, the points in U satisfy the following conditions:

i) For every x, $y \in U$ with $x \neq y$,

$$\limsup_{n\to\infty} \left| f^{(n)}(x) - f^{(n)}(y) \right| > 0,$$

$$\liminf_{n\to\infty} \left| f^{(n)}(x) - f^{(n)}(y) \right| = 0.$$

ii) For every $x \in U$ and periodic point $x_0 \in I$,

$$\limsup_{n\to\infty} \left| f^{(n)}(x) - f^{(n)}(x_0) \right| > 0.$$

Note that the second part of the Period-3 Theorem improves upon Sarkovskii's result by characterizing some aspects of the aperiodic points. Namely, every aperiodic trajectory comes as close as we want to another aperiodic trajectory—but they never intersect. Furthermore, the aperiodic orbits have no point of contact with any of the periodic orbits. We also note that the conditions of the theorem are satisfied by any function f that has a point of period 3, although it's not necessary for f to have such a point for the type of chaos described to emerge.

The Period-3 Theorem provides us with a set of sufficient conditions for a mapping f to display "chaotic" behavior. There are two main criteria: (1) a countable number of periodic orbits, and (2) an uncountable number of aperiodic orbits. We can add to this the requirement that the limiting behavior be an unstable function of the initial state, i.e., that the system is highly sensitive to small changes in the initial state x_0. Before turning our attention to a description of various tests that can be used to identify when chaos is present in processes for which we have only numerical observations, let's consider the case of multi-dimensional maps. In short, we want to consider whether these concepts are peculiar to the one-dimensional setting, or whether we cam extend the notion of chaos to higher-dimensional mappings.

Exercises

1. Let $\mathcal{S}f$ denote the Schwarzian derivative of the function f. Suppose $\mathcal{S}f < 0$ and that f has n critical points. (a) Prove that f can have at most $n + 2$ attracting periodic orbits. (b) In the case of the logistic map, sharpen this result to prove that for each value of the parameter a, the map $x \rightarrow ax(1 - x)$ can have at most one stable periodic orbit.

2. Confirm or disprove the following asserted consequences of the Sarkovskii ordering: (a) The existence of period length $p = 3$ ensures the existence of any other period of length q. (b) If only a finite number of period lengths occur, their lengths must be powers of 2. (c) If a period length p exists that is not a power of 2, then there are infinitely many different periods.

3. Determine all the periodic points of the following maps and classify them as repelling, attracting, or neither. In particular, determine whether or not any of these maps have a point of period 3 (assume all maps are defined on the closed interval $[0, 1]$): (a) $f(x) = x(1 - x)$. (b) $f(x) = x(x^2 - \frac{1}{9})$. (c) $f(x) = x(x^2 - 1)$; (d) $f(x) = e^{x-1}$.

6. Snap-Back Repellors and Multi-Dimensional Chaos

The main ingredient in the scenario of transition to chaos for scalar maps is the shift in stability of a fixed point of the map—a stable point becomes unstable and two new stable fixed points emerge. Continuation of this process eventually results in chaotic motion. Our concern here is with the degree to which such a scenario carries over to a mult-idimensional map $f\colon R^n \to R^n$, $n \geq 2$.

For scalar maps, the central characterizing result is the Period-3 Theorem, which tells us the circumstances under which aperiodic motion will occur. So, for example, the system

$$\begin{aligned} x_{t+1} &= (ax_t + by_t)(1 - ax_t - by_t), \\ y_{t+1} &= x_t, \end{aligned}$$

which has a stable cycle of period 3 when $a = 1.9$ and $b = 2.1$, and no chaotic motion for these parameter values, shows that the Period-3 Theorem does not extend directly to higher-dimensional settings. Our question is whether there is a similar type of result that does apply to this more general situation. We can get some insight into what's involved in answering this question by examining the proof of the Period-3 result.

A key ingredient in proving the Period-3 Theorem is the fact that if I is a compact subset of the real line and f is a continuous map on R such that $I \subset f(I)$, then f has a fixed point in I. This result no longer holds in R^n. So to provide a test for chaos in higher dimensions, we need another concept that implies a period-3 cycle when $n = 1$, but also implies aperiodic orbits when $n > 1$. In short, we need what's called a "snap-back" repellor.

Let $f\colon R^n \to R^n$ have a fixed point z. Further, assume that f is continuously differentiable in some neighborhood B containing z. We call z a *repelling* fixed point if the characteristic values of the Jacobian matrix $[\partial f / \partial x]$ are all outside the unit circle for every $x \in B$. The repelling fixed point is called a *snap-back repellor* if there exists a point $x_0 \neq z \in B$, and an integer M such that $f^M(x_0) = z$ and

$$\det \left[\frac{\partial f(x_0)}{\partial x} \right] \neq 0.$$

It can be shown that the existence of a snap-back repellor and the existence of a period-3 orbit are equivalent when $n = 1$. Consequently, it's natural to conjecture that snap-back repellors might imply the existence of aperiodic orbits for $n > 1$. This conjecture was established by F. R. Marotto in the following important result.

SNAP-BACK REPELLOR THEOREM. *Snap-back repellors imply chaos in R^n.*

Speaking more precisely, the conclusion of the Snap-Back Repellor Theorem is that the existence of a snap-back repellor implies the same conclusions in R^n as the Period-3 Theorem asserts for one-dimensional maps (upon replacing absolute values by the euclidean norm in R^n). That is, there are then periodic orbits of all integer periods, and an uncountable number of exponentially diverging aperiodic orbits. In short, chaos.

Example: The Hénon Attractor

We consider the two-dimensional system

$$\begin{aligned} x_{t+1} &= 1 + y_t - ax_t^2, \\ y_{t+1} &= bx_t, \end{aligned}$$

where $a, b \in R$. This system has the fixed points

$$\begin{aligned} x^* &= (2a)^{-1}\left[-(1-b) \pm \sqrt{(1-b)^2 + 4a}\right], \\ y^* &= bx^*. \end{aligned}$$

To ensure that these points are real, we impose the condition $a > -(1-b)^2/4$. In addition, since we are primarily interested in *unstable* fixed points, we make the additional assumption that $a > a_* = 3(1-b)^2/4$, which implies the instability of all fixed points. For purposes of analysis, we take $b = 0.3$. For values of a in the range $a_* < a < \hat{a} \approx 1.06$, there is a regime of period-doubling. For $\hat{a} < a < a^* \approx 1.55$, x^* is a repelling fixed point. And since $\det[\partial f/\partial x] = -b \neq 0$, this suggests x^* could be a snap-back repellor. Work by Michael Bendicks and Lennart Carleson has provided the analytical fireworks needed to confirm this conjecture. For the sake of completeness, we note that almost all trajectories tend to infinity for values of $a > a^*$.

Extensive analysis of the Hénon mapping shows a number of important features distinguishing its behavior from what's seen for one-dimensional maps. Among these properties are:

1) For a fixed value of a, the long-term behavior depends upon the initial point (x_0, y_0), i.e., the (x, y)-plane is divided into basins, and we may be led to different periodic or aperiodic orbits depending upon which basin of attraction we start in. This contrasts with the situation for one-dimensional maps, which can have at most only a single stable periodic orbit.

2) Some of the attractors of the Hénon map can be self-similar, i.e., they have *fractal* dimensions (we shall return to this property later in the chapter).

3) Hénon's map displays *homoclinic* points, i.e., points of intersection between stable and unstable manifolds. It is exactly the homoclinic points that correspond to the snap-back repellors giving rise to the chaotic behavior.

With the foregoing mathematical weapons at our disposal for the study of dynamical processes, we now turn our attention to the way in which chaos can naturally emerge in a variety of situations in the social and life sciences.

Exercises

1. For the logistic map $x \to ax(1-x)$ with $a > 4$, show that every point in the open interval $(-\infty, \frac{1}{2})$ is a snap-back repellor for the fixed point at $x = 0$. (Such points that tend to a repellor under iteration are often called *homoclinic points.*)

2. Consider the Hénon map. (a) Compute the Jacobian matrix J for the map and show that $\det J = b$. (b) Show that the Hénon map is invertible and compute its inverse. (Answer: $f^{-1}(x,y) = (1 - ax^2 + y, \, bx)$.) (c) If $b = 0$, the Hénon map transforms the entire plane into a parabola P given by $x = a - y^2$. Prove that the restriction of the Hénon map to P is topologically conjugate to the quadratic map $g(y) = a - y^2$. (Hint: Project P onto the y-axis and compute the induced map.)

3. (a) Show that the Hénon map can be decomposed into the composition of three separate maps, i.e., if we write the Hénon map as $T\colon R^2 \to R^2$, then we can find maps T_1, T_2 and T_3 such that $T = T_1 \circ T_2 \circ T_3$, where

i $T_1\colon (x,y) \to (x, 1 - ax^2 + y)$ is an area-preserving, nonlinear bending of the plane;

ii. $T_2\colon (x,y) \to (bx, y)$ is a contraction in the x direction;

iii. $T_3\colon (x,y) \to (y,x)$ maps the bent and contracted area back onto itself.

(b) For the parameter values used in the text, show that the effect of this sequence of transformations on an initial ellipse of points is to produce an attracting set that is a strange attractor, i.e., it displays local stretching and global folding.

4. Set $b = 1$ in the Hénon map. (a) Compute the characteristic values at the fixed points $p_{\pm}$. (b) Show that these characteristic values are complex and of absolute value 1 if the parameter a lies in a range, $a_0 < a < a_1$. (c) Give the explicit form of a_1 and a_2 in terms of a. (This result shows that the fixed points $p_{\pm}$ are not hyperbolic for this range of values of a.)

7. *Tests for Chaos*

Suppose we're given the output $Y = \{x_t\}$ of some scalar dynamical process, where $x_{t+1} = f(x_t, \alpha)$. What kind of mathematical "fingerprints" would we look for to conclude that the system was displaying chaotic behavior? In other words, what types of tests could be applied to the sequence Y to determine whether or not Y is a sample path from a system operating in the chaotic regime? In this section we'll examine several answers that have been proposed to this question.

• *Power Spectrum:* Since a countable number of periodic orbits together with an uncountable number of aperiodic trajectories are a characteristic feature of chaotic phenomena, Fourier analysis is the most direct attack on the question of whether chaos is lurking in the data.

The autocorrelation function of the map f is given by

$$c_j = \lim_{N \to \infty} \frac{1}{N} \sum_{k=1}^{N} x_k x_{k+j} = \langle x_0 x_j \rangle, \qquad j = 1, 2, \ldots,$$

with the Fourier transform of this function being

$$C(\omega) = c_0 + 2 \sum_{j=1}^{\infty} c_j \cos j\omega.$$

If we let $y(\omega)$ denote the Fourier transform of the sequence Y, then we note the important relation $C(\omega) = |y(\omega)|^2$. Thus, $C(\omega)$ defines the *power spectrum* of the sequence of iterates Y.

We have seen that the transition to chaos for the map f involves a cascade of period-doubling bifurcations. At a value of the parameter α for which there is a stable cycle of period 2^n, the power spectrum $C(\omega)$ will consist of a set of "spikes" at the values $\omega = \pi m/2^n$, where m is an integer less than 2^n. As α increases and a bifurcation takes place leading to a new cycle of period 2^{n+1}, new contributions to $C(\omega)$ will appear as additional spikes at $\omega = \pi(2m-1)/2^{n+1}$, where m is an integer such that $2m - 1 < 2^n$. In the range $\alpha > \alpha_\infty$, $C(\omega)$ consists of the same spikes as for the periodic values of α, together with a broad-band component representing the "noisy," or chaotic, aperiodic trajectories.

A consequence of the above observations is that we can expect to see a rather sharp change in the power spectrum of f as α passes from the regions of periodic to period-doubling to chaotic motion. To illustrate this phenomenon, consider the spectrum shown in Fig. 4.4, which is taken from an actual experiment involving the heating of a liquid layer from below that is lighter than the liquid above and, hence, rises by convective currents as the level of heating increases. The top part of the figure shows isolated peaks corresponding to a fundamental frequency and its harmonics; the system is *periodic* in this regime. In the middle we see several independent frequencies; here the motion is *quasi-periodic.* Finally, the bottom part of the figure shows wide peaks on a background of a continuous spectrum, indicating the transition to *aperiodic* motion. Now let's look at another test for identifying chaotic behavior.

• *Lyapunov Exponents:* A principal characteristic of the aperiodic trajectories is that although they remain bounded, each such trajectory diverges exponentially from any other. This observation provides us with another approach to the determination of the onset of chaos. Consider the initial

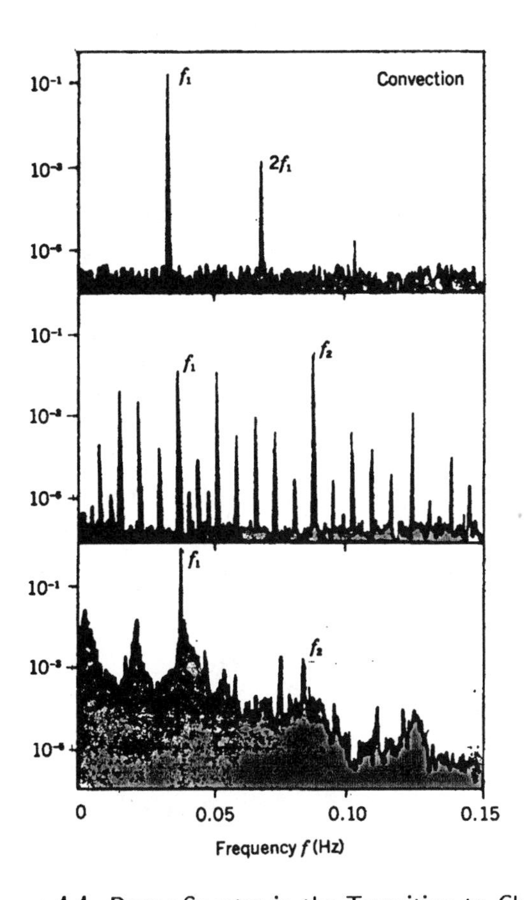

Figure 4.4. Power Spectra in the Transition to Chaos

state x_0 and a nearby initial state $x_0 + \Delta$. These two initial states give rise to trajectories whose difference w_t satisfies the equation

$$w_{t+1} = \mathcal{M}(x_t)w_t, \qquad w_0 = \Delta,$$

where $\mathcal{M} = [\partial f/\partial x]$ is the Jacobian matrix of f. Note that for one-dimensional maps, $\mathcal{M}$ is just the ordinary derivative $f'(x)$. If we introduce a norm $\|\cdot\|$, then the exponential rate of divergence of the trajectories can be expressed as

$$\sigma(x_0) = \lim_{t\to\infty} \frac{1}{t} \log \frac{\|w_t\|}{\|\Delta\|}\ .$$

This can be simplified to

$$\sigma(x_0) = \lim_{n\to\infty} \log |\hat{\lambda}_n(t)|,$$

where $\hat{\lambda}_n$ represents the largest (in magnitude) characteristic value of the matrix

$$A_n = [\mathcal{M}(x_n) \times \mathcal{M}(x_{n-1}) \times \cdots \times \mathcal{M}(x_1)]^{1/n}.$$

Here the points $x_1, x_2, \ldots, x_n$ are any n successive points arising from the iteration of f. In the special case of a one-dimensional map f, this representation simplifies to

$$\sigma = \lim_{n \to \infty} \frac{1}{n} \sum_{i=1}^{n} \log \left| \frac{df(x_i)}{dx} \right|.$$

The *Lyapunov exponent* σ represents the rate at which the two trajectories are diverging when they start close together. When $\sigma > 0$, we have a chaotic orbit. Except for a set of measure zero, σ is independent of the initial state x_0. If $\sigma < 0$, there is a stable cycle and the orbit, after an initial transient, is periodic. It's interesting to note that near the critical value α_∞ at which chaotic motion begins (i.e., where σ becomes positive), it can be shown that

$$\sigma \approx |\alpha - \alpha_\infty|^\eta,$$

where $\eta = \log 2 / \log \delta$, with δ again being the Feigenbaum constant. So the sign of the Lyapunov exponent provides us with another test for the onset of chaos.

- *Correlation Dimension:* The bifurcation diagram in Fig. 4.2 makes it transparently clear that in the chaotic region the trajectory of the logistic map doesn't spend equal amounts of time in all parts of the attractor. Rather, it visits some regions far more frequently than others. Since completely random motion would involve the system's trajectory covering the attractor uniformly, this observation provides the starting point for our next test of chaos.

Let $\mathcal{S} = \{x_i\}$, $i = 1, 2, \ldots, N$ be a finite collection of points lying on the attractor of a dynamical system whose state space is of dimension n, i.e., $x_i \in R^n$. For arbitrary $\epsilon > 0$, define the *correlation integral* $C(\epsilon)$ of the points to be

$$C(\epsilon) = \lim_{N \to \infty} \frac{1}{N^2} \times \{ \text{ number of pairs } (i, j) \text{ such that the distance } |x_i - x_j| < \epsilon \}.$$

Intuitively, we see that this integral measures the probability that the sequence $\mathcal{S}$ and its time-shifted version are within a distance ϵ. Simple geometrical arguments suggest that $C(\epsilon)$ should scale as ϵ^d for some $d > 0$. The number d is called the *correlation dimension* of the system. More specifically,

$$d = \lim_{\epsilon \to 0} \frac{\log C(\epsilon)}{\log \epsilon}.$$

It's not hard to see that if the "signals" $\mathcal{S}$ are pure random noise, then the points of $\mathcal{S}$ will spend the same amount of time in each part of the attractor. In that case, we will have $C(\epsilon) \sim \epsilon^n$. But if $\mathcal{S}$ comes from deterministic mechanism like the logistic map, then d will be strictly less than n. And, in fact, in this chaotic case d will not only be less than n, but also nonintegral. As a result, the correlation dimension d offers a simple, computable way to distinguish between deterministic and stochastic randomness. For example, for the logistic map with $\alpha = 3.57$, the correlation dimension satisfies the bounds $0.492 < d < 0.502$, confirming that this map reflects genuine deterministic rather than stochastic randomness.

The usual situation in practice is that we have a time series of data that is scalar, i.e., $n = 1$. For a variety of technical reasons, some of which we'll discuss in Section 12, it's desirable to create a vector time series from such a data sequence. We do this by creating a window of size m, thereby generating a sequence of m-dimensional vectors from the original scalar series. More precisely, suppose the given time series consists of the measured points $\mathcal{S} = \{x_i\}, i = 0, 1, 2, \ldots, N$, where each x_i is a real number. Then we define the *m-window* at x_i as $X(m,i) \doteq (x_i, x_{i+1}, \ldots, x_{i+(m-1)})$ for each i and for $m = 1, 2, \ldots$. Clearly, $X(m,i) \in R^m$. So we have succeeded in embedding the original scalar series into R^m. The number m is called the *embedding dimension* for the time-series $\mathcal{S}$.

Using the above embedding trick with a fixed value of m, we obtain a sequence of vectors in R^m as we vary i. If the original scalar time series is purely random, then we would expect the set of points $\{X(m,i)\}$ to fill R^m uniformly as we let $i \to \infty$. The correlation dimension d would then approach m. On the other hand, if there is deterministic structure in the data, the correlation dimension d should converge to a number less than m as $i \to \infty$. It's clear, however, that the estimate of the correlation dimension obtained by this embedding procedure depends on the choice of the embedding dimension m (as well as upon ϵ, the tolerance level between two points in R^m). If we take m too small, then we will miss some of the deterministic structure in the data; taking m too large wastes a lot of computer time in searching through a space of too high dimension. We will speak about these matters in more technical detail in Section 12. For now, we ask the reader to take on faith the fact that everything works out just as intuition says it should.

The correlation dimension gives us a yes-or-no answer as to whether a system is chaotic. But sometimes we'd like a more precise estimate of just exactly *how* chaotic the process is. For this, we have another test based upon the loss of information as such processes unfold.

• *The Kolmogorov Entropy:* One of the hallmarks of a chaotic system is that any imprecision in the initial state x_0 is magnified during the course of the system's evolution. This fact suggests still another test for the presence of chaos.

Consider an n-dimensional system moving on a strange attractor. Let's partition state space into boxes of size ϵ^n, and agree to measure the state of the system at time intervals of length τ. Let the quantity $P_{i_0 i_1 \ldots i_n}$ represent the probability that the system state is found in the box i_0 at time $t = 0$, in the box i_1 at time $t = \tau$ and, more generally, in the box i_k at time $t = k\tau$. Thus, we can code any trajectory simply by keeping track of the sequence of boxes it moves through. As a result, the quantity

$$K_N = - \sum_{i_0 \ldots i_N} P_{i_0 \ldots i_N} \log P_{i_0 \ldots i_N}$$

represents the amount of information needed to locate the system on the trajectory $i_0 \ldots i_N$ up to accuracy $O(\epsilon)$. So we see that the difference $K_{N+1} - K_N$ is the *additional* information required in order to predict in which cell the system will be in at time $t = (N+1)\tau$ if we know it was previously in the cells $i_0 \ldots i_N$. Therefore, the quantity $K_{N+1} - K_N$ is a measure of the *loss* of information about the system state in going from a time interval of length N to one of length $N + 1$.

We define the *K-entropy* K to be the average rate of information loss. More formally,

$$K = \lim_{\tau \to 0} \lim_{\epsilon \to 0} \lim_{\nu \to \infty} \frac{1}{\nu \tau} \sum_{i_0 \ldots i_N} P_{i_0 \ldots i_N} \log P_{i_0 \ldots i_N}.$$

The entropy $K = 0$ for regular motions, while $K = \infty$ for purely random systems. Chaotic processes are in the in-between region $0 < K < \infty$. The Exercises show that the K-entropy is proportional to the time interval over which the state of a chaotic system can be predicted. Given the connection between the spreading of trajectories as measured by the positive Lyapunov exponents and the average loss of information as measured by K, it should come as no surprise to the reader to see the connection

$$K = \sum_i \text{positive } \lambda_i.$$

In particular, for scalar chaotic processes, the K-entropy and the Lyapunov exponent coincide.

Note that to compute the K-entropy by the procedure sketech above, it's necessary to know the joint probability functions $P_{i_0 \ldots i_N}$. A direct computation of these quantities may be tedious. However, there are simpler algorithms making use of the same information needed to compute the correlation integral discussed earlier. The reader is invited to consult the chapter Notes and References for citations to this work.

• *Invariant Distributions:* Another approach to the characterization of chaos is via the notion of the invariant distribution of the map f. We say that $P(x)$ is an *invariant distribution* (*invariant measure, probability distribution*) if $P(x) = f(P(x))$, with $\int P(x)\,dx = 1$. Generally speaking, there will be many invariant distributions for a given f; however, we can single out a unique distribution by demanding that the space average over the initial conditions x_0 equal the time average over a trajectory $\{x_t\}$.

Basically, what the invariant distribution measures is the likelihood of a given point $x^* \in I$ appearing during the course of iteration of the map f. So, for instance, if the parameter α corresponds to a stable cycle $\mathcal{C}$ of period T, then the invariant distribution for f at this value of the parameter consists of the function

$$P(x) = \begin{cases} \frac{1}{T}, & \text{for } x \in \mathcal{C}, \\ 0, & \text{otherwise.} \end{cases}$$

Except for a set of measure zero, every x_0 leads to this distribution under iteration of the map f. If α is a parameter leading to aperiodic motion, again almost all initial conditions lead to a unique equilibrium distribution. However, in this case the distribution may be discontinuous in x, and is typically nonzero over a continuum of x values. How can we construct the distribution P?

The construction of $P(x)$ makes use of the fact that for maps f having a single hump, there are always two inverse points for a given value of $P(x)$. Referring to Fig. 4.5, in which the intervals dx_1 and dx_2 inverse to the interval dx are displayed by reflecting dx in the line $f(x) = x$, we see that the number of trajectory points falling into the interval dx equals the number falling into the two intervals dx_1 and dx_2. Thus,

$$P(x)\,dx = P(x_1)\,dx_1 + P(x_2)\,dx_2.$$

Now, by noting that

$$\frac{dx}{dx_i} = \left|\frac{df}{dx}\right|_{x_i}, \qquad i = 1, 2,$$

we obtain the functional equation for $P(x)$ as

$$P(x) = \frac{P(x_1)}{|df/dx|_{x_1}} + \frac{P(x_2)}{|df/dx|_{x_2}}.$$

This equation can be solved by successive approximations to find the invariant distribution P.

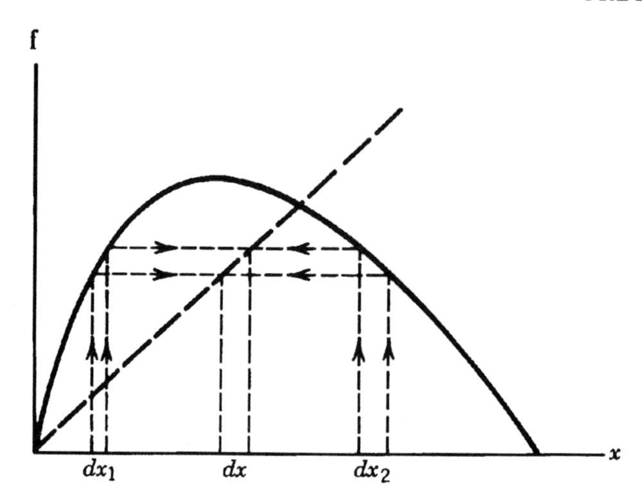

Figure 4.5. Construction of the Distribution P(x)

Example: The "Tent" Map

We consider the piecewise-linear map

$$f(x) = \begin{cases} 2x, & \text{for } 0 \leq x \leq \frac{1}{2}, \\ 2 - 2x, & \text{for } \frac{1}{2} \leq x \leq 1. \end{cases}$$

It's easy to see that for this map the functional equation for $P(x)$ is

$$P(x) = \frac{1}{2}\left[P\left(\frac{x}{2}\right) + P\left(1 - \frac{x}{2}\right)\right],$$

which has the obvious solution $P(x) = 1$. This computation validates our earlier claim that the "tent" map has iterates that are indistinguishable from those obtained by the flipping of a fair coin, since this distribution of states is identical to that obtained by dividing the x-axis into two halves, labeled H or T, and then marking the states in accordance with the outcome of the coin toss. We can conclude chaotic motion from this tent map since the distribution P is continuous.

Here is another view of the problem of computing the invariant distribution, this time making use of integral, rather than functional, equations. The invariant distribution $P(x)$ determines the density of the iterates of a map f, which for simplicity, we take to be unimodular on the unit interval. By definition,

$$P(x) = \lim_{n \to \infty} \sum_{i=0}^{n} \delta\{x - f(x_0)\},$$

where $\delta(\cdot)$ is the usual Dirac delta function. As noted earlier, we assume ergodicity, which means that the invariant distribution does not depend on the initial state x_0.

After one iteration, the initial point x_0 evolves to $f(x_0)$. This means that a delta-function distribution $\delta(x - x_0)$ evolves after one time step to $\delta\{x - f(x_0)\}$, which can be written as

$$\delta\{x - f(x_0)\} = \int_0^1 \delta\{x - f(y)\}\delta\{y - x_0\}\,dy.$$

This expression can be generalized to the evolution of an arbitrary density at time n, yielding the integral relation

$$P_{n+1}(x) = \int_0^1 P_n(y)\delta\{x - f(y)\}\,dy.$$

Letting $n \to \infty$, we obtain the invariant distribution $P(x) = \lim_{n\to\infty} P_n(x)$. In integral equation form, we can express $P(x)$ as the solution of the homogeneous Fredholm integral equation of the second kind

$$P(x) = \int_0^1 \delta\{x - f(y)\}P(y)\,dy.$$

It's also possible to obtain the Lyapunov exponent for f directly from the invariant distribution P as

$$\sigma = \int P(x)\log\left|\frac{df}{dx}\right|\,dx.$$

For the tent map we have

$$\sigma = \int_0^1 \log 2\,dx = \log 2 > 0,$$

again showing that the motion is chaotic.

• *The Brock-Dechert-Scheinkman (BDS) Statistic:* Our last test for hidden structure is a statistical one, making use of a striking property of chaotic processes—their invariance under linear transformations. This means that if one carries out a linear transformation of chaotic data, both the original and the transformed data have the same correlation dimension and the same Lyapunov exponents.

Recognizing that the correlation integral defined above depends on the number of data points N, the tolerance level ϵ and the embedding dimension m, let's write it as $C(m, \epsilon, N)$. By appealing to ergodic theorems, it can be shown that as the number of data points becomes large, the limit

$$\lim_{N\to\infty} C(m, \epsilon, N) = \Pr\,\{\|X(m,t) - X(m,s)\| < \epsilon\}$$

exists. Let's call this limiting quantity $C(m, \epsilon)$. In this expression, the quantities $X(m,t), X(m,s)$ are independent draws from the joint cumulative distribution function Pr $\{x(m,r) < X(m,r)\}$, with $x(m,r) = \{x_r, x_{r+1}, \ldots, x_{r+m-1}\}$. Here x is a random variable and X is a draw from the cumulative distribution of x.

If the sequence $\{X(m,t)\}$ is purely random (i.e., IID), then we have

$$\log C(m, \epsilon) = m \log C(1, \epsilon).$$

Consequently, for purely random processes we must have the correlation dimension

$$d(m, \epsilon) = \left(\epsilon \frac{dC(m, \epsilon)}{d\epsilon}\right) \left(\frac{1}{C(m, \epsilon)}\right) = m$$

for all embedding dimensions m.

These observations suggest the following statistical test for randomness. First form the quantity

$$B(m, \epsilon, N) = N^{1/2}[C(m, \epsilon, N) - C(1, \epsilon, N)^m].$$

Brock, Dechert and Scheinkman have shown that $B(m, \epsilon, N)$ converges to the normal distribution with mean zero and variance V as $N \to \infty$, where the variance V can be estimated from the data. Denote this estimate by the quantity $V(m, \epsilon, N)$. Then if the data is purely random, the statistic

$$W(m, \epsilon, N) = B(m, \epsilon, N)/\sqrt{V(m, \epsilon, N)}$$

converges in distribution to the standard normal distribution $N(0, 1)$ as $N \to \infty$.

We now have an easy test for randomness: Compute the quantity W. If $|W|$ is large, then we can conclude that the data is *not* random at a significance level that can be read off from a standardized normal table.

With these various tests at our disposal for checking the possibly chaotic behavior of dynamical systems, let's look at a couple of examples involving real-world data.

Example 1: Daily Mood Fluctuations

Social psychologist Edward Hannah was interested in the following question: Is it the case that the great variability we commonly observe in people's moods is due simply to inherent randomness and imprecision of measurement? Or could it be the case that there is some underlying chaotic dynamics that generates some of that variability? What Hannah was interested in determining was the nature of the fluctuations of behavioral and psychological phenomena such as the waxing and waning of motivation and effort, the ups and downs of interpersonal relationships and, in general, the changes in human emotions and mood.

To study this issue, Hannah examined the daily mood fluctuations in three male subjects in their early-20s. These subjects were asked to monitor thier moods every 15 minutes (one subject) or every 30 minutes (the other two) during their waking hours over a two-week period. To measure these moods, Hannah used a simple 24-item "mood checklist," enabling him to quantify things like affect and vigor, as well as overall mood. Figure 4.6 shows the time series of mood fluctuations for the three subjects, along with a time series of purely random data given for the sake of comparison.

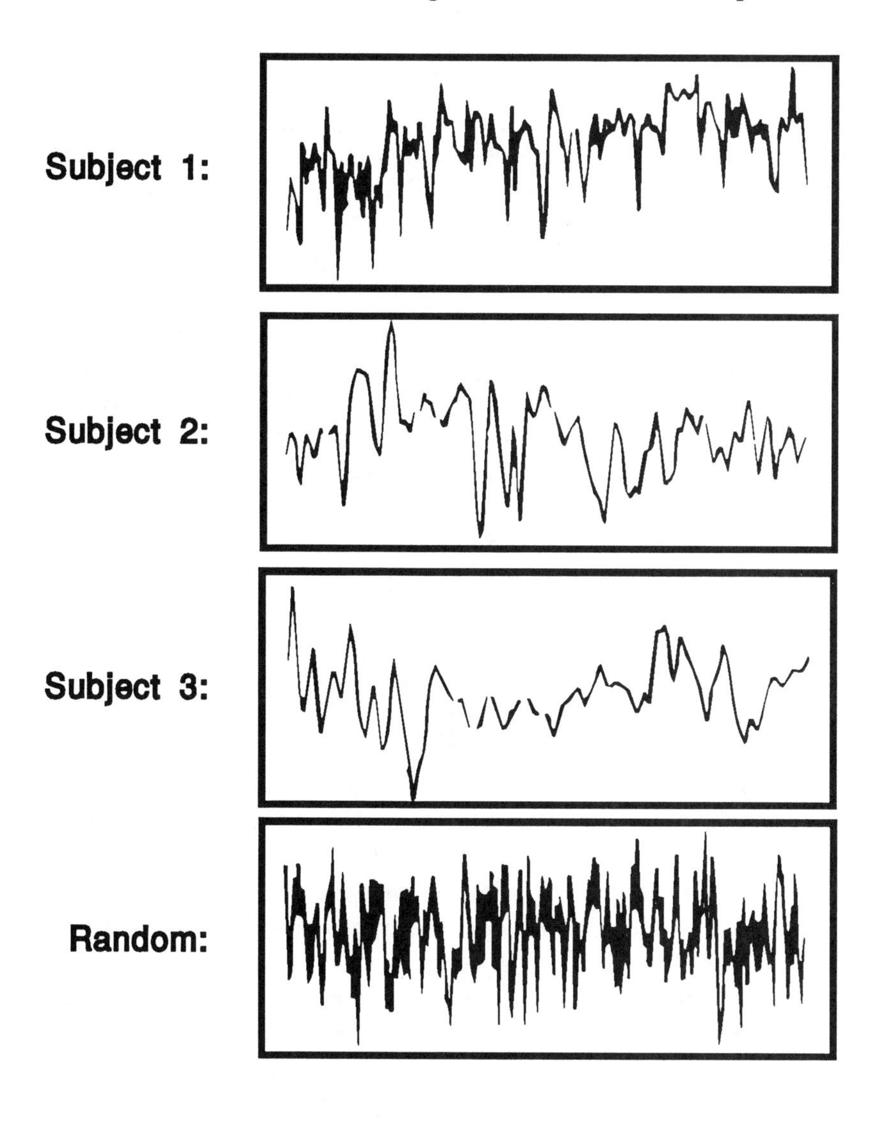

Figure 4.6. Mood Data for Three Subjects

To test this experimental data for hidden structure, Hannah computed the correlation dimension and the Lyapunov exponent for each subject, as well as for the random series. Using an embedding dimension $m = 4$, he obtained the following results:

Table 4.1. Hannah's Experimental Results

	Correlation Dimension d	**Lyapunov Exponent λ**
Subject 1	1.68 ± 0.06	0.966
Subject 2	1.84 ± 0.06	0.522
Subject 3	1.71 ± 0.03	0.527
Random Data	4.01 ± 0.88	0.018

The calculated quantities show just exactly what one would expect for processes governed by deterministic chaos. The correlation dimension estimates are substantially less than the embedding dimension and nonintegral for the subjects—but not for the random data. Moreover, the Lyapunov exponent was always positive for the subjects, as opposed to its being near zero for the random data.

Hannah has argued that it's natural to expect the mood system to display chaotic features, since it's well known that chaotic processes are the most efficient and effective way to dissipate effects from the environment. So, if we want mood systems to be sensitive to the environment *and* effective at dissipating environmental influences, then it's reasonable to think that the governing dynamics would show strongly chaotic behavior—just as the experiment suggests.

Example 2: Gold Price Movements

Mood fluctuations might well be related to another type of fluctuation that humans around the world pay careful attention to each day: the fluctuation of stock and commodity prices on speculative markets. In a 1988 study, Murray Frank and Thanasis Stengos decided to use the fluctuation of the price of gold as a basis for testing one of the most cherished assumptions of financial theorists the world over, the so-called *Efficient Markets Hypothesis (EMH).*

The EMH comes in many flavors, but the one that interests us here is the weakest version, which sometimes masquerades under the rubric of the *Random Walk Hypothesis.* In rough terms, the hypothesis asserts that a time series formed of price *differences* is a completely random sequence. In short, knowledge of the past price history of a stock or commodity gives no information of any kind about what that security will do in the next time period. If the random-walk idea is true, then of course all schemes for beating the market based upon statistical processing of price histories are doomed to failure, at least in the long-run; there can be no such scheme for consistently beating the market. Frank and Stengos devised an experiment to test this contention in the gold market by looking into actual price data for hidden chaotic structure.

Let p_t be the price of gold at period t, and define the relative price difference $r_t = (p_t - p_{t-1})/p_{t-1}$. The Frank and Stengos study revolves about the linear regression equation for predicting the sequence $\{r_t\}$. The regression equation is

$$r_t = \beta_0 + \beta_1 r_{t-1} + u_t,$$

where u_t is a mean-zero, unit variance normal random error term. Statistical theory asserts that if the gold market is efficient, then we should have $\beta_0 = \beta_1 = 0$, i.e., the price differences form a normally-distributed random sequence. Using standard regression techniques on actual gold price data, the estimates the regression coefficients turned out to be $\beta_0^* = 0.0004$ and $\beta_1^* = -0.606$ with a high confidence level. Thus, the standard statistical tests strongly suggest that the gold market is indeed efficient. But not so fast!

The situation changed dramatically when Frank and Stengos used the tests for chaos on the same gold price data. Computing the correlation dimension of the data versus the dimension of a randomly-generated series of numbers, it turned out that the correlation dimension was not only non-integral, but also consistently on the order of half as great as that obtained for the random data. Furthermore, Frank and Stengos estimated the K-entropy as $K = 0.15 \pm 0.07$. Both of these calculations suggest strongly that there is some underlying nonlinear structure in the gold price data that is not being picked up by the linear methods used to estimate the parameters β_0 and β_1 in the regression model.

With this financial example as our cue, the next two sections will look at ways in which chaotic processes can crop up in even the simplest kind of problems in the realm of macroeconomics and human decisionmaking, both within and outside the context of speculative markets.

Exercises

1. Obtain a time series of data from some source, e.g., the stock market quotes in your daily newspaper, and test that series for chaotic behavior using all of the procedures discussed in this section.

2. Let the system Lyapunov exponents be given by $\sigma_1 \geq \sigma_2 \geq \ldots$, with j being the largest integer such that

$$\sum_{i=1}^{j} \sigma_i \geq 0.$$

Prove that the correlation dimension d satisfies the inequality

$$d \leq j + \frac{1}{|\sigma_{j+1}|} \sum_{i \leq j} \sigma_i.$$

3. Show that the invariant distribution for the logistic map in the fully chaotic regime ($\alpha = 4$) is given by

$$P(x) = \frac{1}{\pi \sqrt{x(1-x)}}.$$

4. Prove that for one-dimensional maps, the K-entropy and the Lyapunov exponent coincide.

5. Consider the tent map. (a) Show that after n iterations of this map, an interval originally of length ϵ increases to one of length $L = \epsilon e^{\lambda n}$, where λ is the Lyapunov exponent. (Thus, when L becomes larger than 1 we can no longer locate the trajectory in the unit interval, and can say only that it has a probability $P(x)\,dx$ of being in an interval $(x, x + dx)$, where $P(x)$ is the invariant density.) (b) Show that the preceding result implies that precise prediction about the state of the system is possible only for times n that are less than T_c, where $T_c = \frac{1}{\lambda} \log(1/\epsilon) = \frac{1}{K} \log(1/\epsilon)$. (This shows that the K-entropy is proportional to the time interval over which the state of a chaotic system can be predicted.)

6. (a) Prove that for an n-dimensional Hamiltonian system, the Lyapunov exponents come in pairs and satisfy the relations

$$\sum_{i=1}^{2n} \lambda_i(x) = 0,$$

$$\lambda_i(x) + \lambda_{2n-i+1}(x) = 0.$$

(b) Show that each $\lambda_i(x)$ is a constant of the motion, i.e., $d\lambda_i(x)/dt = 0$.

8. Economic Chaos

The neoclassical theory of capital accumulation provides an explanation of investment cycles that lies exclusively in the interaction of the propensity to save and the productivity of capital—provided enough nonlinearities and a production lag are present in the system. This theory can be used to establish the existence of irregular economic oscillations that need not converge to a periodic cycle. Moreover, because they are unstable, errors in parameter estimation or errors in the initial conditions, however minute, can accumulate quickly into substantial forecasting errors. Such irregular fluctuations can emerge after a period of apparently balanced growth, so that the future behavior of the model cannot be anticipated from its past.

Although it cannot be proved that any real economies are chaotic in the foregoing sense, the example we give here shows that irregular fluctuations of a highly unstable nature constitute one characteristic mode of behavior in dynamic economic models, and that such fluctuations may emerge out of conventional economic theories. It's also of interest to note that the past behavior of a nonlinear system may be no guide whatsoever for inferring even *qualitative* patterns of change in its future, since the type of model we are discussing here may evolve through apparently different regimes even though no *structural* changes have occurred. Now let's give concrete examples to illustrate some of these points.

Example 1: Capital Accumulation

Assume that the production function of the economy is given in the homogeneous Cobb-Douglas form $f(k) = Bk^\beta$, where $B, \beta > 0$. Here k is the capital/labor ratio of the economy. The capital accumulation dynamics are

$$k_{t+1} = \frac{f(k_t) - h(k_t)}{(1+\lambda)},$$

where $h(\cdot)$ is the consumption wealth function and λ is the population growth rate. Per capita consumption depends upon wealth, interest rates and income, but we can use the production function to eliminate income from consideration. We now equate marginal productivity of capital $f'(k)$ with the interest rate to arrive at the function h. These arguments give

$$h(k) = (1-s)f(k),$$

where s is the marginal propensity to save. Putting all these remarks together, the dynamics for capital accumulation become

$$k_{t+1} = \frac{sBk^\beta}{1+\lambda}.$$

For $\beta > 0$, investment cycles cannot occur. Instead growth converges to an equilibrium capital/labor ratio $k^* = [sB/(1+\lambda)]^{1/(1-\beta)}$.

To illustrate how unstable oscillations and chaos might arise in such a model, suppose we introduce a productivity inhibiting effect into the dynamics by changing the production function to

$$f(k) = Bk^\beta(m-k)^\gamma, \qquad \gamma > 0.$$

Since now output falls rapidly as $k \to m$, we can think of the term $(m-k)$ as representing the harmful effects upon output of an excessive concentration of capital (e.g., too "fat and lazy" to work). Keeping a constant savings rate s, the new capital accumulation dynamics are

$$k_{t+1} = Ak_t^{\beta}(m - k_t)^{\gamma},$$

where $A \doteq sB/(1+\lambda)$.

For small values of A and k_0, growth will be monotonic, converging to a stable equilibrium. For A in the range

$$\frac{\beta m}{\beta+\gamma} < A\left(\frac{\beta}{\beta+\gamma}\right)^{\beta}\left(\frac{\gamma}{\beta+\gamma}\right)^{\gamma} m^{\beta+\gamma} \leq m,$$

the system displays period-doubling bifurcations. Let A^* be the value of the parameter A for which equality is attained on the right side of the above expression. Then for $A > A^*$, irregular investment and growth cycles occur and chaotic behavior sets in. In fact, it can be shown that there exists an entire interval of values of the productivity multiplier B for which chaotic behavior will occur.

A somewhat sobering aspect of the above result is that it provides a basis for skepticism about *any* modeling effort that relies upon parameter estimation unless it can be demonstrated beforehand that the true parameter values do not lie in the chaotic region. If this is not the case, then there is little hope that observations on the past behavior of the system will provide a basis for identifying the parameter values, and a model based upon such a spurious identification will almost certainly be quite useless.

The foregoing example has been based upon purely theoretical considerations, quite independent of any empirical evidence from measured data on investment cycles. Let's now look at a closely related kind of economic phenomenon, the business cycle, from an empirical perspective, and employ some of the chaos tests developed earlier to see if there is any interesting nonlinear structure lurking in the data.

Example 2: Chaos in Monetary Aggregates

The Federal Reserve Bank keeps track of many different types of monetary aggregates: currency in circulation, traveler's checks, Eurodollars, money market deposits, savings deposits T-bills, commercial paper and so forth. The Fed uses four different simple-sum aggregate indexes, M1, M2, M3 and L as a basis for many of its decisions on interest rates and money supply. A study by Ping Chen tested 12 time-series of data involving different monetary indexes, including simple-sum aggregates (SSM), as well as Divisia monetary demand aggregates (DDM) and Divisia monetary supply aggregates (DSM). Each of these series yielded about 800 data points

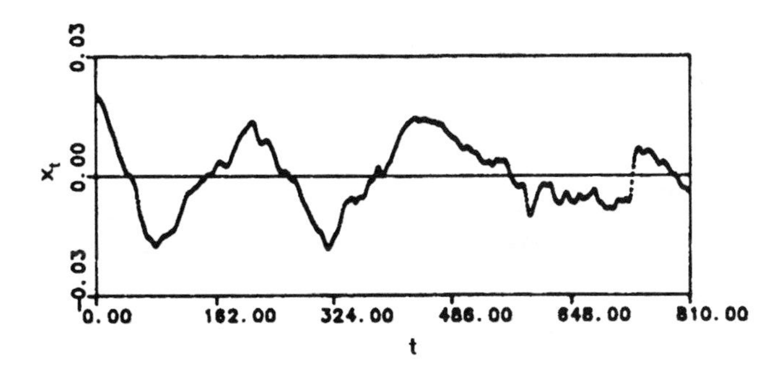

Figure 4.7. Log-Linear Detrended SSM2 Data

over the years 1969-1984. For brevity, here we report only the results Chen obtained from the SSM2 and DDM2 data.

Since the monetary indexes all display growth trends associated with the general growth of the economy, we want to factor out this part of the change in indexes before searching for deterministic randomness in the data. In order to preserve long-term correlations in the economic fluctuations, log-linear detrending is often employed. If S_t is the raw data at time t, then the log-linear detrended data is

$$X_t = \log S_t - (k_0 + k_1 t),$$

where k_0 is a constant representing the intercept point (i.e., $k_0 = \log S_0$), while k_1 is the constant growth rate. Chen's study looked for structure in the detrended SSM2 and DDM2 data, which is shown above in Fig. 4.7 for the SSM2 data when k_1 is four percent.

To test for chaos in the SSM2 and DDM2 data, Chen used an embedding dimension $m = 5$ and calculated the largest Lyapunov exponent over a 180-week interval. The results are shown in Fig. 4.8, where we see chaos indicated by the fact that this largest Lyapunov exponent consistently remained greater than zero.

The next test involved computing the correlation dimension of the data for various embedding dimensions and for different values of the tolerance parameter ϵ. The reason for experimenting with a range of ϵ values is that if the tolerance is taken too large, then the correlation integral becomes too saturated at the total number of data points. But if ϵ is chosen too small, then the computation picks up too much of the experimental noise in the data. Figure 4.9 shows a log-log plot of the correlation integral $C(\epsilon)$ versus ϵ for embedding dimensions $m = 2, 3, \ldots, 6$ using the SSM2 data. The plots rotate downward and to the right as m increases. The correlation dimension is taken to be the slope of these curves at their point of convergence, which appears to be when $m \approx 5$. This slope is about 1.5, which being both non-integral and substantially less than the embedding dimension again strongly

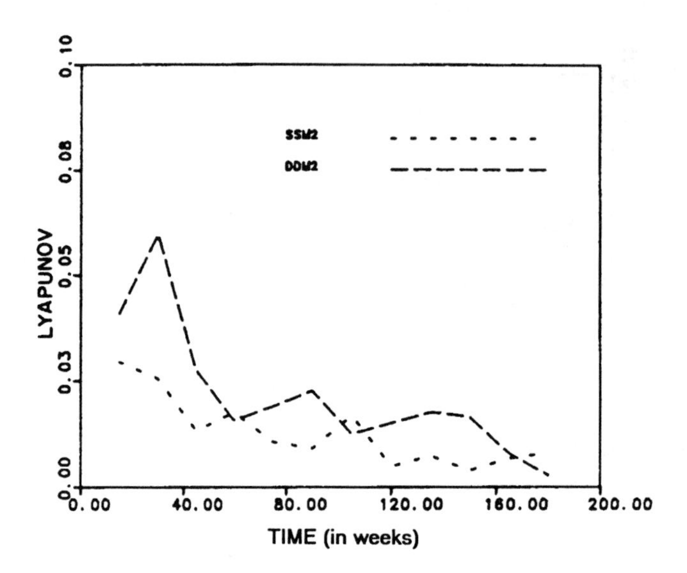

Figure 4.8. The Largest Lyapunov Exponent for the SSM2 and DDM2 Data

suggests the possibility of deterministic randomness in the data. There are many more results of this type in Chen's work, which the interested reader can find cited in the Notes and References.

The results presented in this section have shown that chaotic behavior can naturally arise in economic processes. Nevertheless, one should be cautious in drawing any general conclusions from this work. Other studies on other kinds of economic processes have found scanty evidence of deterministic randomness in the data, or at least have not found evidence convincing enough to cause mainline economists to throw out the existing models. So in our quest for the truly chaotic, let's look into some other areas of human endeavor where deterministic randomness appears to play a role.

Exercises

1. Suppose the first N points of an economic data set with $2N$ points are independent and uniformly distributed in $(-\frac{1}{2}, \frac{1}{2})$, while the second N points are independent and uniformly distributed in $(-\alpha, \alpha)$. Let $\frac{1}{2} > \epsilon > 2\alpha$, and write

$$p_1 = \Pr \{|x_i - x_j| < \epsilon,\ j < i \leq N\},$$
$$p_2 = \Pr \{|x_i - x_j| < \epsilon,\ i < N < j\},$$
$$p_3 = \Pr \{|x_i - x_j| < \epsilon,\ N < i < j\}.$$

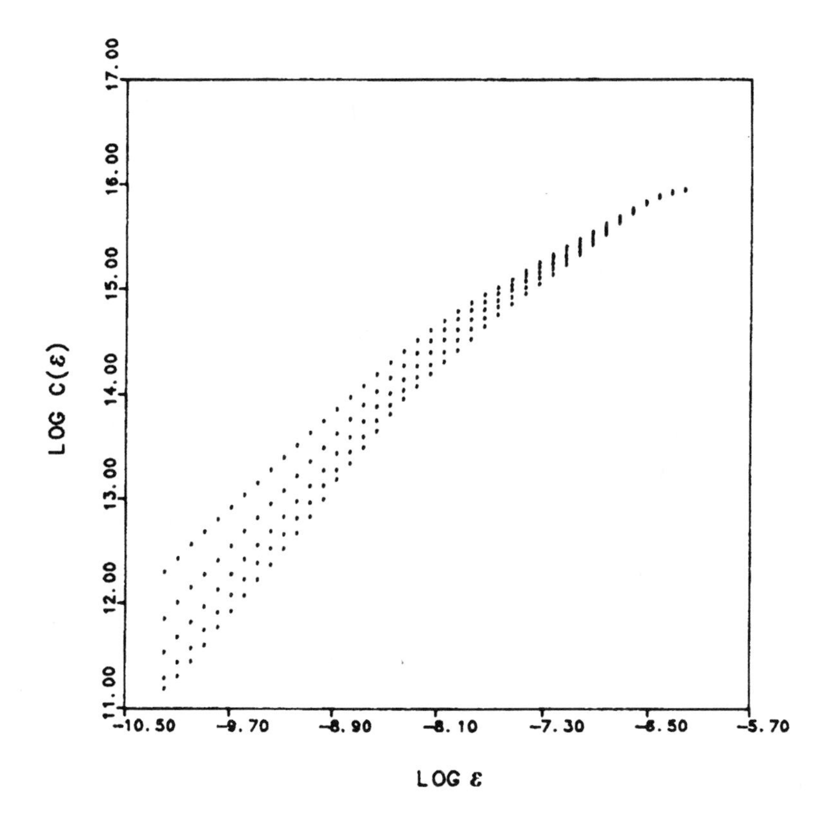

Figure 4.9. The Correlation Integral log $C(\epsilon)$ vs. log ϵ for the SSM2 Data

(a) Show that $p_1 = 2\epsilon - \epsilon^2, p_2 \approx 2\epsilon, p_3 = 1$ and that the correlation integrals are

$$C(1, 2N, \epsilon) \approx \frac{1}{4}(p_1 + 2p_2 + 1) \approx \frac{1}{4}(6\epsilon - \epsilon^2 + 1),$$
$$C(2, 2N, \epsilon) \approx \frac{1}{4}(p_1^2 + 2p_2^2 + 1) \approx 3\epsilon^2 + \epsilon^4 - \frac{1}{4}\epsilon^3 + \frac{1}{4}.$$

(b) Thus, prove that the quantity $C(2, 2N, \epsilon) - [C(1, 2N, \epsilon)]^2$ is greater than zero for all $\epsilon \in (0, 1)$. Hence, conclude by the BDS test that there might be interesting nonlinearities lurking at the heart of this set of data. In short, the data cannot be explained by linear models.

9. Beer, Bears, Bulls and Chaos

One of the characteristic features of human activity distinguishing it from the doings of nature is the need to continually take conscious decisions. Whether it's in the corporate boardroom, on the battlefield or in the local supermarket, we're called upon to make choices about almost every aspect of our daily lives. And in a large number of cases, the effect of those decisions is not determined solely by the individual's choice, but rather by how that choice interacts with the choices made by many others. Profit or loss on our decisions about stock transactions or romantic attachments are good examples of this kind of interactive effect. So we may legitimately ask if these kinds of interactions can give rise to the sort of weird, wonderfully wacky chaotic behavior we've seen so amply displayed in earlier examples in this chapter.

To address this question, in this section we'll examine two fascinating studies in human decisionmaking—one in the lab, the other on the street—each of which illustrates the boundless human capacity for turning order (or perhaps I should say *orders)* into chaos. Our first case involves the famous Beer Game, which has been played at the Sloan School of Management at MIT by generations of MBA students over the past thirty years.

Example 1: The Beer Game

In order to reach a widespread market, it's customary for breweries (and other industries, as well) to employ a hierarchical distribution system with dealers at several different levels. The basic hierarchy consists of the following levels:

- A *distributor,* who receives the beer from the factory and ships it to the main markets;
- Regional *wholesalers,* who receive the beer from the distributor and allocate it to local outlets such as liquor stores and bars;
- *Retailers,* who then disperse the products to the end-consumers.

To guard itself against irregularities in demand and supply, each link in the distribution chain maintains a suitable level of inventory. Besides ensuring the availability of beer at the consumer level, the hierarchical distribution system is meant to facilitate swift restocking if a dealer's inventory runs low. The chain should also function as a buffer to protect the production line from fluctuations in consumer demand. Thus, seasonal and other low-frequency components of the demand variations should propagate towards the factory in a damped fashion.

Figure 4.10 shows the basic structure of the simplified beer distribution system we'll consider here. We assume there is only one inventory at each level. Orders for beer propagate from right to left, while products are shipped from stage to stage in the opposite direction. Since the processing of orders and the production and shipment of beer involve time delays, we assume a communication delay of one week (one time-period) from one stage to the next. In the same way, we assume it takes one week to ship

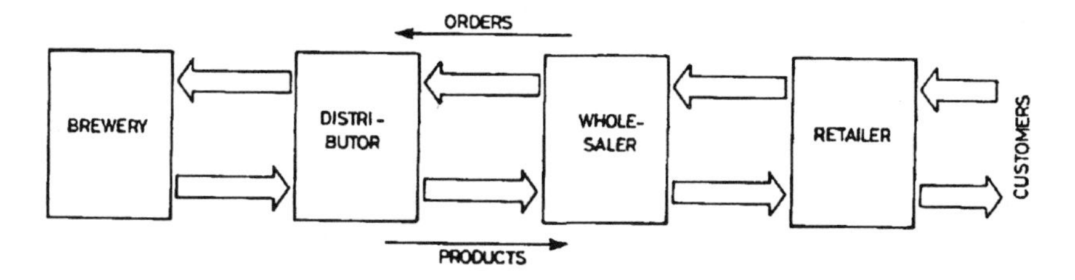

Figure 4.10. The Production-Distribution System in the Beer Game

beer between two sectors. The production time is taken to be three weeks. Finally, we take the production capacity of the brewery to be unlimited.

Each week customers order beer from the retailer, who ships the requested quantity out of inventory. Customer demand is exogeneous. In all the results reported here, customer demand is at a constant level of four cases of beer per week until week 5, at which time demand increases to eight cases per week. The demand is then maintained at this level for the rest of the process. The system is initialized with 12 cases of beer in each dealer's inventory.

In response to variation in customer demand and to other pressures, the retailer adjusts the order for beer that he places with the wholesaler. As long as the inventory is sufficient, the wholesaler ships the beer requested. Orders that cannot be met are kept in backlog until delivery can be made. Similarly, the wholesaler orders and receives beer from the distributor, who in turn orders and receives from the brewery. The rules of the game stipulate that orders must always be filled if they are covered by available inventory, and orders that have already been placed cannot be cancelled nor can deliveries be returned.

The goal of the participants is to minimize their cumulative costs over the duration of the game (usually 40 weeks). Due to the costs of holding inventory, stock levels should be kept as low as possible. On the other hand, failure to deliver immediately may force customers to look to alternative suppliers. For this reason, penalty costs are assessed for accumulating a backlog of unfilled orders. Therefore, each stock manager must attempt to keep his inventory at the lowest possible level, while at the same time avoiding a "stockout." If the inventory begins to fall below the desired level, extra beer must be ordered to rebuild the inventory. If stocks begin to accumulate because of a slackening in demand, the order rate must be reduced. In the experiments discussed here, inventory holding costs are taken to be $0.50 per case per week, while the costs of having backlogs are set at $2 per case per week.

The decision variable each week in all sectors is taken to be the amount of beer to be ordered from the immediate supplier. The participants can base their ordering decisions on all information locally available to them, i.e., the current value of their inventory/backlog, previous values of these variables, expected orders, anticipated deliveries, and so forth. In addition, the participants may utilize their overall conception of the way in which the distribution chain functions.

Figure 4.11 displays a typical result of a play of the Beer Game, showing the variation in the effective inventory for the different sectors, i.e., the actual inventory minus the backlog. (Note: A negative effective inventory represents a backlog of unfilled orders.)

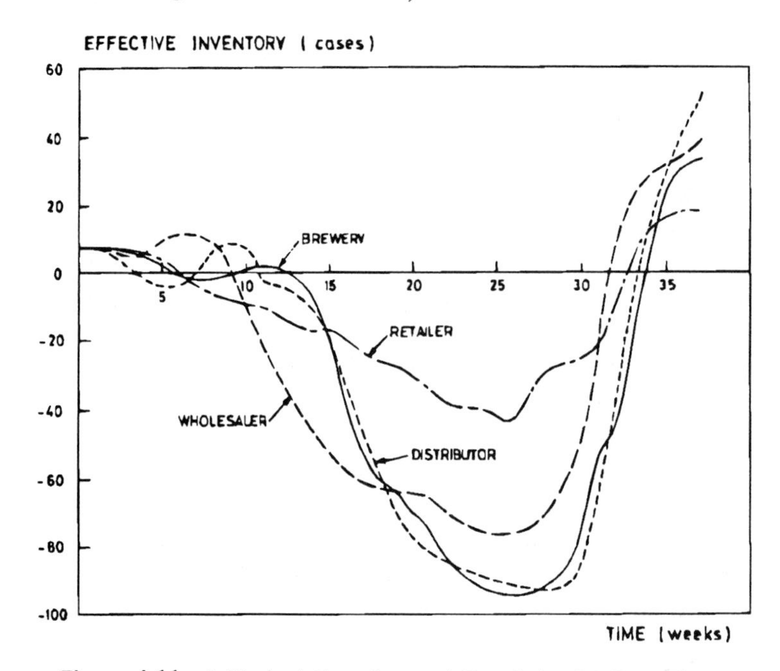

Figure 4.11. A Typical Experimental Result in the Beer Game

The usual results of a play of the Beer Game are characterized by large-scale oscillations growing in amplitude from retailer to wholesaler and from wholesaler to distributor. So by the time the original stepwise increase in customer orders reaches the brewery, it typically leads to an expansion of production by a factor of more than six. Another feature of these games is the increase in orders, which propagates in a wavelike fashion down the chain, depleting the inventories one by one until it's finally reflected at the brewery when the large surplus of orders placed during the out-of-stock period is produced. These features show clearly that there is a strong amplification mechanism at work in the system. At the same time, the behavior is restricted by various nonlinearities. For instance, we have a pronounced nonlinear relationship between the nonnegativity of orders and shipments. Together with the relatively high number of state variables, these nonlinearities admit an extraordinary variety of complex dynamic behaviors.

The amplification process observed in the beer distribution chain is connected with the built-in delays. Assume that a particular sector suddenly experiences a significant increase in demand. To discover whether the change in demand is of a more permanent character, players usually hesitate a little before adjusting their own orders by a similar amount, since the very purpose of an inventory is to absorb high-frequency components in the demand fluctuations. However, on account of this hesitation, and by virtue of the built-in communication and shipping delays, demand then exceeds inventory replacements for several weeks. So, during this period the inventory decreases. Thus, to build the inventory back up to its desired level, the players must increase the orders they place with their suppliers beyond the level of the immediate incoming orders. As the players come to realize that the increase in demand is of a more enduring nature, they generally increase their orders even more with an eye toward rebuilding their inventory.

The amplification phenomenon is a direct consequence of the structure of the distribution chain, in much the same way that the macro structure of an economy amplifies small fluctuations in the demand for consumer goods into significant variations in the demand for capital goods via the well-known accelerator effect. On the other hand, it's important to realize that the beer distribution chain can be operated in a stable manner. In fact, experience shows that many players are capable of doing just that, and only about a quarter of the participants end up using ordering policies that lead to deterministic chaos. Large-scale oscillations are always observed in the transient behavior, however, and in all cases the ordering decisions are highly suboptimal, incurring costs in excess of the minimal possible costs by a factor of more than 400 percent. Now let's look at a dynamical model that formalizes much of the empirically-observed features of the Beer Game.

The stock management problem can be thought of as a stock and flow structure involving ordering, acquisition, and shipment of stock units in accordance with decision rules used by the manager at each stage in the system. In addition to the inventory, the supply line of orders is a relevant stock variable. Thus, just as the inventory accumulates differences between acquisitions and shipments, the supply line of orders accumulates the differences between ordering and acquisition. Hence, the supply line represents orders that have been placed for products that have not yet been delivered. Shipments may depend on variables that are both internal and external to the system. These variables are also usually influenced by the stock itself. The acquisition rate is determined by the supply line and by the average acquisition lag, which we'll term AL. In general, AL will depend on the supply line itself, as well as on a number of other variables.

The following equations formalize what's been learned from the results of the Beer Game over the past thirty years. Firstly, we have the obvious condition that the orders must be nonnegative, i.e.,

$$O_t = \max(0, IO_t),$$

where IO_t denotes the indicated order rate at time t, i.e., the actual size of the order placed by the dealer.

We express the expected demand as the participant's forecast of incoming orders:

$$ED_t = \theta \cdot ORR_{t-1} + (1 - \theta) \cdot ED_{t-1}.$$

Here ED_t and ED_{t-1} are the expected demand at times t and $t-1$, respectively, ORR is the order receiving rate, and $\theta\,(0 \leq \theta \leq 1)$ is a parameter controlling the rate at which expectations are updated.

Stock adjustments are assumed to be linear in the discrepancy between the desired and the actual inventory. Thus, letting INV_t be the inventory level at time t, the adjusted stock level AS_t satisfies the relation

$$AS_t = \alpha_S(DINV - INV_t).$$

Here the stock adjustment parameter α_S is the fraction of the discrepancy ordered in each round. Because the participants lack the time and information to determine the optimal inventory level, the desired inventory $DINV$ is taken to be constant throughout the course of the process, although $DINV$ may vary from manager to manager. In a similar manner, we can express the supply line adjustments as

$$ASL_t = \alpha_{SL}(DSL - SL_t),$$

where DSL is the desired supply line, and α_{SL} is the fractional supply line adjustment rate.

Defining $\beta = \alpha_{SL}/\alpha_S$ and $Q = DINV + \beta \cdot DSL$, the expression for the indicated order rate becomes

$$IO_t = ED_t + \alpha_S\,(Q - INV_t - \beta \cdot SL_t).$$

Since $DINV$, DSL, and β are all non-negative, $Q \geq 0$. Further, it is unlikely that participants place more emphasis on the supply line than on the inventory itself. The supply line does not directly affect the costs, nor is it as directly relevant as the inventory. Therefore, we assume $\alpha_{SL} \leq \alpha_S$ and $\beta \leq 1$. The quantity β may be interpreted as the fraction of the supply line taken into account by the decisionmakers. If $\beta = 1$, the participants fully recognize the supply line and do not double order; if $\beta = 0$, the orders placed are forgotten until the beer arrives. With these dynamical equations in hand, let's look at some results.

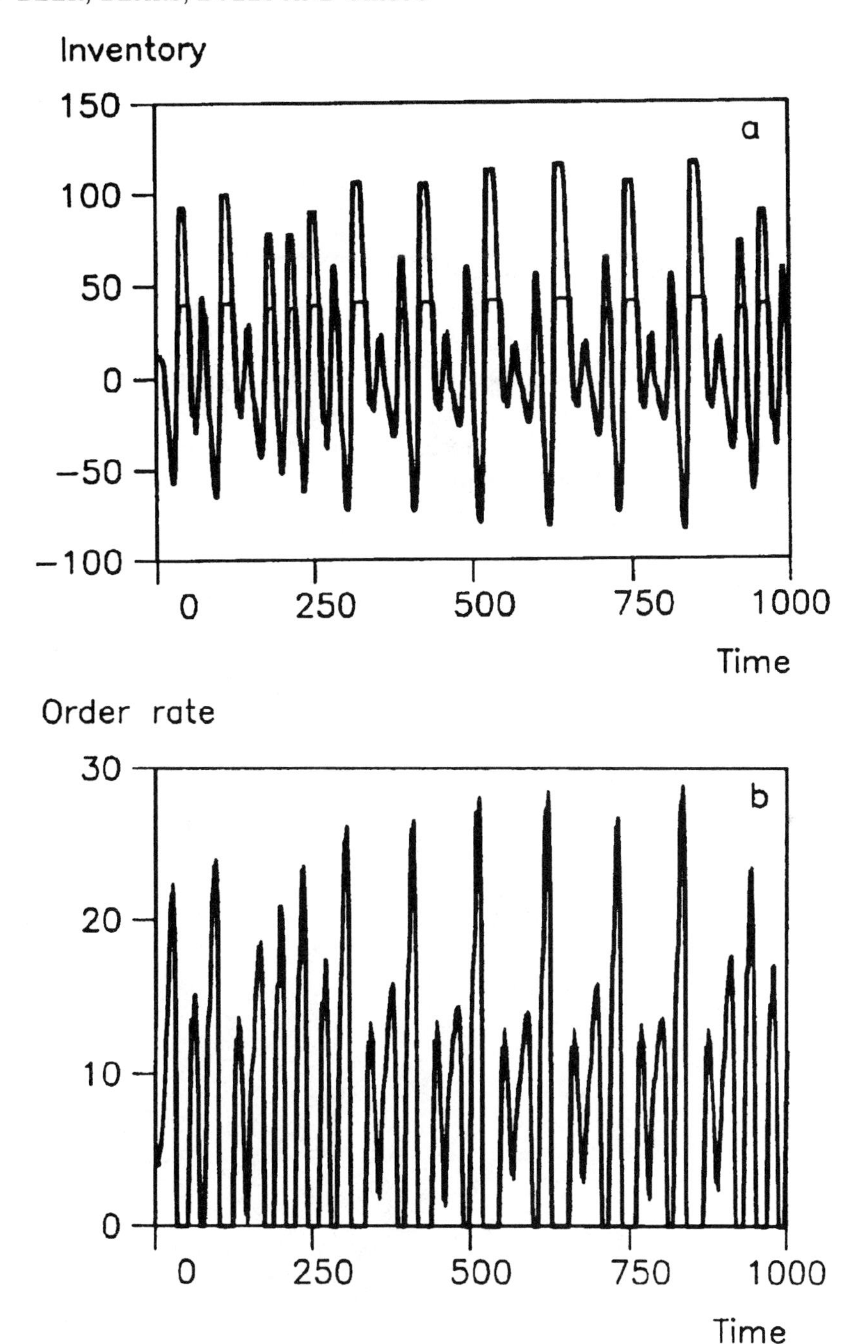

Figure 4.12. Simulation of the Beer Distribution Chain Dynamics

Figure 4.12 shows a typical example of a simulation of the Beer Game dynamics with the parameters having the values $\theta = 0.00$, $\alpha_S = 0.30$, $\beta = 0.15$, and $Q = 17$. These values were obtained by statistically estimating the parameters, using the empirical results of 44 runs of the Beer Game that involved 176 participants over a period of four years. The figure shows the results of the computer simulation to a time horizon of 1,000 weeks. The model behavior is clearly aperiodic; the system never repeats itself, but continues to explore new regions in state space. Calculations show that the largest Lyapunov exponent is positive, suggesting that the system is sensitive to the initial conditions and that the behavior is chaotic.

The Beer Game dynamics contains 27 state variables. Compared with real managerial systems, the model is a vast simplification; compared with most physical systems investigated in the nonlinear dynamics world, however, the model is very complex. In certain regions of parameter space the distribution system has three positive Lyapunov exponents. Therefore, we might expect the system to display an unusually complicated spectrum of behaviors. Figure 4.13 shows the distribution of behavior types in the (α_S, β)-policy plane when $\theta = 0.25$ and $Q = 17$. Here the steady-state solutions of 200 × 200 simulations in the (α_S, β)-plane are plotted using a gray-scale code: light gray indicates stable behavior, dark grey represents aperiodic behavior, and black denotes periodic behavior. (Observe that here we have not distinguished between quasi-periodic and chaotic behavior.)

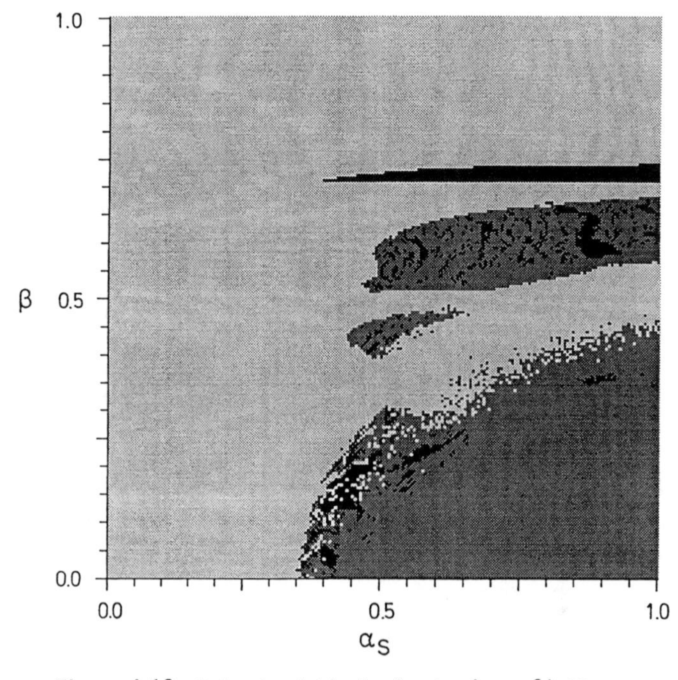

Figure 4.13. Behavioral Modes in the (α_S, β)-Plane

A closer inspection of Figure 4.13 shows that the policy plane contains several regions of unstable behavior, separated by "fjords" of stable behavior.

For instance, in the regions around $\beta = 0.50$ and $\beta = 0.70$ the model is stable for all values of α_S, while the narrow peninsula near $\beta = 0.72$ contains only small-amplitude periodic (and quasi-periodic) solutions. The other regions of unstable behavior are dominated by large-amplitude fluctuations. The occurrence of unstable behavior is most clearly seen in the lower-right corner, where α_S is large and β relatively small. Therefore, to stabilize the distribution chain it's necessary to use an ordering policy in which inventory discrepancies are adjusted relatively slowly and a significant fraction of the supply line is taken into consideration. However, β should not be too large, since a large value of β increases the costs. This is because the system will then stabilize in a state for which the inventories are negative.

The foregoing account has only scratched the surface of the many fascinating lessons to be learned about complex dynamics from the Beer Game. We encourage the reader to consult the work cited in the Notes and References for the details. Now let's turn our attention away from microeconomics in the laboratory to the making of hard cash on the floor of the exchange, looking at another example of decisionmaking leading to chaos. In an earlier section, we saw how speculation on the gold market could lead to price fluctuations that showed all the hallmarks of chaos. Now let's look at the New York Stock Exchange through the same pair of spectacles, seeing if there's any merit to the idea that stock price fluctuations follow a random walk—just as efficient-markets theorists always claim.

Example 2: Chaos in Stock Prices

Consider the following simple model of stock price changes

$$r_t = F(I_{t-1}, u_t),$$

where r_t is the logarithmic return, i.e., the difference $\log P_t - \log P_{t-1}$, where P_t is the stock price at time t. The quantity I_{t-1} is the information available at time $t-1$, while the sequence $\{u_t\}$ is an independent, identically distributed (IID) stochastic process independent of $I(t-1)$ at each time t. The conditional mean is given by $m = E[r_t \,|\, I_{t-1}]$, while the conditional variance is $E[(r_t - m)^2 \,|\, I_{t-1}]$. The kth conditional moment is, by definition, $E[(r_t - m)^k \,|\, I_{t-1}]$. In particular, the conditional skewness is the third conditional moment divided by the cube of the conditional standard deviation, while the conditional kurtosis is the fourth conditional moment divided by the fourth power of the conditional standard deviation minus 3. The conditional moment generating function is $M(s, I_{t-1}) = E[\exp(sr_t) \,|\, I_{t-1}]$.

In plain English, the conditional mean is a measure of the likelihood that a stock price will increase or decrease as a function of the past information I_{t-1}. So the term "conditional" here means conditioned on the past information I_{t-1}, which could be thought of as a set of regression coefficients if one were to estimate the conditional mean from the actual data. In standard efficient markets/random walk theories, the conditional mean is zero or, at least, constant. In other words, the EMH states that past patterns (heads and shoulders, distances between successive peaks and troughs of chartist constructed curves from Horsey charts, etc.) predict nothing about future movements in the stock price.

The EMH has been under attack in recent years. For example, in addition to higher returns in January, lower returns from Friday's close to Monday's close, and systematic return differences over the month, academic finance has found evidence for mean reversion in stock price data. Thus there may be linearly predictable 3–5 year swings in stock returns. Such evidence contradicts simple versions of the random walk theory. However, the mean reversion evidence is controversial. It is sensitive to the Great Depression years, and seems to disappear in large firms in the post-WWII period. So it may just be due to chance.

Earlier, we spoke about conditional means. So let's turn now to a discussion about the variance, or volatility, of returns as well as the role played by higher moments such as the skewness.

Conditional variance is a measure of the volatility of the stock price as a function of past information. Past patterns of the stock price movements may have predictive power for forecasting the future *volatility* in the movement of the stock price, even though the past patterns have no predictive power insofar as assessing whether the stock price itself will go up or down. Before considering this hypothesis, let's make a short digression to look at measures of central tendency and dispersion.

Skewness is a measure of symmetry. For example, it's zero for the bell-shaped normal curve, which is perfectly symmetric about its mean. Kurtosis is a measure of the peakedness, or fat-tailedness, of a distribution. Thus, it measures the likelihood of extreme values. For the normal distribution, the kurtosis is 3. Many people adjust things so that the kurtosis is zero for the normal distribution. That way kurtosis is measured *relative* to the normal. Leptokurtic distributions have an adjusted kurtosis greater than zero (larger than the normal), while platykurtic distributions have an adjusted kurtosis less than zero (less than the normal). Leptokurtosis is very common for stock returns, which are usually measured as we have done here by the difference of the logarithms of the stock's price in consecutive time periods. This means that it's very likely that stock returns will experience greater extremes than if they were normally distributed. This is especially true for high-frequency returns, while returns look more normally distributed at lower frequencies. Now let's return to our main theme, testing of the EMH.

The overall modeling procedure to test for hidden structure in the stock price data is first to use basic economic theory—the received wisdom in academic finance—and fundamental analysis to guide us in choosing a forecasting model. We then use some nonstandard statistical tests to see if this model is adequate, i.e., represents all the structure in the data. We do this by fitting the model to the actual data, thereby obtaining forecast errors from it. Next we test the forecast errors for the presence of "extra nonlinear structure" in the following manner: First, we create a totally random series that has the same mean, variance, skewness, kurtosis, and so on as the original series of forecast errors. Secondly, we compute a measure (the Brock, Dechert, Scheinkman (BDS) statistic discussed in Section 7) of the difference between the original series of forecast errors and this randomly-constructed comparison series. If the measure is zero, the series of forecast errors is random and there is no extra structure to be discovered and exploited. If the measure is not zero, we calculate the probability that the nonzero difference is due only to chance. If the probability of this difference being just a chance fluctuation is less than an error tolerance level of, say, 5 percent, then we tentatively accept the conclusion that there is extra structure that the original forecasting model has missed. In this case, we then use economic theory, fundamentalist analysis, and nonlinear techniques to improve the original forecasting model. Finally, we repeat the preceding steps until we obtain a series of forecast errors that satisfy the test for being IID. Now let's look at some results.

William Brock, Blake LeBaron and their colleagues at the University of Wisconsin have applied the foregoing approach to study returns on the equal-weighted and value-weighted NYSE stock indexes for both weekly and monthly data. They found that both these time series displayed extra structure in the forecast errors of fitted linear models, even after known regularities such as the January effect and the monthly effect and turn-of-the-week effects were taken out of the data. Furthermore, they found that there has been a shift in the structure of stock returns between the period 1962–1974 and the period 1975–1985. This shift remains even after accounting for systematic movements in volatility over each period. Finally, they discovered that forecast errors from standard (in academic finance) forecasting models of the conditional variance (i.e., the volatility) still contain extra, potentially forecastable, structure. This is true for both stock returns on the equal- and value-weighted indexes. The calculations further revealed that the probability that the above results are due solely to chance is very low.

Thus, there does indeed appear to be predictable structure in stock returns. It's related to business-cycle movements, interest rate changes, tax policy, demographics, and so on. We encourage the interested reader to consult the works cited in the Notes and References for the gory details of these studies, as well as for related work on other financial instruments like T-bills, interest rates and foreign currency exchange rates. Now let's turn our attention from the flow of money to a quite different kind of flow—the movement of populations.

Exercises

1. In Figure 4.13 there are regions of unstable (aperiodic) behavior surrounded completely by stable or periodic regions. Interpret this structure in the policy plane in terms of the Arnold "tongues" discussed in Problem 11.

2. Chaotic behavior always involves the competing forces of stretching and folding. Discuss how these essential ingredients might show themselves in a chaotic model for stock price fluctuations. (Hint: Consider things like information congestion and volume bunching.)

10. Population Fluctuations

To illustrate the way in which chaotic oscillations can arise in simple models of population growth, consider a population composed of individuals of two age classes. If x_t and y_t represent the levels of the two groups at generation t, then we take the dynamics governing the population growth to be

$$\begin{aligned} x_{t+1} &= b_1(N)x_t + b_2(N)y_t, \\ y_{t+1} &= sx_t, \end{aligned}$$

where $N_t = x_t + y_t$. The quantities $b_1(N)$ and $b_2(N)$ are birthrate functions, while s is the fraction of the younger population x that goes into the next generation y.

For the sake of straightforward analysis, let's assume that the birthrate functions are the same for both populations and are given by

$$b_1(N) = b_2(N) = b_0 \exp(-\alpha N),$$

where b_0 and α are positive constants. Furthermore, assume that all cohorts from the younger generation survive into the older group ($s = 1$). With these assumptions, the dynamics become

$$\begin{aligned} x_{t+1} &= b_0(x_t + y_t)e^{-\alpha(x_t+y_t)}, \\ y_{t+1} &= x_t. \end{aligned}$$

Let's fix the "crowding factor" α at the level $\alpha = 0.1$ and study the behavior of the populations as a function of the joint birthrate b_0.

The behavior of this two-dimensional system is fundamentally different from what's seen in the scalar case. First of all, for $b_0 < 8.95$ there is a unique, globally attracting fixed point at $x^* = y^* = [\log(2b_0)]/(2\alpha)$. When $b_0 \approx 8.95$, a pair of 3-point cycles appear surrounding the equilibrium. One of these cycles is stable; the other unstable. However, the equilibrium point remains locally attracting until $b_0 = e^3/2$. Thus, in the range $8.95 < b_0 < e^3/2$, a stable 3-point cycle coexists with the stable equilibrium point.

Note how this situation contrasts with the one-dimensional case. There a 3-point cycle would automatically imply chaos by the Period-3 Theorem. As $b_0 \to e^3/2$, the unstable inner cycle decreases in amplitude until at $b_0 = e^3/2$ it coalesces with the stable equilibrium point leaving an unstable equilibrium.

Further increases in b_0 merely increase the amplitude of the 3-point cycle, which is now globally attracting until $b_0 \approx 14.5$. At this point, a period-doubling bifurcation occurs in which each of the endpoints of the 3-point cycle splits into 2 points, creating a stable 6-point cycle. As b_0 is further increased, the 6-point cycle gives way to further period-doubling bifurcations creating 12-, 24-, 48-, ... point cycles. The parameter intervals corresponding to these $3 \cdot 2^k$-point cycles monotonically decrease just as in the one-dimensional case. A new kind of bifurcation occurs near $b_0 \approx 17$. At this point the orbit of the system appears more and more chaotic. However, if b_0 is further increased to $b_0 \approx 24$, the chaos is replaced by a stable 4-point cycle. This 4-point cycle then generates a sequence of $4 \cdot 2^k$-point cycles, which eventually terminate in another chaotic regime. Figure 4.14 shows the pattern of alternating bands of chaos and $n \cdot 2^k$-point cycles. Note that the size of the periodic bands progressively decreases in width, so that for sufficiently high birthrates chaotic dynamics is virtually assured.

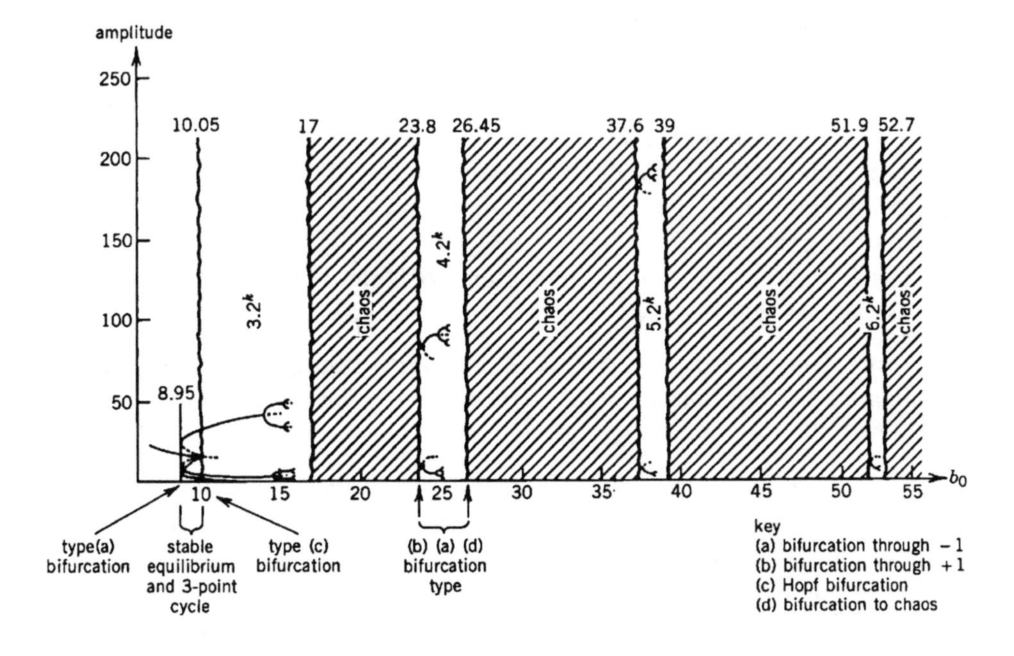

Figure 4.14. Periodic and Chaotic Bands of Population Growth

There are two remarks worthy of note about the dynamical behavior displayed by this elementary model. First, the emergence of "pattern out of chaos" as shown by the continual reappearance of the periodic cycles for ever-increasing birthrates. By way of contrast, in the one-dimensional case once we passed into the chaotic regime we had cycles of *all* periods together with aperiodic motion. Here the situation is entirely different. Now we have *either* periodic cycles or aperiodic orbits, but not both for a given level of the parameter b_0. A second macroscopic feature is the appearance of periodic "outbreaks." Since each successive bifurcation creates a new fixed point near the origin, the orbit tends to spend more and more time at low population levels, punctuated by occasional brief excursions away from the origin culminating in rapid "crashes."

This example illustrates as forcefully as possible the major differences between the behavior of scalar processes and what can happen in even the simplest sort of multi-dimensional setting. We shall see more evidence of this dimensional dichotomy as we go along.

In closing this example, we note that the situation is even more complicated if the two cohorts do not have identical birthrates. In this case, there exists the possibility of quasi-periodic motion, and the emergence of chaos is far more complex when the chaotic orbits come out of the quasi-periodic orbits than when they emerge out of periodic cycles.

11. Continuous Chaos and the Lorenz Attractor

Up to now our attention has been directed exclusively to discrete-time systems whose dynamical behavior is characterized by the iteration of an n-dimensional map. But many processes are described by differential equations (vector fields), not difference equations (maps). The issue we want to explore in this section is to what degree the phenomena of period-doubling bifurcations and the emergence of aperiodic trajectories carry over to continuous-time systems.

The first point to note in connection with the attractors of continous flows is the classical result of Poincaré and Bendixson.

POINCARÉ-BENDIXSON THEOREM. *A nonempty, compact attractor of an autonomous planar flow must be either an equilibrium point or a closed orbit.*

Consequently, no planar, autonomous dynamical system can display the aperiodic trajectories characteristic of chaotic motion. Two-dimensional systems can have only point equilibria and/or limit cycles (periodic orbits) as their long-run behavior. So in looking for chaos in continuous-time systems, we must concentrate our search on systems of dimension $n \geq 3$.

The first such system to display nonperiodic behavior was discovered in theoretical meteorology by Edward N. Lorenz and presented in a paper published in 1963. Briefly, the physical background of Lorenz's problem involved the heating of a fluid layer of uniform depth H, the temperature difference between the top and the bottom of the layer being kept at a constant level ΔT. This problem represents a simplified version of atmospheric heating by the Sun. Such a system has a steady-state solution that varies linearly with depth. So if this solution is unstable, convection currents should develop. If we confine all motions in the fluid to the (x, z)-plane, allowing no variations in the y-direction, the equations of fluid motion can be written as

$$\frac{\partial}{\partial t}\nabla^2\psi = -\frac{\partial(\psi, \nabla^2\psi)}{\partial(x, z)} + \nu\nabla^4\psi + g\alpha\frac{\partial\theta}{\partial x},$$

$$\frac{\partial\theta}{\partial t} = -\frac{\partial(\psi, \theta)}{\partial(x, z)} + \frac{\Delta T}{H}\frac{\partial\psi}{\partial x} + \kappa\nabla^2\theta.$$

Here ψ is a stream function for the two-dimensional motion, while θ is the departure of the temperature from the linear temperature gradient seen in the case of no convection. The constants g, α, ν and κ denote the acceleration of gravity, the coefficient of thermal expansion, the kinematic viscosity and the thermal conductivity, respectively. Assume that the upper and lower boundaries are free, so that both ψ and $\nabla^2\psi$ vanish at both boundaries.

It can be shown that convection will occur if the *Rayleigh number*

$$R_a = g\alpha H^3\Delta T\nu^{-1}\kappa^{-1},$$

exceeds the critical value

$$R_c = \pi^4 a^{-2}(1 + a^2)^3,$$

where a is a parameter expressing the period of the convective oscillation. The smallest value of R_c occurs when $a^2 = 1/2$, and is given by $R_c = 27\pi^4/4$.

Let's expand the above dynamics into a double Fourier series in x and z, the coefficients involving the time t alone. Introduce the terms

$$a(1 + a^2)^{-1}\kappa^{-1}\psi = X\sqrt{2}\sin(\pi a H^{-1}x)\sin(\pi H^{-1}z),$$

$$\pi R_c^{-1}R_a\Delta T^{-1}\theta = Y\sqrt{2}\cos(\pi a H^{-1}x)\sin(\pi H^{-1}z) - Z\sin(2\pi H^{-1}z),$$

where X, Y and Z are functions of t alone. Substituting these expressions into the original equations for ψ and θ and equating coefficients of like terms, we obtain the following system for the functions X, Y and Z:

$$\dot{X} = -\sigma X + \sigma Y,$$

$$\dot{Y} = -XZ + rX - Y,$$

$$\dot{Z} = XY - bZ.$$

Here $\sigma = \kappa^{-1}\nu$ is the Prandtl number, $r = R_c^{-1}R_a$ and $b = 4(1+a^2)^{-1}$. These are the famous *Lorenz equations.* The variable X is proportional to the intensity of the convective motion, Y is proportional to the temperature difference between the ascending and descending currents, and Z is proportional to the distortion of the vertical temperature gradient from linearity. Similar signs of X and Y means that warm fluid is rising and cold fluid is descending, whereas a positive value of Z indicates that the strongest temperature gradients occur near the boundaries. We will examine the behavior of these equations for various ranges of values of the parameters σ, r and b.

Since it's of greatest physical interest to consider variations in the Rayleigh number, we fix the parameters σ and b and consider only variation in r. For definiteness, let $\sigma = 10, b = 8/3$. In this range stable convection rolls lose their stability for $r \approx 25$, and are replaced by another kind of large amplitude motion. This was the kind of chaos that Lorenz set out to study. Let's follow his analysis by setting $r = 28$ and take an initial state near the saddle point at $p = (0, 0, 0)$. Lorenz found that the solutions rapidly approached the branched surface shown in Fig. 4.15. The boundary of this surface is part of the unstable manifold $W^u(p)$ of the saddle point p. Figure 4.15 shows the first 50 loops of one "side" of the unstable manifold of p, the surface being shaded and the branch indicated.

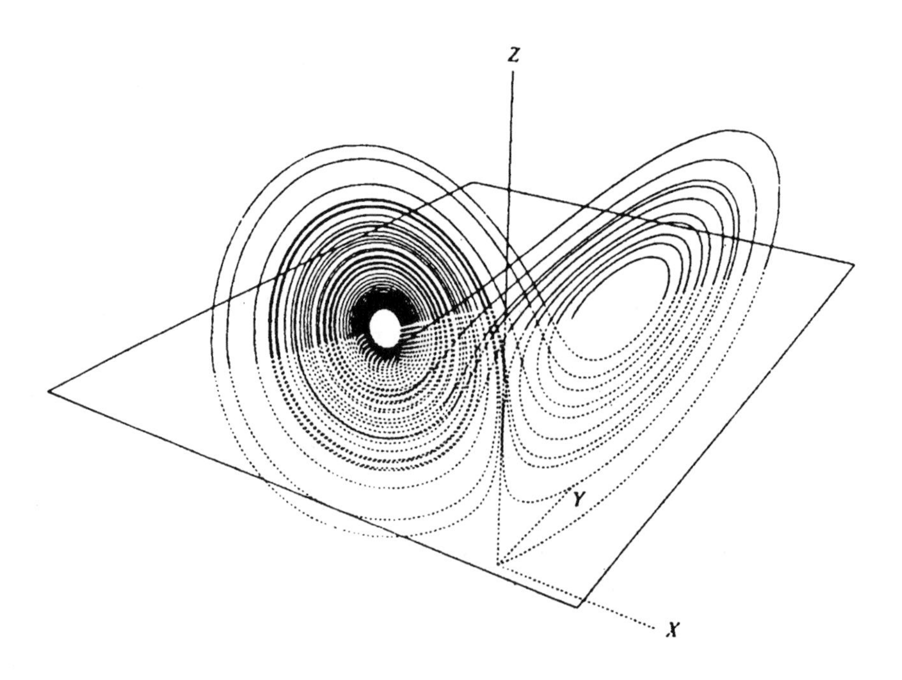

Figure 4.15. The Lorenz Attractor

The presence of the branch in the Lorenz attractor shows that the attractor has infinitely many sheets. Furthermore, solutions do not intersect within the attractor but move from sheet to sheet as they circulate over the branch. This kind of motion is exactly analogous to the type of aperiodic behavior displayed by one-dimensional maps, and leads to the conjecture that the Lorenz attractor is indeed an example of continuous chaos. Since the geometric and analytic study of this particular system is by now rather well-advanced, we refer the reader to the papers and books cited in the Notes and References for details, and pass on to the more general question of *strange attractors* and their identification in actual practice.

Exercises

1. The Lorenz system has a fixed point at the origin. (a) Show that the Jacobian matrix for the system at this point is

$$J(0) = \begin{pmatrix} -\sigma & \sigma & 0 \\ r & -1 & 0 \\ 0 & 0 & -b \end{pmatrix}$$

(b) Compute the characteristic values of the system at the origin. (c) Prove that the system undergoes a change of stability at $r = 1$.

2. In the Lorenz equations, eliminate y and solve for z in terms of x^2. (a) Show that these operations reduce the Lorenz system to

$$\ddot{x} + (1+\sigma)\dot{x} + x\sigma \left\{ 1 - r + \frac{1}{2\sigma^2}x^2 + \left(1 - \frac{b}{2\sigma}\right) \times \int_0^\infty [x(t-\tau)]^2 e^{-b\tau}\, d\tau \right\} = 0.$$

(b) Assume $b > 0$, and split the term $e^{-b\tau}$ into a delta function and the deviation from it. By substituting this into the expression from part (a), establish that

$$\ddot{x} + (1+\sigma)\dot{x} + \frac{dU}{dx} + x\left\{ (\sigma - b/2) \int_0^\infty [x^2(t-\tau) - x^2(t)] e^{-b\tau}\, d\tau \right\} = 0,$$

where

$$U(x) = \sigma \left\{ \frac{1-r}{2}x^2 + \frac{1}{4b}x^4 \right\}.$$

(This form shows that the Lorenz system describes the motion of an overdamped particle in a potential well $U(x)$ under the action of an additional force that depends on the *history* of the motion.)

3. Referring to the preceding exercise, (a) show that when $r < 1$ the potential well has one stable minimum at $x = 0$, while there is a bifurcation at $r = 1$ at which the state $x = 0$ becomes unstable. In this case, show that for $r > 1$ the system has two new stable minima at $x = \pm[b(r-1)]^{\frac{1}{2}}$. (b) Show that the cumulative memory term has a destabilizing effect on the two equilibria if $\sigma > b/2$ and a stabilizing effect if $\sigma < b/2$. (c) How can you interpret the term $(1+\sigma)\dot{x}$?

4. Prove that the divergence of the Lorenz system is given by the expression $\nabla \cdot f = -(\sigma + b + 1)$, where f stands for the right-hand side of the Lorenz equations. Thus, conclude that the phase volume is contracted everywhere by the Lorenz dynamics; hence, the attracting set for the system must have zero volume in R^3.

12. Poincaré Maps and Strange Attractors

Consider a set of differential equations

$$\dot{x} = V(x), \qquad x \in R^n.$$

The phase space for this system is n-dimensional, having coordinates x_i, $i = 1, 2, \ldots, n$. The *Poincaré section* for the system is found by selecting a surface Σ_R in the phase space that intersects the trajectory of the system transversally. Roughly speaking, this means that the tangent to the system trajectory does not lie in the surface Σ_R at the point of intersection (see Fig. 4.16). The *Poincaré map* is then obtained by choosing a point $x_k \in \Sigma_R$ and integrating the dynamics forward in time to find the next intersection x_{k+1} of the orbit with Σ_R. In this way we construct the map

$$x_{k+1} = f(x_k), \qquad x_k \in R^{n-1}.$$

If the vector field V is smooth and Σ_R is everywhere transverse to V, then it can be shown that the Poincaré map f is also smooth.

Geometrically, it's easy to see that a periodic orbit of the original system corresponds to a fixed point of the Poincaré map, whereas an aperiodic orbit corresponds to discrete chaos in the Poincaré map. Thus, transferring attention to the Poincaré map enables us to shift much of the analysis of the nature of the attracting sets of continuous flows to the simpler (and lower-dimensional) setting of maps.

Since the phase-space volume must contract for dissipative systems, the stable, steady-state motion for an n-dimensional system must lie on a "surface" of dimension less than n. Loosely speaking, we call such a surface an *attractor* for the flow. Let's recall from Chapter 2 that for A to be an attractor,

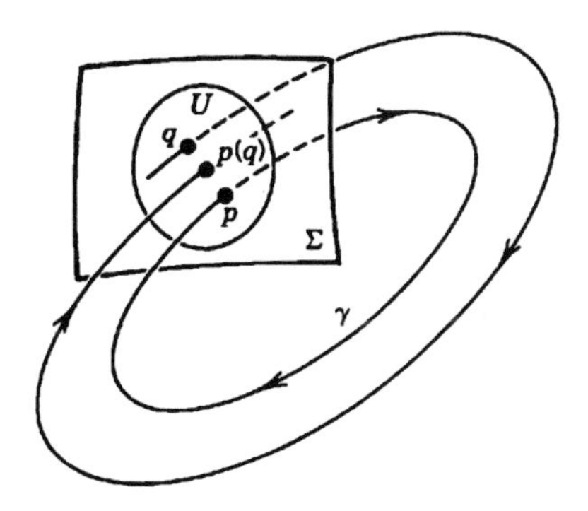

Figure 4.16. The Poincaré Section and Map

1) A must be invariant under the flow of the system,

2) There must be an open neighborhood U around A such that all points $x_t \in U$ approach A as $t \to \infty$,

3) No part of A can be transient, i.e., only the steady-state motions of the system lie in A, and

4) A cannot be decomposed into disjoint invariant components.

For scalar dynamics, an attractor A is called a *strange attractor* if in addition to the above properties, A contains periodic orbits of all periods, as well as an uncountable number of aperiodic orbits. More generally, an attractor is strange if it contains a transversal homoclinic orbit. We refer the reader to the literature cited in the Notes and References for the painstaking (and painful!) details associated with these higher-dimensional situations. Our goal here is only to show a little of the structure of the attractor set A when A is strange.

At first glance, it may appear that to study the properties of A we have to consider the entire n-dimensional phase portrait of the system. However, by appealing to deep embedding theorems, David Ruelle and Floris Takens have shown how to construct the multi-dimensional phase portrait of the system using only observations on a *single* variable. The method follows the procedure we saw earlier for embedding a scalar time series into R^m. Only now we go in the opposite direction. The Ruelle-Takens results show that for almost every coordinate function $x_i(t)$, and almost every time delay T, the m-dimensional portrait constructed from the set of vectors $\{x_i(t_k), x_i(t_k + T), \ldots, x_i(t_k + (m-1)T)\}$, $k = 1, 2, \ldots, \infty$, yields a proper embedding of the original manifold if $m \geq 2n+1$; i.e., if we take a sufficiently large number of observations of $x_i(t)$, we will obtain a manifold having the same geometrical properties as the original phase portrait.

Figure 4.17 illustrates the the embedding result. Here the first part of the figure shows measurements taken on a single variable in the regime of a

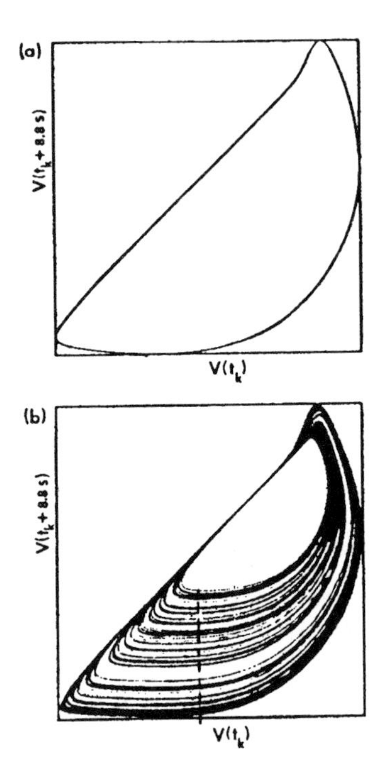

Figure 4.17. Embedded Phase Portraits for Periodic and Strange Attractors

periodic orbit, while the second half of the figure shows the same variable for parameters in the chaotic regime. Note that in the second part of the figure, the third axis, $x(t_k + 17.6)$, is normal to the page.

The embedding trick discussed in Section 7 for scalar time series can also be used as the basis of a procedure to forecast future values of a scalar chaotic processes $\{x_t\}, t = 0, 1, 2, \ldots, N-1$. Here's how. First we choose an embedding dimension m and form the point $X(m,t) = (x_t, x_{t+1}, \ldots, x_{t+m-1})$ in R^m. Suppose we want to predict the value x_{t+1}. The forecasting scheme involves finding the $m+1$ closest neighbors $X(m,s)$ in R^m. That is, from the N possible m-windows that can be formed from the sequence $\{x_t\}, t = 0, 1, 2, \ldots, N$, we find the $m+1$ points in R^m that are closest to $X(m,t)$. We then use these $m+1$ points as the vertices of a simplex in R^m. This simplex contains $X(m,t)$ as an interior point. The predicted value of x_{t+1} is now obtained by projecting the vertex points of the simplex forward one time step. The predicted value x_{t+1} itself is then taken to be the last component of the vector $X(m, t+1)$, a quantity found by exponentially weighting the projected values of the new vertex points. The weights in this scheme are determine by the original distance of $X(m,t)$ from the vertices of the simplex. And if we wanted to predict p time steps into the future, then all we would have to do is keep track of where the vertices of the simplex ended up p steps later and apply the same weighting rule. Now let's see how this procedure works in practice.

Example: The Tent Map

To test the above prediction scheme, Robert May and George Sugihara constructed a time series of 1,000 data points by iterating the tent map

$$x_{t+1} = \begin{cases} 2x_t, & 0 < x_t < 0.5, \\ 2 - 2x_t, & 0.5 < x_t < 1. \end{cases}$$

These data points are shown in Fig. 4.19(a). Part (b) of the figure shows the predicted versus actual data values at $p = 2$ time steps into the future. Part (c) shows the degradation of the prediction when the method is used to predict values $p = 5$ steps into the future. In both cases, an embedding dimension $m = 3$ was used. Finally, in part (d) we see the correlation coefficient between the actual and predicted values as a function of the number of time steps into the future the prediction covers. As expected, for a chaotic attractor the prediction accuracy decreases exponentially as the predictions are extended further into the future.

We can also use the Poincaré section to study the attractors displayed in Fig. 4.17. The orbits for the strange attractor lie essentially along a sheet. Thus, intersections of this sheetlike attractor with a plane lie, to a good approximation, along a curve and not on a higher-dimensional surface. The Poincaré map for this attractor is shown in Fig. 4.18, from which we can see the "one-hump" structure familiar from our earlier discussion. Analysis of this map shows that the largest Lyapunov exponent is positive, confirming that the set A is indeed a strange attractor.

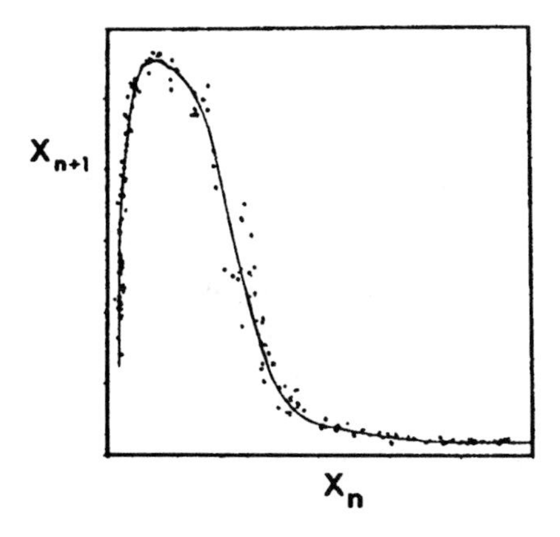

Figure 4.18. The Poincaré Map for a Strange Attractor

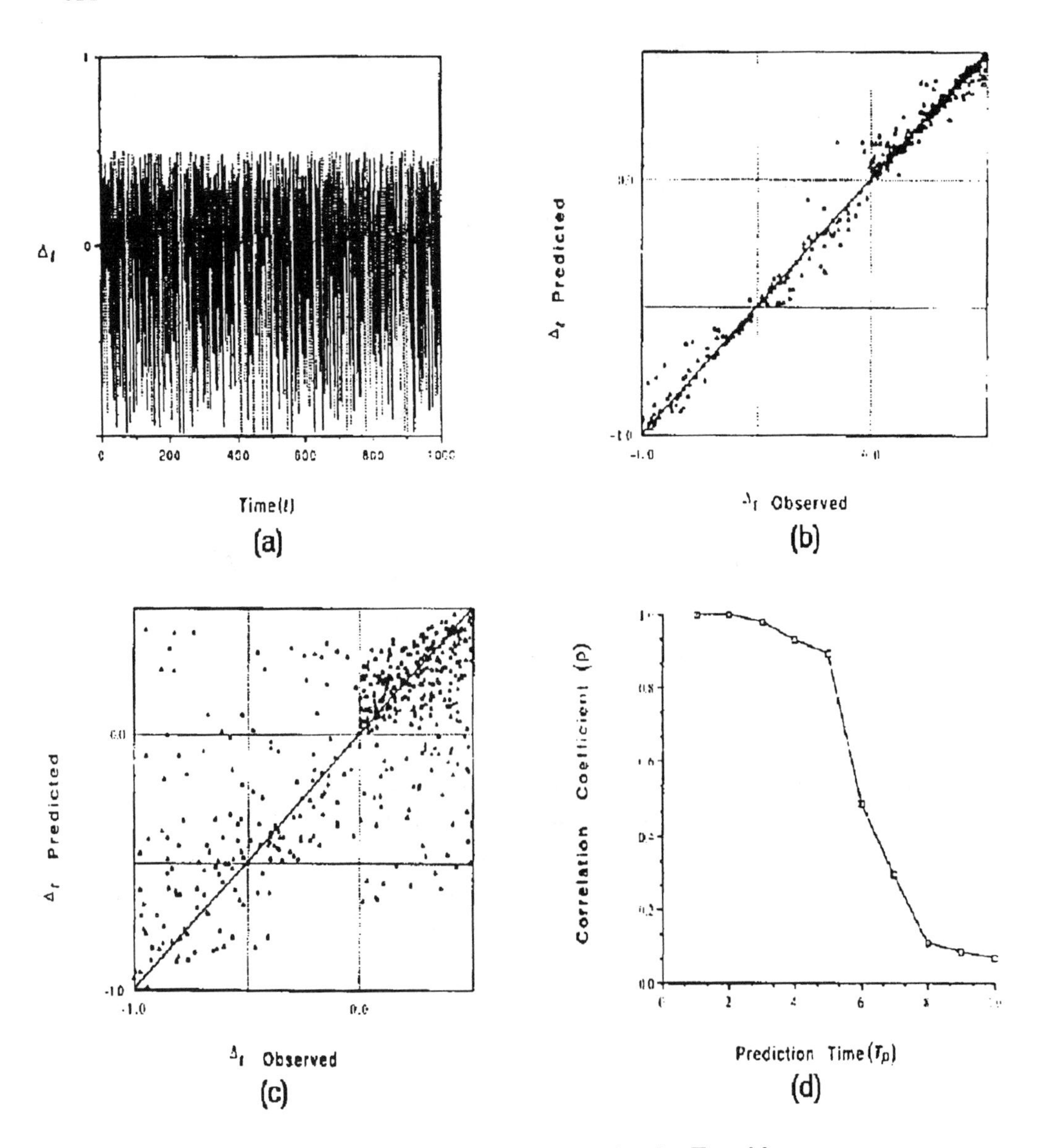

Figure 4.19. Forecasting Results for the Tent Map

Using any reasonable definition of dimension, the dimension D of simple attractors can easily be seen to be integers: $D = 0$ for fixed points; $D = 1$ for limit cycles; $D = n$ for an n-torus. However, for strange attractors D turns out to be nonintegral. The Lyapunov exponents give us a way to classify attractors, both simple and strange. Here's how.

Consider the initial state x_0 for the flow. Let's build a local coordinate system at x_0 by choosing at random n linearly independent vectors originating at x_0. Let these vectors be labeled $\{e_0^{(i)}, i = 1, 2, \ldots, n\}$. Now we orthogonalize this set and integrate the system forward to a state x_1 using the linearized equations of motion

$$\dot{e}_0^{(i)} = \sum_{i=1}^{n} M_{ij}(x_0) e_0^{(j)}, \qquad e_0^{(i)} = x_0^{(i)}, \qquad i = 1, 2, \ldots, n.$$

Here $M(x_0) = [\partial f(x_0)/\partial x]$ is the system Jacobian matrix. Applying this procedure again yields a new set of vectors $\{e_1^{(i)}\}$ originating at x_1. Now consider the following set of geometrical ratios:

$$\begin{aligned} C_1 &= \text{ length ratios } |e_1^{(i)}|/|e_0^{(i)}|, \\ C_2 &= \text{ parallelogram area ratios } |e_1^{(i)} \wedge e_1^{(i)}|/|e_0^{(i)} \wedge e_0^{(i)}|, \\ &\vdots \\ C_n &= n\text{-dimensional hyperparallelopiped volume ratio } \frac{\bigwedge_{i=1}^{n} e_1^{(i)}}{\bigwedge_{i=1}^{n} e_0^{(i)}}. \end{aligned}$$

Here "$\wedge$" is the symbol for the exterior vector product operation, the generalization to higher dimensions of the standard two-dimensional vector cross product familiar from elementary vector analysis. In the set C_k, there are exactly $\binom{n}{k}$ real numbers representing the various geometrical ratios. If we continue this process at each step of the trajectory and take the long-term averages of each of these sets of numbers, we obtain a global collection of numbers characterizing the attractor A. It can be shown that if the vectors $\{e_0^{(i)}\}$ are chosen randomly, then the average of the numbers C_1, i.e., the length ratios, converges to the largest Lyapunov exponent σ_1 for the system. Similarly, the average of the area ratios (the numbers C_2) converges to the sum of the two largest Lyapunov exponents, $\sigma_1 + \sigma_2$, and so on. Finally, the last quantity C_n converges to the contraction rate of a volume element along the trajectory,

$$\sum_{i=1}^{n} \sigma_i = -\operatorname{div} V.$$

The Lyapunov exponents can be ordered as

$$\sigma_1 \geq \sigma_2 \geq \cdots \geq \sigma_n.$$

It has been suggested that the sign pattern of the above ordering can be used as a scheme for the qualitative classification of attractors for dissipative flows. For example, if we have the sign pattern $(\sigma_1, \sigma_2, \sigma_3) = (-, -, -)$ when $n = 3$, this would indicate an attractor consisting of a fixed point with dimension $D = 0$. Similarly, the signature $(0, -, -)$ corresponds to a limit cycle having dimension $D = 1$. On the other hand, the pattern (+, 0, −) is characteristic of a chaotic attractor. Since the signature changes with changes in whatever parameters may be in the system, this kind of classification scheme can be extremely useful in identifying parametric changes leading to qualitative shifts from one type of attractor to another.

As we have repeatedly emphasized, the Lyapunov exponents are a measure of how sensitive the system is to changes in the initial conditions. However, there is another, closely related, kind of sensitivity for chaotic processes that we need also take into account. This is the problem of computational error. The basic question here is the following: In a numerical study of a chaotic process, in what sense do the numerical results reflect the true dynamics of the actual system? For example, we might integrate numerically the same chaotic system with the same initial conditions, using single-precision arithmetic in one experiment, double-precision in another. What kind of confidence can we place in the two end results when it could easily happen that after just a few time steps, the two trajectories differ by the same order of magnitude as the quantities being computed? Put more specifically, if truncation errors are of the order 10^{-8} for a system whose variables are of magnitude $O(1)$, and if distances between two neighboring trajectories double at each iteration, on the average, then two trajectories starting at a distance 10^{-8} apart will be separated by one unit in less than 20 iterations. Consideration of this problem leads to what has been termed "shadowing."

The basic idea underlying shadowing is the same as the process that spies are supposed to be regularly engaged in—staying close to some other person. The equivalent in the world of chaos is telling whether a set of noisy data generated by computer iternation with round-off and/or truncation errors, can be followed within predefined limits by some "true" trajectory. In other words, while a noisy trajectory might diverge rapidly from the true trajectory with the same initial conditions, there might exist a different true trajectory (one with a different initial condition) that stays near the noisy trajectory for a long time. In this case, we say that the different true trajectory "shadows" the noisy one, and in this case we could justify taking the noisy trajectory as a good approximation to the actual chaotic process. The problem is to understand those systems for which this type of shadowing occurs. While as yet there appear to be no completely general results of this type, methods developed by Celso Grebogi and his co-workers enable us to prove shadowing for several prototypical types of chaotic processes. It can be hoped that the techniques used in these studies can be extended to a more general theory of when such shadowing can be expected to occur. Let's look at one of their examples.

Example: The Standard Map

About the simplest nontrivial, nonlinear Hamiltonian system is the so-called *standard map,* given by the equations

$$I_{n+1} = I_n + (K/2)\sin 2\pi\theta_n \mod 1,$$
$$\theta_{n+1} = \theta_n + I_{n+1} \mod 1.$$

The greater the value of K, the more nonhyperbolic and chaotic the system is. So to make things interesting, we look at the shadowing distance for values of $K > 0$. The table below shows the shadowing distance δ_x as a function of K. In all cases, the number of time steps was taken to be one million.

Table 4.2. Shadowing Distances

K	δ_x
1.0	2.9×10^{-7}
1.05	5.4×10^{-8}
2.25	6.1×10^{-10}
2.5	2.6×10^{-10}
3.0	2.9×10^{-9}
4.0	1.7×10^{-10}
5.0	3.5×10^{-10}
7.5	9.5×10^{-11}

The conclusion to be drawn from this table of shadowing distances is that it is possible to very accurately shadow the standard map for all values of K up to $K \approx 0.97$, where the last KAM surface is broken and where the islands of stability become large. Grebogi and his co-workers estimate that for most two-dimensional Hamiltonian maps having chaotic trajectories, if the noise amplitude is n_f, then we can expect to find a shadowing distance $\delta_x \leq \sqrt{n_f}$ for a trajectory of length $N \approx 1/\sqrt{n_f}$.

Now let's put some of these theoretical and computational ideas to use in consideration of what some have called *the* unsolved problem of classical physics: turbulent fluid flow.

Exercises

1. Consider the dynamical system

$$\dot{r} = ar(1 - r),$$
$$\dot{\theta} = 1,$$

where a is a real parameter. (a) Verify that the solution of this system is $r(t) = r_0/[r_0 - (r_0 - 1)e^{-at}]$, $\theta(t) = \theta_0 + t$. (b) Show that the circle $r = 1$ in the plane is an attracting (repelling) limit cycle of period 2π for $a > 0$ ($a < 0$). (c) Consider the Poincaré section given by the line $\theta = 0$ (i.e., the points $y = 0, x > 0$ in rectangular coordinates). Determine the Poincaré map for this section. (Answer: $P(x) = x/[x - (x - 1)e^{-2\pi a}]$.)

2. Consider the Ikeda map

$$f(x, y) = (1 + \mu(x \cos t - y \sin t), \mu(x \sin t + y \cos t)),$$

where $t = 0.4 - 6/(1 + x^2 + y^2)$. To ensure that the system has a strange attractor, take the parameter value $\mu = 0.7$. (a) Calculate the first 100 iterates of this map with the initial state $x_0 = y_0 = 0$. Then use the first 90 data points as the basis upon which to predict the rest, using the Embedding Theorem. (b) Do the same thing for some of the other maps we've discussed in this chapter, such as the Lorenz system or the logistic map.

3. Discuss some of the practical computational difficulties that have to be overcome in using the embedding technique to forecast noisy and/or chaotic time series.

4. Consider the equation for the forced pendulum given by $\ddot{x} + \sin x = 2.4 \cos t$, with the initial condition $x_0 = \dot{x}_0 = 0$. (a) Consider whether the trajectory of this system is chaotic. (b) Compute a shadowing trajectory for the system. (c) Estimate how closely the shadowing trajectory follows the true system trajectory and for how long.

13. Chaos and the Problem of Turbulence

The great triumphs of nineteenth-century theoretical physics, such as Maxwell's theory of electromagnetism and Boltzmann's theory of statistical mechanics, are widely touted as illustrations of the seemingly onward and upward, monotonic progress of science, in general, and physics in particular. What is usually left unsaid is an account of the equally great *failures* of science, failures that most scientists fervently wish would simply curl up into a little ball, roll off into a corner and disappear, much like the now mythical ether. Perhaps the greatest failure of this sort in classical physics is the inability to give any sort of coherent account of the puzzling phenomenon of turbulence. The laminar flow of fluids was already well-accounted for by the classical Navier-Stokes equation characterizing the stream function. But the commonly observed transition from smooth laminar flow to periodic eddies and then on into fully-developed turbulence was completely beyond the bounds of the conceptual framework and techniques of nineteenth-century physics. In fact, it was not until the discovery of aperiodic orbits, strange attractors and the like, that even twentieth-century physicists began to get a glimpse of what is beginning to look like an actual *theory* of turbulent flow. In this section we present a skeletal outline of this newly-emerging paradigm for turbulence, primarily as an introduction to the actual physical interpretations that can be attached to the purely mathematical phenomenon of chaos.

The central difficulty in giving a mathematical account of turbulence is the lack of any single scale of length appropriate to the description of the phenomenon. Intuitively—and by observation—turbulent flow involves nested eddies of all scales, ranging from the macroscopic down to the molecular. So any mathematical description of the process must take all these different scales into account. This situation is rather similar to the problem of phase transitions, where length scales ranging from the correlation length, which approaches infinity at the transition temperature, down to the atomic scale all play an important role in the overall transition process.

Before entering into an account of turbulence using the ideas of chaos, we emphasize the point that the kind of machinery developed here is relevant only to the problem of the *onset* of turbulence, and has nothing to do with fully-developed turbulence of the type studied in mechanical and civil engineering. Furthermore, mathematical chaos is concerned primarily with temporal random behavior, while real-world turbulence involves stochastic behavior in space, as well. Moreover, the mathematical models built upon chaotic dynamics generally involve only planar geometries rather than the three-dimensional geometry of real life. Consequently, until these kinds of gaps are filled, the theories given here can only be regarded as initial approximations to real turbulent flow. Nevertheless, they serve as just the sort of first step needed to understand the mechanism by which smooth fluid flow begins to turn turbulent.

The classical Navier-Stokes equations describing the stream velocity and pressure of fluid flow are given in dimensionless units by the expression

$$\frac{\partial v}{\partial t} + v \cdot \nabla v = -\nabla p + \frac{1}{R}\nabla^2 v,$$

together with the incompressibility condition

$$\nabla \cdot v = 0.$$

The Poisson equation for the pressure is

$$\nabla^2 p = -\nabla \cdot (v \cdot \nabla v).$$

These equations are subject to whatever boundary conditions are appropriate for the problem under study.

The only dimensionless number that appears explicitly in these equations is R, the *Reynolds number* of the fluid. It has been known for many years that when $R \gg 1$ turbulence sets in, with the typical thresholds being values of R between 1,000 and 2,000. At these levels of R, the velocity field v becomes highly disordered and irreproducible, so that from a practical point of view it becomes essentially random. This is an observational conclusion, and has never been established theoretically with anything like the kind of rigor and generality that theoreticians would like. The constant R measures the ratio of the two nonlinear terms in the velocity field equation to the linear term, and in practical situations involving the atmosphere and the oceans it is always a few thousand or more. So if you're dealing with real fluids, you're dealing with turbulence.

In view of the preceding remarks, the mathematical characterization of the onset of turbulence essentially reduces to providing a convincing description of how the constant velocity flow fields of laminar flow in the region of small Reynolds number R progress to turbulent flow as R is increased. Over the past several years, a number of mathematical "scenarios" have been presented as to how this transition might take place. Here is a brief summary of the major competing positions.

- *The Landau-Hopf Scenario*—This scenario views the emergence of turbulence as being attributable to the appearance of an increasing number of quasi-periodic motions resulting from successive bifurcations in the system.

When the Reynolds number is small, the fluid motion is laminar and stationary, corresponding to an attractor consisting of a stable *fixed point* in the phase space. As R is increased, this fixed point loses its stability and begins to repel all nearby trajectories. Since a small change in R cannot cause a global disruption in the flow, the fixed point may become repelling locally. But it will remain attracting for regions located far enough away. Thus, the local repulsion and global attraction causes the formation of a *limit cycle*, which is a periodic motion of the flow, i.e., a sort of "whirlpool." This process of generation of a limit cycle from a stable fixed point is exactly the *Hopf bifurcation* discussed in Chapter 2.

Further increasing R leads to a value at which the limit cycle loses its stability and becomes repelling. At this point, an attracting 2-torus surrounding the unstable limit cycle appears. The motion becomes quasi-periodic if the two frequencies on the torus are incommensurable, i.e., rationally independent. Landau and Hopf then assume that this process continues indefinitely, and identify the final state with an infinite number of incommensurable frequencies representing turbulence.

The Landau-Hopf scenario suffers from a number of serious difficulties of both a theoretical and an experimental nature. First of all, the scenario requires the successive appearance of new incommensurable frequencies in the power spectrum. In laboratory experiments, turbulent spectra do develop a few independent frequencies. But they soon turn into broad noisy bands. Furthermore, the Landau-Hopf picture ignores the important physical phenomenon of phase-locking. In fact, in nonlinear systems new incommensurable frequencies cannot continue to appear indefinitely without interacting with each other. Nearby frequencies then tend to get locked, thereby diminishing the number of independent frequencies. Finally, the Landau-Hopf scenario contains no provision for sensitive dependence on the initial conditions. For all of these reasons, the Landau-Hopf road to turbulence is unlikely to be nature's mechanism for the onset of turbulent flow.

• *The Ruelle-Takens Scenario*—In 1971 David Ruelle and Floris Takens showed that the Landau-Hopf picture is unlikely to occur in nature, and provided an alternate mechanism that can be compactly summarized as: fixed point $\rightarrow$ limit cycle $\rightarrow$ 2-torus $\rightarrow$ strange attractor; i.e., quasi-periodic motion on a 2-torus loses its stability and gives birth to turbulence directly. In short, the scenario is based upon the mathematical fact that if a system undergoes two Hopf bifurcations as the Reynolds number is increased from a stationary initial state, then it is likely (generic) that the system has a strange attractor after the second bifurcation.

Experimental evidence involving the velocity field of a fluid confined between rotating cylinders and convective flow of the Rayleigh-Bénard type can be interpreted as being consistent with the Ruelle-Takens scheme.

• *The Feigenbaum Scenario*—This picture of turbulence is based upon using the Poincaré map P_R associated with the flow of the Navier-Stokes equation, and invoking the period-doubling $\rightarrow$ aperiodic orbit transition we have already seen for maps. Under some smoothness assumptions on the map P_R, it can be shown that the bifurcation pattern associated with the Feigenbaum scenario involves a stable fixed point giving way to a stable periodic orbit as R is increased. However, in this scenario the bifurcation is of the *pitchfork* type rather than the Hopf type, as was the case in the previous scenarios. Thus, as R increases, the linear part of P_R has a real root that moves out of the unit circle at -1 rather than a pair of complex conjugate roots that leave the unit circle away from the real axis.

The Feigenbaum scenario claims that if subharmonic bifurcations are observed at values R_1 and R_2 of the Reynolds number, then one can expect another bifurcation near

$$R_3 = R_2 - \frac{R_1 - R_2}{\delta} \,,$$

where $\delta = 4.66920\ldots$ is the universal Feigenbaum number. These predictions, as well as several others involving the shape of the power spectrum, have been verified in the lab on a variety of physical problems, including heat transport by convection in liquid helium.

• *The Pomeau-Manneville Scenario*—This view of the onset of turbulence involves the phenomenon of transition to turbulence through *intermittency.* Here the laminar phase is not lost entirely, but rather is "interlaced" with intermittent bursts of chaotic behavior. Mathematically, the nature of this scenario involves a *saddle-node* bifurcation, in which a stable and an unstable fixed point collide. Both then disappear (into complex fixed points). The phenomenon underlying this scenario is illustrated by the map

$$x_{t+1} = f(x_t) = 1 - \mu x_t^2, \qquad x_0 \in [-1, 1], \qquad \mu \in [0, 2].$$

The third iterate $f^{(3)}$ can be shown to have a saddle-node when $\mu = 1.75$. For $\mu > 1.75$, $f^{(3)}$ has a stable periodic orbit of period 3, and an unstable one nearby. These orbits collide at $\mu = 1.75$. If $\mu - 1.75$ is $O(\epsilon)$, then a typical orbit will need $O(\epsilon^{-\frac{1}{2}})$ iterations to cross a fixed, small interval around $x \approx 0$. As long as the orbit is in this small interval, an observer will have the impression of seeing a periodic orbit of period 3; as soon as the system leaves this small interval, iterations of the map look chaotic.

We can summarize the Pomeau-Manneville scenario as follows. Assume the Poincaré map of the system has a saddle-node at a critical value of the Reynolds number R_c. Then as R is varied, intermittently turbulent behavior of random duration appears, with laminar phases of mean duration approximately $|R - R_c|^{-\frac{1}{2}}$ in between.

A major difficulty with this scenario is that it has no clear-cut precursors, since the unstable fixed point that is going to collide with the stable one may not be visible. There are theoretical ways for getting out of this difficulty, but the degree to which they are physically plausible remains in doubt. As far as experimental evidence is concerned, although intermittent transition to turbulence can be seen in a number of experiments, only a small subset of these experiments seem to fit the scenario portrayed above. It should be noted that observation of the power spectrum is of little use in support of this scenario, and one must look instead at the time behavior of the system. Figure 4.20 displays a summary of the three main scenarios (excluding the Landau-Hopf model).

With the preceding mathematically distinct pictures in mind, now let's take a look at a natural system outside the realm of physics and engineering where phenomena suggestive of turbulence also seem to occur.

Exercises

1. Give a convincing set of arguments for why we cannot base a theory of turbulence on stochastic differential equations or equations having only quasi-periodic motions.

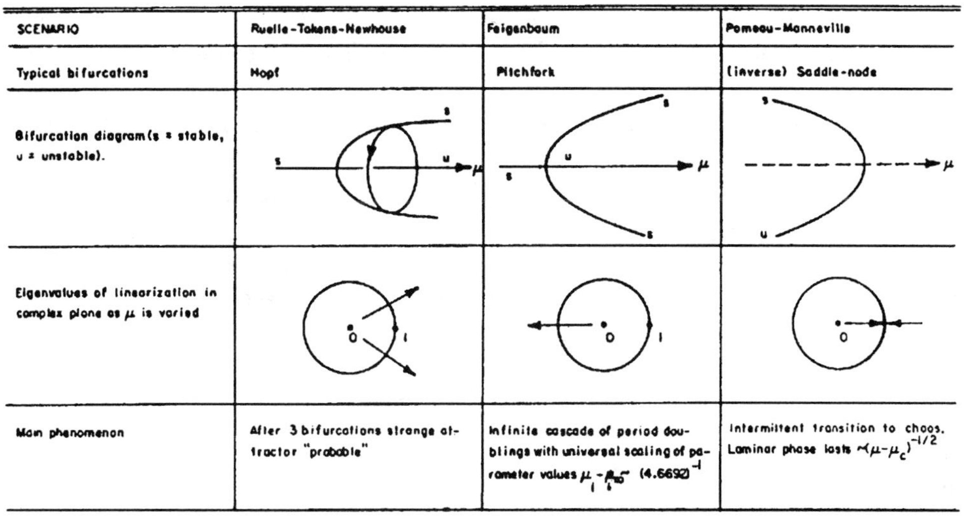

SCENARIO	Ruelle-Takens-Newhouse	Feigenbaum	Pomeau-Manneville
Typical bifurcations	Hopf	Pitchfork	(inverse) Saddle-node
Bifurcation diagram (s = stable, u = unstable).			
Eigenvalues of linearization in complex plane as μ is varied			
Main phenomenon	After 3 bifurcations strange attractor "probable"	Infinite cascade of period doublings with universal scaling of parameter values $\mu_i - \mu_\infty \sim (4.6692)^{-i}$	Intermittent transition to chaos. Laminar phase lasts $\sim(\mu-\mu_c)^{-1/2}$

Figure 4.20. Scenarios for the Onset of Turbulence

2. In laser physics, the following set of equations describe the field strength E, the polarization P and the inversion D (in properly-chosen units):

$$\begin{aligned}\dot{E} &= \chi(P - E),\\ \dot{P} &= \gamma(ED - P),\\ \dot{D} &= \alpha(\lambda + 1 - D - \lambda EP),\end{aligned}$$

where χ, γ, α and λ are parameters. (a) Show that by a change of variables in the state and parameter spaces, these equations can be transformed into those of the Lorenz system for fluid flow. (Answer: The relevant transformations are: $t \to t'\sigma/\chi$, $P \to a\eta$, $D \to \zeta$, $\alpha \to \chi b/\sigma$, $\gamma \to \chi/\sigma$, $\lambda \to r-1$, $E \to a\psi$, where $a = [b(r-1)]^{-\frac{1}{2}}$.) (b) In view of the fact that the Lorenz system describes the onset of turbulence in the atmosphere for certain values of the parameters σ, r and b, discuss the appropriate values of the laser parameters that would lead to turbulent behavior. (c) Try to interpret physically the meaning of turbulence in the laser context. (d) Which of the various scenarios for turbulence seem most appropriate for lasers?

14. Turbulent Behavior in the Spread of Disease

Classical models for the spread of disease assume that infection is caused by contact sufficiently close that the infectious agent is spread from the infected individual to one who is susceptible. Let's assume that, on the average, each individual will have the same number of contacts with every other individual over a given period of time. Further, assume that the population is constant, and that the unit of time is taken to be the time that an individual is infectious. In susceptible-infectious-susceptible, or what are called S–I–S)-type models, the individual does *not* gain immunity from the disease at the end of the infectious period, but rather goes back into the pool of susceptibles with no immunity. Diseases like the common cold and malaria follow this pattern.

Under the foregoing assumptions, if we let $I(t)$ be the number of infectives at time t, with $S(t)$ being the number of susceptibles, then the dynamics of the spread of disease can be plausibly given as

$$I(t+1) = S(t)[1 - e^{-\alpha I(t)}],$$
$$S(t+1) = I(t) + S(t)e^{-\alpha I(t)},$$

where $e^{-\alpha}$ is the probability of no contact between two individuals in one time period. In addition, let the total population be given by $N = I(t)+S(t)$. If we normalize the variables by the population as

$$x_1(t) = \frac{I(t)}{N}, \qquad x_2(t) = \frac{S(t)}{N}, \qquad A = \alpha N,$$

we obtain normalized dynamics as

$$x_1(t+1) = x_2(t)[1 - e^{-Ax_1(t)}],$$
$$x_2(t+1) = x_1(t) + x_2(t)e^{-Ax_1(t)},$$

with $x_1(t) + x_2(t) = 1$ for all t. In view of this constraint, we can write $x_1 \doteq x$ and study the single equation

$$x_{t+1} = (1 - x_t)(1 - e^{Ax_t}).$$

For the above one-dimensional model, it's easy to show that there is a unique fixed point x^* satisfying

$$x^* = (1 - x^*)(1 - e^{-Ax^*}).$$

This point is globally stable for all $A \geq 0$. So if there is to be turbulent behavior in the spread of disease, the above model must be extended.

The most straightforward way to extend the model is to assume that an infective individual spends one time unit in the infected state, then two time units in a removed (i.e., isolated) state and then returns to the susceptible state. If we let $R(t)$ represent the number of individuals in the removed state at time t, arguments of the type given above yield the normalized equations

$$\begin{aligned} x_1(t+1) &= [1 - x_1(t) - x_2(t) - x_3(t)][1 - e^{-Ax_1(t)}], \\ x_2(t+1) &= x_1(t), \\ x_3(t+1) &= x_2(t). \end{aligned}$$

For $A > 1$, this system has a unique fixed point at $x_1 = x_2 = x_3 = x^*$ satisfying

$$x^* = (1 - 3x^*)(1 - e^{-Ax^*}).$$

The stability of this fixed point is lost at

$$A = \frac{4e - 3}{e - 1} \approx 4.58,$$

at which point the system enters into a stable periodic cycle. At $A = 10$, the solution appears to be asymptotic to a stable 5-cycle, at $A = 30$ there appears to be convergence to a 10-cycle, while at $A = 40$ the limiting behavior looks to be a 20-cycle. This numerical evidence strongly suggests that for large values of the contact rate A, the period-doubling regime will terminate in the kind of transition to aperiodic motion characteristic of a strange attractor. It's tempting to speculate that should such behavior be analytically confirmed, the chaotic state would be interpretable as a random outbreak of the disease, seemingly uncorrelated with the number of initial infectives. Furthermore, such a pattern of outbreak would bear strong spatial and temporal resemblance to the onset of turbulent fluid flow, in the sense that there would be transitions between infectives and susceptibles at all scales of time and space. Operationally, such a situation would imply that the usual procedures of quarantine of the infected could no longer be expected to stem the disease, and other methods would have to be taken to reduce the contact rate A to a level below the turbulence threshold.

Exercises

1. Consider the system $\dot{x} = f(x)$, where $x \in R^n$ and

$$\begin{aligned} f_1(x) &= (1 - x_1 - x_2 - \cdots - x_k)(1 - e^{-Ax_1}), \\ f_j(x) &= x_{j-1}, \qquad j = 2, 3, \ldots, k. \end{aligned}$$

Let S be the subset of R^k defined by

$$S = \left\{ x \in R^k : 0 \leq x_j \leq 1 \text{ for all } j, \sum_{j=1}^{k} x_j \leq 1 \right\}.$$

(a) Show that $f(S) \subset S$. (b) Prove that if $0 < A \leq 1$, then the sequence $x^*, f(x^*), f^{(2)}(x^*), \ldots$ remains in S for any initial state $x^* \in S$ and converges to the zero vector, i.e., the origin is globally stable. (c) Now let $k = 2$. Prove that if $A > 1$, the system has a unique nonzero equilibrium point $\hat{x} = (p, p)$, with $p = (1 - 2p)(1 - e^{-Ap})$. (d) Show that the number p satisfies the inequalities $0 < p < \frac{1}{3}$. Is this equilibrium stable?

2. Consider the system of the preceding exercise, but now with $k = 3$. (a) Show that for $A > 1$ there is again an equilibrium $\hat{x} = (q, q, q)$, where $q = (1 - 3q)(1 - e^{-Aq})$. (b) Prove that $0 < q < \frac{1}{4}$. (c) Show that the local stability of this equilibrium is lost at a value $A^* = (4e - 3)/(e - 1) \approx 4.58$.

15. Self-Similarity and Fractals

Experiencing the world ultimately comes down to the recognition of boundaries: self/non-self, past/future, inside/outside, subject/object, and so forth. And so it is in mathematics, too, where we are continually called upon to make distinctions: solvable/unsolvable, computable/uncomputable, linear/nonlinear and other categorical distinctions involving the identification of boundaries. In particular, in geometry we characterize the boundaries of especially important figures by giving them names like circles, triangles, ellipses, and polygons. But when it comes to using these kinds of boundaries to describe the natural world, these simple geometrical shapes fail us completely: mountains are not cones, clouds are not spheres, and rivers are not straight lines.

To illustrate the point, Fig. 4.21 shows the coastline of Norway. It's patently obvious that the boundary of Norway is not any kind of simple geometrical curve, but rather is a very complicated, twisty, jagged kind of line. Moreover, even though the square grid in the figure is about $\delta = 50$ km on a side, the profile of the coastline doesn't get any smoother if we refine the scale. For example, taking $\delta = 10$ km (or even $\delta = 1$ km) on a side yields exactly the same kind of jagged boundary. To see this, note that if the coastline were a regular geometric shape like a series of connected straight lines, then the overall length of Norway's border should remain the same, regardless of the length of the "yardstick" we employ to measure it. But as shown in Fig. 4.22, when we refine our scale by reducing δ, the overall length of the border systematically increases. This is a sure-fire tipoff that there's something going on here that lies beyond the bounds of classical geometry. In this section we'll have a look at what this "something" is.

The kind of curve describing the coastline of Norway is what in 1975 Benoit Mandelbrot christened a *fractal.* Fractals are curves that are irregular all over. Moreover, they have exactly the same degree of irregularity at all scales of measurement. So it doesn't matter whether you look at a fractal from far away or up close with a microscope—in either case you'll see exactly

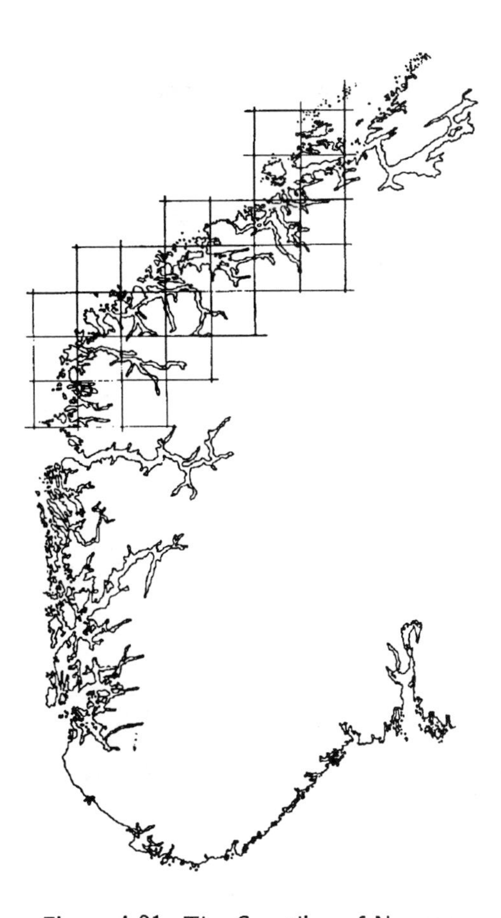

Figure 4.21. The Coastline of Norway

the same picture. If you start looking from a distance, then as you get closer small pieces of the curve that looked like formless blobs before turn into well-defined objects, whose shape is the same as that of the whole object seen previously.

There are many examples of fractals in nature: ferns, clouds, lightning bolts, coastlines, river basin networks and galaxies, to name a few. We'll look at some of these later. For now, let's look at a simple example of a fractal pattern that we saw in Chapter 3. There it arose as the output of the cellular automaton Rule 90. Here we'll dress up this pattern by giving it it's proper name: The *Sierpinski gasket.* The basic idea underlying the construction of this strange object is illustrated in Fig. 4.23. The gasket is formed by starting with the black equilateral triangle at the top. The white triangle is then cut out of the center, leaving three smaller black equilateral triangles. This excision process is then repeated on the three black triangles, obtaining nine new black triangles. This process is then

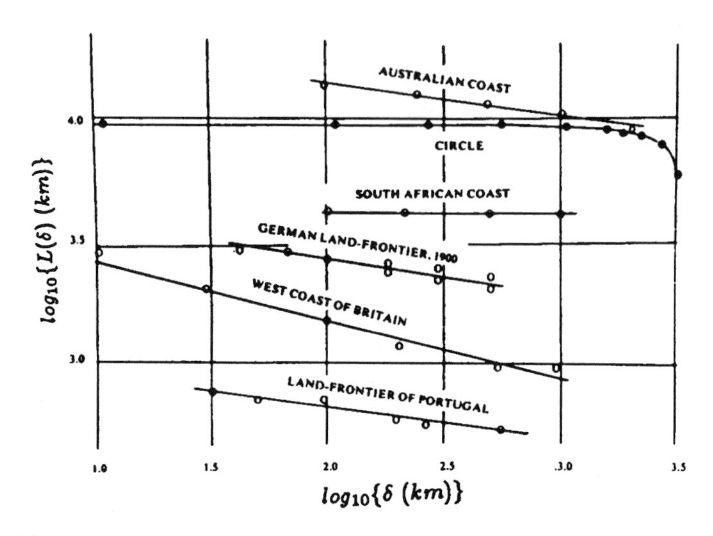

Figure 4.22. The Measured Length of the Coastline for at Different Scale Lengths

Figure 4.23. The Sierpinski Gasket

repeated indefinitely, doubling the detail at each stage. The final pattern as $k \to \infty$ is the famous Sierpinski gasket. The reader should compare the resultant figure with the evolution of the cellular automaton Rule 90 shown in Fig. 3.4 and Fig. 3.5.

The Sierpinski gasket shows what we call *linear self-similarity,* in the sense that the parts of the object are exactly like the whole. But most important fractals are not linearly self-similar. They describe nonlinear chaotic systems or even totally random processes. Let's look at an example of this in the context of a system showing the kind of chaos we've been examining in this chapter.

Example: The Boundary of Spiral Chaos

Consider the dynamical system

$$\begin{aligned}
\dot{x} &= 2.2x - 4.4(y+z)/(x+0.01) + 22, \\
\dot{y} &= 1.2x - y, \\
\dot{z} &= 14x - 140z/(z+0.05), \\
\dot{w} &= ax + 0.5w - (2+a)w/(w+0.05) - 0.1w^2.
\end{aligned}$$

The first three variables of this system represent Michaelis-Menten-type chemical reactions, while the last variable w is a kind of chemical "switch." When a, the forcing parameter of the switch, in greater than zero, the system displays chaotic behavior. But if $a = 0$ the system moves to one of two stable steady-states, depending only on w_0, the initial value of w. On the other hand, if $a > 0$ then the value of w_0 at which the system switches from the upper to the lower steady-state depends on x_0, as well.

To find out more about the nature of the switching curve in the (x_0, w_0)-plane, Otto Rössler conducted a set of numerical experiments fixing the values $a = 0.25$, $y_0 = 6$, $z_0 = 0.1$, and letting w_0 and x_0 varying in the regions $0 \leq x_0 \leq 15$, $0.5 \leq w_0 \leq 2.5$. The results are displayed in Fig. 4.24, where the white points are initial conditions leading to one steady-state while the black points lead to the other. The point to note is that the boundary between the two steady-state solutions is no longer a horizontal straight line, but looks more like some kind of "moonscape." Figure 4.25 further reinforces this feeling. Part (a) shows a five-fold blowup of the part of Fig. 4.24 near the "trough," while part (b) is a ten-fold magnification of the peak inside the trough of part (a). Just as with the Siepinski gasket, in this example we find the boundary separating the domain of attraction of the two steady states being some kind of irregular curve defying description in classical geometrical terms.

As the preceding examples show, for complicated geometrical objects the ordinary notion of dimension may vary with scale. For fractals, the counterparts of the familiar integer dimensions of things like points, lines and squares are values that are not usually whole numbers. The simplest kind of fractal dimension is what's called the *similarity dimension D*. If we

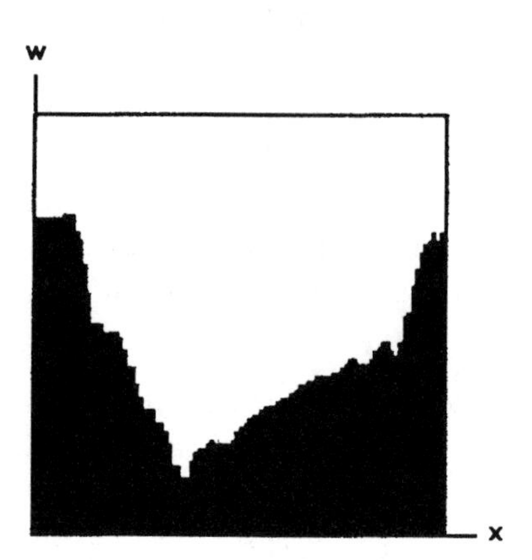

Figure 4.24. Boundary of Two Domains of Attraction

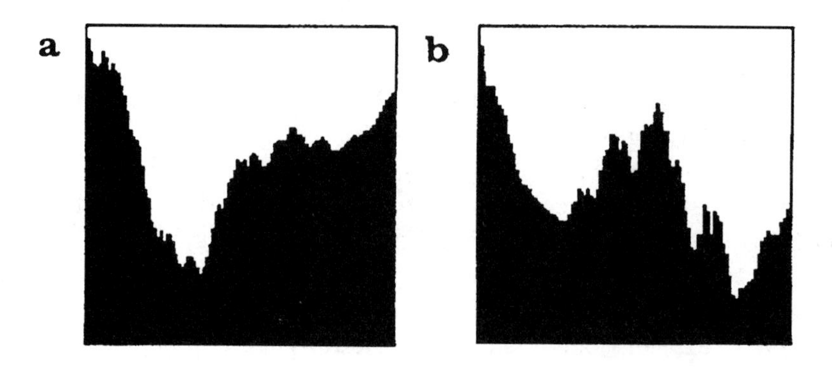

Figure 4.25. (a) Five-Fold, and (b) Ten-Fold Magnifications of the Trough

have a linearly self-similar curve, it could be anything from an almost smooth one-dimensional line to a nearly plane-filling curve that visits virtually every part of the plane. In the first case, D will be only slightly greater than one; in the second, D will be very nearly two. Intuitively speaking, the similarity dimension measures the degree of "roughness" of a fractal shape. Let's see how to calculate this quantity.

If we have a line segment of unit length, then we can cover the line by N nonoverlapping lines of length $1/N$, for any positive integer N. We then say that the line is (linearly) self-similar with scaling ratio $r(N) = 1/N$. Similarly, a rectangle in the plane may be covered by scaled-down rectangles if we change the length scale by $r(N) = (1/N)^{1/2}$. For cubes, the appropriate scaling factor is $r(N) = 1/N^{1/3}$, while for a general self-similar shape we use a scale factor $r(N) = (1/N)^{1/D}$, where D is the similarity dimension discussed above. We can rewrite this expression to obtain D as

$D = -\log N / \log r(N)$. So, for example, in our construction of the Sierpinski gasket the filled triangles are replaced by $N = 3$ triangles, each of which has been scaled down by a factor $r = \frac{1}{2}$. Therefore, the similarity dimension of the Sierpinski gasket is $D = \log 3 / \log 2 \approx 1.58$.

But how do we compute the similarity dimension in practical cases like the Norwegian coastline shown earlier in Fig. 4.21? The procedure is rather straightforward. First of all, cover the region with a set of squares having edge length δ. Let $N(\delta)$ be the number of such squares needed to cover the coastline. It follws from the discussion above that in the limit of small δ, we must have $N(\delta) \sim 1/\delta^D$. Since on a log-log scale the relationship $N(\delta)$ versus δ is linear, we can determine the fractal dimension of the coastline by finding the slope of the the quantity $\log N(\delta)$ viewed as a function of $\log \delta$. We find this straight line simply by computing $N(\delta)$ for smaller and smaller values of δ. Now let's use this same general idea in another context, that of river basin networks.

Example: River Basin Networks

Consider the drainage area A shown in Fig. 4.26, where L denotes the length of the longest river in the network. It's clear from the figure that for rivers with a constant drainage width d, the relationship between L and A would be simply $L = A/d$. In that case we would conclude that the similarity dimension is $D = 2$. On the other hand, if the drainage basin kept the same shape as we magnified it, then we would obtain $D = 1$. In 1982 Mandelbrot suggested the following length-area relationship for real drainage networks:

$$L(\delta) = k\delta^{1-D} A(\delta)^{D/2},$$

where L is the length of the perimeter bounding the area A, δ is the "yard-stick size" used to measure the length, and k is a constant of proportionality. Note here that the area A of the region is also a function of δ, and is assumed to be measured by covering it with little squares of area δ^2.

Studies of drainage networks in the northeastern United States using topographic maps and aerial photographs have shown that empirically we have

$$L = 1.4A^{D/2},$$

with $D = 1.2$. Consequently, the fractal dimension of most rivers and streams is $D = 1.2$ and, as a result, there is good evidence for concluding that in this region the stream length is, on the average, proportional to the $D/2 = 0.6$ power of its drainage area—regardless of the geological or topographic features of the land area.

The preceding examples have shown natural phenomena with something happening on all scales of length. And what seems to make many

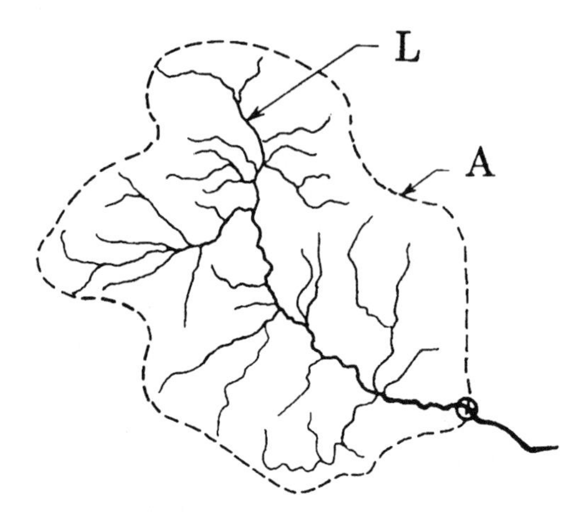

Figure 4.26. A Drainage Area

self-similar natural objects interesting is just this fact that they possess significant features on many length scales. But spatial lengths are not the only intervals that we care about in life. Temporal lengths and frequency intervals are two others. Music is a good example of this, as no matter what scale the music is written on, it must have pitch changes on many frequency scales and rhythm changes on more than one time scale in order to be appealing. Let's illustrate this point by considering some of the music written by Johann Sebastian Bach and how it might be "downsized" using our ideas of self-similarity and fractals.

Example: Bach and Fractal Music

If an electrical engineer were to compute the power spectrum (the squared magnitude of the Fourier transform) $f(x)$ of the relative frequency intervals x between successive notes in Bach's *Brandenburg Concerto,* it would be found that over a large range $f(x) = c/x$, where c is some constant. Thus, Bach's music is characterized by the kind of "noise" that engineers call *1/f noise.* This hyperbolic functional relationship between the frequency and the intervals between successive notes shows up in the spectrum of amplitudes, too, as displayed in Fig. 4.27 showing the amplitude spectrum for Bach's *First Brandenburg Concerto.* The question, of course, is why Bach and so many other composers created music that seems to obey this $1/f$ rule.

Part of the answer to this puzzle lies in the simple observation that in order for any piece of music to be "interesting," it should be neither too regular (like most rock and C&W tunes) nor too unpredictable (like a lot of avant garde compositions). In power spectrum terms, this means that the relationship between the Fourier transform and the frequency should not

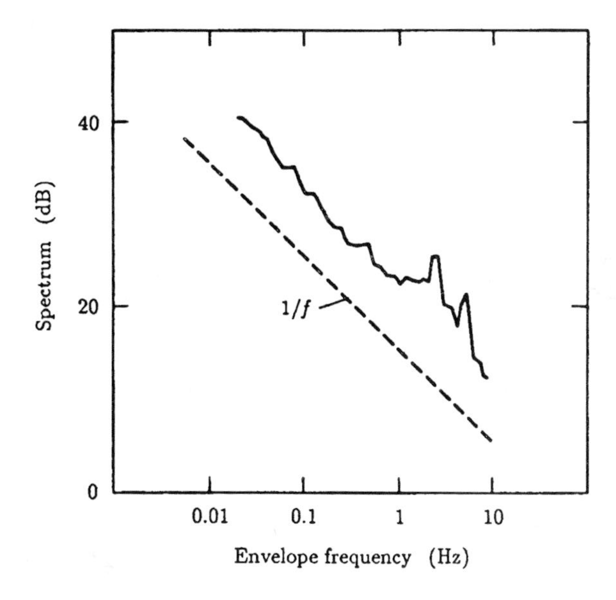

Figure 4.27. The Amplitude Spectrum for the *First Brandenburg Concerto*

behave like monotonic "brown" noise, with its $1/f^2$ frequency dependence, or like unpredictable "white" noise and its $1/f^0$ dependence. The intermediate "pink" noise case corresponds to our $1/f$ relationship. Figure 4.28 shows examples of music composed according to each of these frequency patterns. Part (a) shows "white" music produced from independent notes, part (b) is "brown" music composed of notes with independent frequency increments, while part (c) consists of "pink" music whose notes have a frequency and duration determined by $1/f$ noise.

The homogeneous power laws obeyed by fractal objects suggests a way of "compressing" the music of a Bach or a Mozart down to its irreducible essence. This idea, in turn, opens up the possibility of perhaps using this distilled Bach "essence" to compose new "Bach-like" music. Kenneth Hsu took up this challenge. Figure 4.29 shows the results of some of his work. In the figure, the notes are plotted using the frequency and amplitude spectrums of Bach's music to determine each note's relationship to its neighbors. By removing notes from several of Bach's inventions, Hsu found that basic patterns persisted in the fractal reductions of the music, even if what remained contained as little as 1/64th of the original notes. Thus, music recognizable as Bach survives fractal reduction. And, in fact, to some ears this "reduced Bach" gives the impression of an economy of frills and ornamentation. For those of a musical orientation, Fig. 4.29 shows one of these reductions in the case of Bach's *Invention 5.*

Figure 4.28. (a) White, (b) Brown and (c) Pink Music

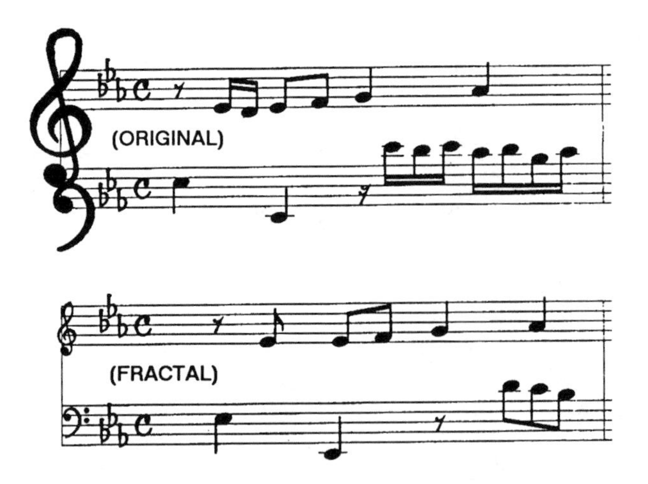

Figure 4.29. Bach's *Invention 5* in Original and Fractal Form

With these basic ideas of fractals in nature under our belts, let's turn to a more thorough consideration of the most common way these weird objects are created in computers—by iteration.

Exercises

1. Construction of the Sierpinski gasket started with a black equilateral triangle and proceeded by sequentially cutting similar triangles out of the "seed." (a) Start with a single black square and apply the same excision procedure, cutting out a smaller square from the center of the black squares at each stage. By this procedure, you will obtain what's called the *Sierpinski carpet.* (b) Compute the similarity dimension D of the carpet. (Answer: $D = \log 8/\log 3 \approx 1.89$.) (c) Starting with a cube instead of a square, apply the same procedure to obtain the *Menger sponge.* Show that the sponge has fractal dimension $D = \log 7/\log 3 \approx 1.77$.

2. Using the scheme suggested in the text, compute the similarity dimension D of the Norwegian coastline. (Answer: $D \approx 1.52$.)

3. In the river basin example, why is the average stream length proportional to half the similarity dimension instead of to D itself?

4. Define a metric on R^2 by the rule $d(x, y) = |x_1 - y_1| + |x_2 - y_2|$. (a) Can you see why this is called the *Manhattan metric?* (b) Let A be a compact subset of R^2. Suppose A has fractal dimension D_1 when evaluated using the Euclidean metric, while it has fractal dimension D_2 when evaluated using the Manhattan metric. Show that $D_1 = D_2$.

16. Fractals and Domains of Attraction

We have seen in Chapter 2 that the most important thing one can know about a differential or difference equation is its set of attractors. Associated with each attractor, there is a set of initial states of the dynamical system that end up in that attractor under iteration of the system's vector field. These points are called the *domain of attraction* of the attractor. In our spiral chaos example of the last section, we saw that the boundary of the domain of attraction can be a very complicated curve or surface. In this section we'll show that even for the simplest kind of nonlinear dynamics, the structure of the domain of attraction is most often a fractal. Let's see why.

Fix a point c in the complex plane. Now form a sequence of complex numbers by the rule

$$z_k = z_{k-1}^2 + c, \qquad z_0 = 0, \qquad k = 1, 2, \ldots . \tag{$*$}$$

So for each choice of the parameter c, we have a different quadratically nonlinear dynamical system. It's clear that for some points c, the sequence $\{z_k\}$ will diverge to infinity. The choice $c = 2$ is such a point. But for other points like $c = i$, the sequence remains bounded. We now form a set M in the complex plane in the following way: We put the number c in M if the sequence $\{z_k\}$ does **not** go off to infinity; otherwise, c is not in M. So, for

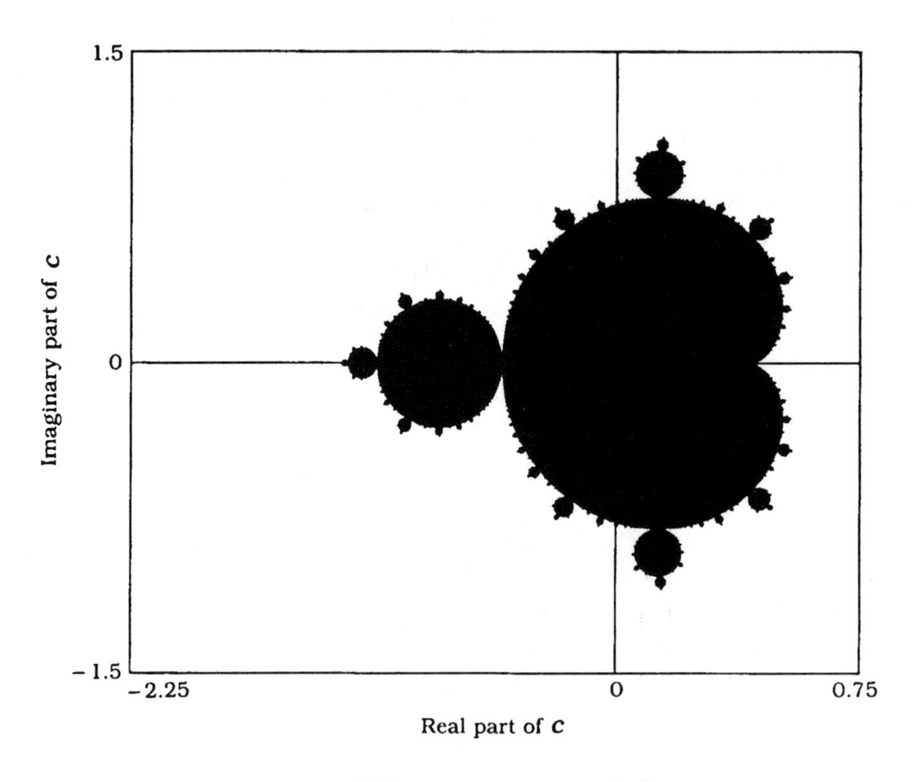

Figure 4.30. The Mandelbrot Set

instance, the point $z = 2$ is not in M, while the point $z = i$ is. The set M is called the *Mandelbrot set* and is shown above in Fig. 4.30.

It turns out that when the starting point of our iteration is in the interior of M, the quadratic map $(*)$ yields an orbit that is orderly and well-behaved. But if z_0 is outside M, the orbit, while perfectly deterministic, is wild and disorderly, wandering all over the complex plane. The boundary of M, separating the orderly and disorderly behaviors, turns out to be unbelievably messy. As one might surmise from inspecting Fig. 4.30, new copies of the entire Mandelbrot set continue to "bud off" from the original at all scales. Furthermore, there are many additional structures that appear at various magnification levels. For instance, Fig. 4.31 shows a magnified view of part of the boundary of the set M. New structures continue to appear as one goes to higher and higher magnifications, as the set M remains forever intricate on any scale. It's this kind of never-ending sequence of new patterns that leads many to claim that the Mandelbrot set is the most complicated object known to humankind.

To add further ammunition to this claim, in 1991 the Japanese mathematician Mitsuhiro Shishikura proved that the dimension of the boundary

Figure 4.31. Magnified View of Part of the Boundary of the Mandelbrot Set

of the Mandelbrot set is 2. This means that not only is the boundary of M indeed a fractal, but that it "wiggles" as much as any curve in the plane possibly can. In short, if you measure the complexity of a curve by its "wiggliness," then no curve can be more complex than the boundary of M. However, this does **not** mean that the boundary of M is a space-filling curve. Merely having dimension 2 is not enough for this as is shown by the case of a two-dimensional sponge, which is all "holes." Whether or not the boundary of the Mandelbrot set fills up a two-dimensional continuum in the plane is, at present, still an open question.

The Mandelbrot set can be thought of as characterizing the domain of attraction of the starting point $z_0 = 0$ for the *family* of quadratic maps $z \to z^2 + c$. But this is not usually what we mean when we speak of the domain of attraction. The usual idea is that we have a *single* system with one or more attractors, and we want to characterize those initial states that lead to the different attractors. This was the situation, for example, with the spiral chaos example discussed in the last section. And just as with that particular example, it turns out that if the system has a strange attractor, the boundary of that kind of attractor has remarkable fractal properties.

Let's go back to our earlier quadratic mapping

$$z_k = z_{k-1}^2 + c, \qquad k = 1, 2, \ldots,$$

and fix the parameter value at $c = -1$. Now we consider the set of initial points z_0 such that the sequence $\{z_n\}$ remains bounded as $k \to \infty$. The boundary of the black part of Figure 4.32 shows these points, while the black interior is the strange attractor for this system. The set of boundary

Figure 4.32. The Julia Set of the Quadratic Map $z \longrightarrow z^2 - 1$

points is called the *Julia set* of the system, while the filled-in black part is called the *filled-in Julia set.* For the sake of completeness, let's note that the complement of the Julia set (the white part) is called the *Fatou set.* These names owe their origin to the French mathematicians Gaston Julia and Pierre Fatou, who studied this kind of set in the early part of this century.

It's interesting to see what happens to the Julia set for our quadratic map when $c = 0$. In that case, the iterates are just the powers of z, z^2, z^4, z^8 and so on. Clearly, if the initial point z_0 lies inside the unit circle, then its orbit remain bounded; starting outside the unit circle, it flies off to infinity. And if $|z_0| = 1$, the orbit moves round the circle indefinitely. Therefore, for this system the Julia set is simply the unit circle $|z| = 1$. Amazing things now happen to the Julia set when we perturb c away from zero.

For even the slightest nonzero value of c, the Julia set gets distorted—but not into a smooth, roughly circular curve. No. What happens is that the curve becomes a fractal. Figure 4.33 shows a sequence of distortions of the original unit circle as we gradually increase the magnitude of c, going from the "almost circle" in part (a) to the so-called Douady rabbit in part (b) and on to a set of disconnected islands in (c).

This quadratic map, elementary as it is, shows us that the boundary of the domain of attraction of a strange attractor can be just about as complicated as a mathematical object can get. And, in fact, it's rather easy to see that the Julia set for this map and the Mandelbrot set are intimately related. Some of these relationships are brought out in the Exercises and Problems. But to get a real feel for the depth and beauty of these objects, the more detailed literature cited in the Notes and References is must reading.

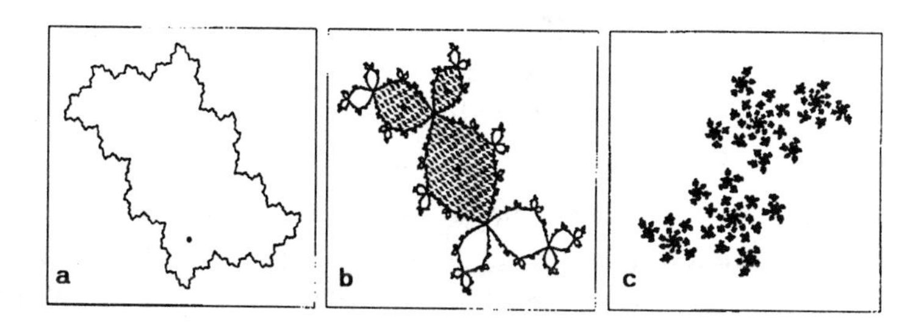

Figure 4.33. The Julia Set for Various Nonzero Values of c

Exercises

1. Let $W = \{w_1, w_2, \ldots, w_n\}$ be a set of contraction mappings of the plane, i.e., each w_i moves pairs of points in the plane closer together. Choose an arbitrary starting point $z_0 = (x_0, y_0)$ and randomly pick a transformation from the set W to use at each iteration. (a) Show that with probability 1, the sequence $z_0, z_1, z_2, \ldots$ obtained in this way will converge to a limiting set of points P. (b) Prove that P is the unique set satisfying the relation

$$P = w_1(P) \cup w_2(P) \cup \cdots \cup w_n(P).$$

(c) Experiment with this process using the transformations

$$w_i(x, y) = \begin{bmatrix} a_i & b_i \\ c_i & d_i \end{bmatrix} \begin{bmatrix} x \\ y \end{bmatrix} + \begin{bmatrix} e_i \\ f_i \end{bmatrix}, \qquad i = 1, 2, 3, 4,$$

where the elements of the matrices are given by

i	a	b	c	d	e	f	p
1	0	0	0	0.16	0	0	0.01
2	0.85	0.04	-0.04	0.85	0	1.60	0.85
3	0.20	-0.26	0.23	0.22	0	1.60	0.07
4	-0.15	0.28	0.26	0.24	0	0.44	0.07

Here the entries in the column labeled p denote the probability of selecting the corresponding transformation w_i.

2. Newton's method for finding a root of the equation $f(z) = 0$ is given by the iterative scheme

$$z_{n+1} = z_n - \frac{f(z_n)}{f'(z_n)}.$$

Consider the function $f(z) = z^2 - 1$. (a) Show that if the starting point z_0 has a nonnegative real part, then Newton's method converges to either the root $+1$ or -1. (b) Prove that if the starting point is purely imaginary, i.e., $z_0 = ir_0$, where r_0 is real, then Newton's method does not converge at all. (c) Show by a change of variable that in this latter case, the iterative scheme is equivalent to the Circle-2 map $x_{t+1} = 2x_t \mod 1$.

3. Consider Newton's method for the cubic equation $f(z) = z^3 - 1$. (a) Show by counterexample that from a given z_0, the iterates do not necessarily converge to the root closest to z_0. (b) Prove that the domains of attraction of the three roots form a "cloverleaf" with 3 leaves intersecting at the origin.

4. Each of the "bulbs" of the Mandelbrot set has a specific dynamical meaning. For example, the region inside the main cardioid consists of all c values for which the quadratic map $z \to z^2 + c$ has an attracting fixed point. What dynamical interpretation can you give to: (a) the circular region immediately to the left of the cardioid, (b) the small bulbs budding off from the main cardioid, and (c) the bulbs emerging from the circular region of part (a)?

5. Consider the quadratic polynomial $p(z) = z^2 - 2$. (a) Show that the Julia set of p is the closed interval of the real line $-2 \leq x \leq 2$. (b) Prove that the polynomial $p(z)$ is topologically conjugate to the logistic map $z \to 4z(1 - z)$. (This is an alternate way of showing that the logistic map has a dense set of periodic points in the unit interval).

Discussion Questions

1. In a by now famous series of "debates" over the physical *interpretation* of the state vector in quantum mechanics, Niels Bohr championed the so-called Copenhagen position that nature is intrinsically random, and that it is impossible—*in principle*—to measure simultaneously and with complete accuracy two conjugate attributes of a particle like its position and momentum. Albert Einstein, on the other hand, took the point of view expressed in his famous aphorism, "God does not play dice with the Universe." He claimed that the Copenhagen interpretation was incomplete, and that the Uncertainty Principle underlying the measurement problem was an artifact of limitations on our ability to measure and not an inherent property of nature. Do the results we have seen on deterministic chaos shed any light on this debate?

2. In Chapter 6 we shall consider in detail the concept of a *completely observable* system. Rougly speaking, this is a dynamical process whose measured output contains sufficient information for us to determine exactly what initial state the system began in. By extending the linear arguments

given in Chapter 6 to polynomial systems, it's straightforward to show that the "chaotic" logistic equation $x_{t+1} = 4x_t(1 - x_t)$ is completely observable in the system-theoretic sense, i.e., by observing the output, we can identify the initial state. How can you reconcile this result with the fact that the logistic equation displays random-looking behavior?

3. In his kinetic theory of gases, Ludwig Boltzmann argued that molecular motion should be thought of as random, and that every molecule explores the entire region of phase space that is accessible to it, subject to energy constraints. This assumption is called the *ergodic hypothesis,* and serves as a central focus in the theory of statistical mechanics. What does the KAM Theorem say about this hypothesis? In particular, does the KAM Theorem provide a tool for confirming or refuting Boltzmann's hypothesis?

4. In the period-doubling route to chaos for one-dimensional maps, the range of parameter values for the successive cycles decreases monotonically and rapidly. Thus, small changes in the parameter can result in the emergence of a cycle of much longer length, or even a cycle of infinite length (an aperiodic orbit). Discuss the implications of this fact for practical identification of chaotic behavior from *real* data. In particular, do you see any practical way of deciding whether a sequence of observed data points is from an aperiodic orbit or is from a long stable cycle?

5. We know that the behavior of the logistic map is equivalent mathematically to a sequence of flips of a fair coin (Bernoulli trials with $p = \frac{1}{2}$). This fact leads to the description of the behavior of the system in terms of the Bernoulli probability distribution. Conversely, given a Bernoulli distribution, we can always use the logistic map as a mechanism for realizing this distribution. What about other probability distributions? That is, given an arbitrary probability distribution $P(x)$, can we always find a deterministic mechanism whose dynamics unfold in accordance with that distribution? What bearing does this question have upon the issues presented in Discussion Question 1? (Hint: Consider computer random number generators.)

6. Intuitively, it seems that a fixed point is a "simpler" kind of attractor than a periodic orbit (limit cycle) which, in turn, seems simpler than an aperiodic orbit. Can you think of any good scheme that would formalize this intuition into a definite procedure for characterizing the *complexity* of an attractor? In this regard, discuss the Kolmogorov-Chaitin concept of complexity (see Chapter 3, Discussion Question 4 and Chapter 9 for more information about these matters).

7. Consider a measuring instrument with a uniform scale of resolution ϵ. Thus, the measurement of a variable yields one of $1/\epsilon$ numbers. If we have an n-dimensional dynamical system and assign such an instrument for the

measurement of each of the n variables, then the phase space of the system is partitioned into ϵ^{-n} boxes. Let $N(\epsilon)$ be the number of boxes needed to cover the attractor of the system. Let $P_i(\epsilon)$ be the "probablility density" of occurrence of the attractor in the ith box, i.e., the fraction of the time that the trajectory is in the ith box. The average information contained in a single measurement is then

$$I(\epsilon) = -\sum_{i=1}^{N(\epsilon)} P_i(\epsilon) \log_2 P_i(\epsilon).$$

Define the *information dimension* D_I of the attractor to be

$$D_I = \lim_{\epsilon \to 0} \left(\frac{I(\epsilon)}{|\log_2 \epsilon|} \right).$$

The information dimension achieves its maximal value when the probability of all boxes is equal, in which case $I(\epsilon) = \log_2 N(\epsilon)$.

Discuss the desirability of trying to reduce the observational resolution ϵ to zero. In particular, since the dynamic storage capacity of the attractor is characterized by the number D_I, is there any *a priori* reason to believe that the maximal storage capacity of the attractor occurs when $\epsilon = 0$? Consider this question within the context of the survivability of biological organisms.

8. Reliable information processing requires the existence of a good code or language, i.e., a set of rules that *generate* information at a given hierarchical level, and then *compress* it for use at a higher cognitive level. To accomplish this, a language should strike an optimum balance between variety (stochasticity) and the ability to detect and correct errors (memory). Consider the degree to which the type of chaotic dynamical systems of this chapter could be used to model this dual objective. In particular, discuss the notion that entropy (randomness) increases when the volume in the state space occupied by the system flow expands, while it is compressed (thereby revealing information) when the volume contracts, i.e., when the system moves toward the attractor.

9. The C^0-Density Theorem asserts that if a vector field is structurally stable, the only attractors of the field are fixed points and closed orbits. Further, the Stability Dogma claims that good mathematical modeling practice dictates that we use structurally stable models as a hedge against the inevitable uncertainty present in our knowledge of the true dynamics and in the imprecision of our observations. In short, we can't put much faith in predictions made using a structurally unstable model. Consider these arguments in light of the claim that structurally unstable chaotic models serve as a good mathematical metaphor for characterizing things like turbulent fluid flow, the outbreak of epidemics and the growth of insect populations.

10. A topic of considerable interest in some economic circles is the *Kondratiev wave,* which describes what its proponents claim is a periodic cycle in global economic affairs. Assuming such a wave is not just an artifact of the economist's imagination, discuss the plausibility and relative merits of the following mathematical "explanations" of the wave:

- The wave is a stable limit cycle arising from the Hopf bifurcation mechanism.
- The wave is one of the periodic orbits emerging during the period-doubling bifurcation of a logistic-type map.
- The wave is a quasi-periodic orbit on an n-torus.

How would you go about using real data and economic relationshiops to test the foregoing hypotheses?

Assume that the dynamical process generating the wave contains a parameter μ representing the rate of technological development. What kind of behavior would you expect to see for small, medium and large values of μ?

11. The human heart provides an example of a physical process whose proper functioning demands a very stable attracting periodic orbit. One of the characteristic signs of heart failure is the loss of this stability, together with the ensuing aperiodic phenomenon of ventricular fibrillation. Consider how you might use chaotic dynamics to model this process. What quantity might serve as the bifurcation parameter?

12. The characteristic feature of dissipative systems is that the phase volume contracts over time, ultimately leading to the attractor. However, we have also seen that such systems may have strange attractors, which implies orbits that *diverge* exponentially fast. How can you reconcile the contracting volume property with the exponentially divergent aperiodic orbits? Try to relate this observation to the property of *equifinality* in biological systems, whereby many organisms tend to develop the same final forms quite independently of their very different genotypes.

13. In most physical experiments we expect the data to be well approximated by a deterministic model. Differences between the predictions of the "best" deterministic model and the data are generally attributed to stochastic effects. When the appropriate deterministic models are chaotic, how does one go about separating the deterministic and the stochastic components of the data? Do you see this kind of issue as possibly constituting a general form of the Uncertainty Principle of quantum mechanics, whereby scientists are forced to accept an irreducible level of uncertainty, or "fuzziness," in their representations of the world?

14. By a variety of straightforward arguments, it's easy to see that the Schrödinger equation governing the wave function of quantum mechanics

does not display the kind of deterministic randomness we've discussed in this chapter—at least, not for finite, bounded, conservative systems. And, in fact, all existing evidence suggests that this equation will not display chaotic behavior even for infinite, unbounded and/or time-dependent systems.

This fact about the Schrödinger equation leads to a paradox: classical deterministic Newtonian dynamics, like the logistic equation, generally exhibit randomness in their temporal evolution; quantum systems do not. Thus, deterministic Newtonian systems have uncomputable trajectories, while quantum systems are computably deterministic. How could we resolve this paradox? In particular, consider the notion that perhaps quantum mechanics, as it's traditionally interpreted, is just not random enough.

15. Consider a throw of the dice at the craps table. By virtue of the dissipation of energy, the dice must always come to rest within a finite time. But the kind of chaotic systems with fractal domains of attraction we've discussed in this chapter all relate to motions that can be sustained forever. Therefore, we must conclude that the toss of a die cannot show the sensitive dependence on initial conditions found in chaotic systems. Yet dice-throwing certainly seems to be **exactly** the kind of process that displays sensitive dependence on the initial conditions. Consider this kind of *finite* sensitivity to initial conditions, and discuss possible connections with standard chaotic behavior.

16. In the book *Chaos Bound,* N. Katherine Hayles discusses parallels between contemporary literature and critical theory and the science of chaos, using works like *The Education of Henry Adams* and Stanislaw Lem's science-fiction classic *His Master's Voice.* Discuss her assertion that the paradigm of chaos includes elements that were evident in literary theory and literature long before they became prominent in the sciences. Here she has in mind the shift of the focus of literary studies toward local, fragmentary modes of analysis in which texts are no longer regarded as deterministic or predictable. Consider her claim that such similarities between the natural sciences and the humanities are the result not so much of direct influence, but due more to having roots in a common cultural matrix.

17. *Zipf's Law* is an empirically-observed relationship between the word rank and the word frequency for many natural languages. Here a word has rank r if it is the rth word when the words of the language are listed according to their frequency. More formally, Zipf's Law says that the relative word frequency function f can be expressed by the relation

$$f(r) \approx \frac{1}{r \log(1.78R)},$$

where R is the number of different words in the language. Thus, if we take $R = 12,000$ for English, then the relative word frequency of the most common words like "the," "of," and "and" are 0.1, 0.05 and 0.033, respectively. The foregoing relation for $f(r)$ is again a $1/f$ law, just as we saw in the text for the "language" of nontrivial music. Discuss what this might mean in the context of spoken and/or written language. By regarding the states of a discrete-time chaotic system as symbols making up the words of a language, think about how Zipf's Law might be obtained as a consequence of the iteration of such a system.

18. Morse's Theorem (see Chapter 2) says that the typical smooth function we can expect to encounter is locally quadratic near a critical point. Discuss this genericity result within the framework of the quadratic maps leading to the Mandelbrot set and its many relatives. In particular, do you think that the genericity guaranteed by Morse's Theorem and the ubiquitousness of systems with strange attractors having fractal domains of attraction have anything to do with each other? Or are they just mathematical coincidences? (For more information on this matter, see Problem 20.)

Problems

1. Consider the dynamical system

$$\dot{x} = x^2,$$
$$\dot{\theta} = 1,$$

which has the solution (flow)

$$\phi_t(x_0, \theta_0) = \frac{tx_0}{1 - tx_0} + \theta_0.$$

a) Using the section

$$\Sigma = \{(x, \theta) : \theta = 0\},$$

show that the Poincaré map for this flow is given by

$$P(x_0) = \frac{x_0}{1 - 2\pi x_0}, \qquad x_0 \in (-\infty, 2\pi).$$

(Remark: This illustrates the fact that the Poincaré map may not always be globally defined.)

b) For the system

$$\dot{r} = r(1 - r^2),$$
$$\dot{\theta} = 1,$$

use the section $\Sigma = \{(r, \theta)\colon r > 0,\ \theta = 0\}$ to conclude that the Poincaré map is given by

$$P(r_0) = \left[1 + \left(\frac{1}{r_0^2} - 1\right) e^{-4\pi}\right]^{-\frac{1}{2}}.$$

Prove that the closed orbit is stable by showing that the derivative dP/dr_0 at $r_0 = 1$ is equal to $e^{-4\pi} < 1$.

2. Consider the system

$$x_{t+1} = f(x_t) \doteq (1 - c)x_t + p(x_t),$$

where $p(x) = bx^s \exp(-xs/r)$. Assume $0 < c \leq 1$, $s > 1$ and the line $y = x$ intersects the graph of p.

a) Sketch the graph of f and show there are two positive fixed points. Denote the smaller by u_c and the larger by v_c.

b) Show that u_c is unstable for all c and that v_c may be either stable or unstable, but that it's stable if c is sufficiently small. Prove that for $c \in (0, 1]$, we have $v_1 \leq v_c < \infty$. Moreover, show that v_c is monotonically decreasing as a function of c, with $v_c \to \infty$ as $c \to 0^+$, and that $v_c \to v^*$ as $c \to 1^-$.

c) Assume that $p'(v^*) < -1$. Prove that there then exists a unique value of c, call it c_a, in the unit interval such that $f_c'(v_c) = -1$. Show that for $c \in (0, c_a)$ the point v_c is stable, while for $c \in (c_a, 1]$ it is unstable.

d) Show that for any $c > c_a$, the system has points of period-2. (Hint: Use the fact that when $c > c_a$, the derivative of f_c^2 is greater than 1 at u_c and at v_c.)

e) For any $c \in (0, 1]$, let y_c be the point at which f_c attains its local maximum. Moreover, assume that $p^2(y_1) < u_1$, where u_c is the smaller of the positive fixed points of f. Under these conditions, prove that there exists a point $c_b \in (0, 1)$ such that for any $c \in (c_b, 1]$, f_c has a point of period 3. Hence, conclude that in this case there are uncountably many aperiodic orbits.

3. Consider the system

$$\dot{x} = -\zeta x - \lambda y + xy,$$
$$\dot{y} = \lambda x - \zeta y + \frac{1}{2}(x^2 - y^2).$$

a) Show that for $\zeta = 0$, $\lambda > 0$, the system is integrable with Hamiltonian

$$H(x, y) = -\frac{\lambda}{2}(x^2 + y^2) + \frac{1}{2}\left(xy^2 - \frac{x^3}{3}\right).$$

This system has three saddle points at

$$p_1 = (\lambda,\, \sqrt{3}\lambda), \qquad p_2 = (\lambda,\, -\sqrt{3}\lambda), \qquad p_3 = (-2\lambda,\, 0),$$

as well as a center at the origin. Draw the phase portrait of the system showing that the saddle points are connected. What happens to these connections if $\zeta > 0$?

b) Prove that when $\lambda = 0$ this is a gradient dynamical system with potential function

$$V(x,\, y) = \frac{\zeta}{2}(x^2 + y^2) + \frac{1}{2}\left(\frac{y^3}{3} - x^2 y\right).$$

Show that in this case the origin is a sink (for $\zeta > 0$), and that there are saddle points at

$$q_1 = (\sqrt{3}\zeta,\, \zeta), \qquad q_2 = (-\sqrt{3}\zeta,\, \zeta), \qquad q_3 = (0,\, -2\zeta).$$

What does the phase portrait look like?

4. The map

$$x_{t+1} = \begin{cases} +1 + \beta x_t, & x_t \in [-1,\, 0), \\ -1 + \beta x_t, & x_t \in (0,\, 1], \qquad 1 < \beta < 2, \end{cases}$$

has the points $x = \pm\frac{1}{\beta}$ as the first two preimages of zero, i.e., points that map to zero. Show that the kth preimages of zero are given by the points

$$x = \sum_{j=1}^{k} \left(\pm\frac{1}{\beta^j}\right).$$

Hence, conclude that those points constituting the stable manifold of $x = 0$ are dense in the interval $[-1,\, 1]$.

5. Suppose we are given the system

$$\begin{aligned} \dot{x} &= Ax + f(x,\, y), \\ \dot{y} &= By + g(x,\, y), \qquad f,\, g \in C^\infty, \end{aligned} \tag{\dagger}$$

where $x \in R^n$, $y \in R^m$ with A and B constant matrices. Suppose further that B is a stability matrix (i.e., $\operatorname{Re}\lambda_i(B) < 0$, $i = 1, 2, \ldots, m$), and A has all of its roots on the imaginary axis ($\operatorname{Re}\lambda_j(A) = 0$, $j = 1, 2, \ldots, n$). In addition, let $f(0) = g(0) = f'(0) = g'(0) = 0$, where $'$ denotes the Jacobian matrix.

If $y = h(x)$ is an invariant manifold for the system and h is smooth, we call h a *center manifold* for the system if $h(0) = h'(0) = 0$. Note that if $f = g = 0$, all solutions tend exponentially fast to solutions of $\dot{x} = Ax$. So, the behavior of the system on the center manifold determines the asymptotic behavior of the entire system, up to exponentially decaying terms. The Center Manifold Theorem enables us to extend this argument to the case when f and/or g are not zero.

CENTER MANIFOLD THEOREM.

i) If $|x|$ is sufficiently small, then there exists a center manifold $y = h(x)$ for the system (†). The behavior of (†) on the center manifold is governed by the equation

$$\dot{u} = Au + f(u, h(u)). \qquad (*)$$

ii) The zero solution of (†) has exactly the same stability properties as the zero solution of $(*)$. Furthermore, if the zero solution of $(*)$ is stable and if $x(0)$, $y(0)$ are sufficiently small, then there exists a solution $u(t)$ of (†) such that as $t \to \infty$

$$\begin{aligned} x(t) &= u(t) + O(e^{-\gamma t}), \\ y(t) &= h(u(t)) + O(e^{-\gamma t}), \qquad \gamma > 0. \end{aligned}$$

iii) Define the operation

$$[M\phi](x) \doteq \phi'(x)\{Ax + f(x, \phi(x))\} - B\phi(x) - g(x, \phi(x)).$$

If $\phi\colon R^n \to R^m$ is a smooth map with $\phi(0) = \phi'(0) = 0$, then if $[M\phi](x) = O(|x|^q)$, $q > 1$, as $|x| \to 0$, we have $|h(x) - \phi(x)| = O(|x|^q)$ as $|x| \to 0$.

Note that part (iii) enables us to approximate the center manifold h by the function ϕ up to terms $O(|x|^q)$.

a) Use the Center Manifold Theorem to show that the system

$$\begin{aligned} \dot{x} &= \ \ xy + ax^3 + by^2x, \\ \dot{y} &= -y + cx^2 + dx^2y, \end{aligned}$$

has a center manifold $h(x) = cx^2 + cdx^4 + O(|x|^6)$. Hence, conclude that the equation governing the stability of the original system is

$$\dot{u} = (cd + bc^2)u^5 + O(|u|^7).$$

So if $a + c = 0$, the origin is stable for the original system if $cd + bc^2 < 0$ and unstable if $cd + bc^2 > 0$. What happens if $cd + bc^2 = 0$?

b) Show how to extend the Center Manifold Theorem to the case of maps, as well as to the situation in which the linearized part of (†) may have characteristic roots in the right half-plane.

6. Consider the general two-dimensional quadratic map

$$\begin{pmatrix} u_{n+1} \\ v_{n+1} \end{pmatrix} = \begin{pmatrix} \Lambda_{11} & \Lambda_{12} \\ \Lambda_{21} & \Lambda_{22} \end{pmatrix} \begin{pmatrix} u_n \\ v_n \end{pmatrix} + \begin{pmatrix} \Gamma_{11} & \Gamma_{12} & \Gamma_{13} \\ \Gamma_{21} & \Gamma_{22} & \Gamma_{23} \end{pmatrix} \begin{pmatrix} u_n^2 \\ u_n v_n \\ v_n^2 \end{pmatrix}.$$

a) Show that this system can always be transformed to the standard form

$$x_{n+1} + Bx_{n-1} = 2Cx_n + 2x_n^2,$$

where B is the determinant of the Jacobian of the original system and $C = (\lambda_1 + \lambda_2)/2$, with λ_1 and λ_2 being the characteristic roots of the linear part of the original system (assuming $|B| < 1$).

b) Show that there is a stable fixed point at the origin for

$$|C| < \frac{1+B}{2},$$

which becomes unstable when

$$|C| < -\frac{1+B}{2}.$$

At this point, a stable 2-cycle appears.

c) Prove that the above system undergoes period-doubling bifurcations as C and B decrease until chaotic motion sets in when $B_\infty = 0$, $C_\infty = (1 - \sqrt{17})/4 \approx -0.781$.

d) Would this same type of behavior take place for a Hamiltonian map, i.e., one for which $B \equiv 1$?

7. Show that the noninvertible map

$$\begin{aligned} x_{n+1} &= 2x_n \mod 1, \\ y_{n+1} &= \alpha y_n + \cos 4\pi x_n, \end{aligned}$$

has Lyapunov exponents $\sigma_1 = \log 2$, $\sigma_2 = \log|\alpha|$. Hence, conclude that the system displays sensitivity to initial conditions.

8. A simple model of host–parasite dynamics is given by

$$\begin{aligned} x_{t+1} &= x_t e^{[r(1-x_t)-\gamma y_t]}, \\ y_{t+1} &= x_t(1 - e^{-\gamma y_t}), \end{aligned}$$

where x represents the host population and y is the parasite. Here the parameters r and γ represent the host population's net rate of increase and the parasites' ability to locate and reproduce in the host's larvae, respectively.

a) Show that the (r, γ)-plane can be partitioned into four regions in which the system behavior can be one of the following stable modes: (i) a fixed point; (ii) a periodic orbit; (iii) an almost-periodic orbit, or (iv) chaos.

b) By regarding the parameters r and γ as "strategies" selected by the host and parasite, respectively, it can be shown that the above system leads to a competitive equilibrium (r^*, γ^*) that is a *Nash equilibrium* for the system, i.e., a point from which neither "player" can deviate without decreasing his utility—in this case, the average population. It turns out that this Nash equilibrium can very easily lie in the cyclic or even chaotic region of the (r, γ)-plane. What interpretation can you attach to such a strategy?

9. Probably the simplest continuous-time system displaying a strange attractor is

$$\begin{aligned}\dot{x} &= -(y+z),\\ \dot{y} &= \quad x + \frac{1}{5}y,\\ \dot{z} &= \quad \frac{1}{5} + z(x-\mu),\end{aligned}$$

where μ is a real parameter. This system is called the *Rössler attractor.*

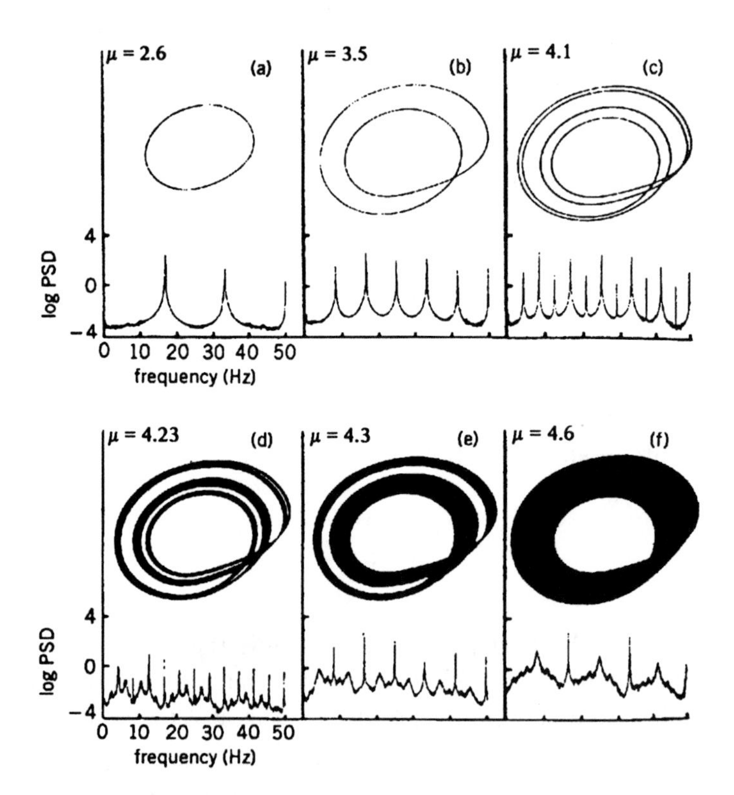

Behavior of the Rössler Attractor

a) The behavior of this system in the (x, y)-plane, as well as the power spectral density, is shown above for increasing values of the single parameter μ. At what level of μ (i.e., (a)–(e)) do you think chaotic motion sets in?

b) The Rössler attractor is sometimes called a *baker's transformation.* Can you see why?

10. Consider the following integral equation used to model the spread of infectious disease:

$$x(t) = \frac{b}{L_2} \int_{t-L_1-L_2}^{t-L_1} x(s)[1 - x(s)]\, ds,$$

where b, L_1 and L_2 are parameters.

a) Show that for $L_2 = 0$ this equation reduces to the difference equation

$$x(t) = bx(t - L_1)(1 - x(t - L_1)),$$

which displays chaotic motion for $b > 3.57$.

b) The above result suggests that for L_2 "small"—but not zero—the original equation may display chaotic motion as well. How would you go about testing this hypothesis? (Hint: Consider the linear approximation to the original equation in the neighborhood of the fixed point $x^* = (b-1)/b$ with $L_2 = 0$.)

11. The following mapping takes the circle S^1 to itself

$$\theta_{n+1} = \theta_n + \Omega - \frac{\mu}{2\pi} \sin 2\pi\theta_n \mod 1, \qquad \mu,\, \Omega \geq 0.$$

a) Prove that for $\mu = 0$ the only orbits are periodic when Ω is rational, and quasi-periodic when Ω is irrational.

b) Show that when $\mu > 0$ there are "tongues" in the (Ω, μ)-plane that emerge from each rational point on the $\mu = 0$ axis, and that the values of μ and Ω within each tongue correspond to periodic orbits for the map. Further, show that these tongues get wider and wider as μ increases, and that the points outside the tongues correspond to quasi-periodic orbits.

c) The quasi-periodic regions do not approach each other closely, even when $\mu \to 1$ from below. Thus, in the Ruelle-Takens scenario there are gaps where the transition from quasi-periodic motion to chaos can take place. Show that this can occur *only* if both parameters μ and Ω are varied together. In other words, the transition from quasi-periodic motion to chaos is a "codimension two" bifurcation.

12. Consider the *linear* control system

$$x_{t+1} = \alpha x_t + u_t,$$

with α real.

a) Show that if there is no control ($u_t \equiv 0$), the solutions to this system do not oscillate.

b) Let the control law be given in feedback form as

$$u_t = \beta |x_t| - 1,$$

with β real. Prove that if the parameters satisfy the conditions

$$\beta > 0, \qquad \beta^2 - \alpha^2 \geq \alpha + \beta + 1,$$

the controlled system has at least N_{n-2} periodic trajectories of period n, where N_k = the kth Fibonacci number.

c) Show that if strict inequality holds in part (b), there are an uncountable number of bounded aperiodic trajectories such that if X and Y are two such trajectories, then

$$|X_t - Y_t| \geq \frac{2(\beta^2 - \alpha^2 - \alpha - \beta - 1)}{(\alpha - \beta)^2(\alpha + \beta)^2} > 0,$$

for sufficiently large t.

13. The generalized baker's transformation of the square is given by

$$x_{n+1} = \begin{cases} a x_n, & \text{if } y_n < \alpha, \\ \frac{1}{2} + b x_n, & \text{if } y_n > \alpha, \end{cases}$$

and

$$y_{n+1} = \begin{cases} \frac{1}{\alpha} y_n, & \text{if } y_n < \alpha, \\ \frac{1}{1-\alpha}(y_n - \alpha), & \text{if } y_n > \alpha, \end{cases}$$

where $0 \leq x_n,\ y_n \leq 1$. Assume the parameters satisfy a, b, $\alpha \leq \frac{1}{2}$ and $b \geq a$.

a) Show that the Jacobian matrix of this system is diagonal and depends only upon y.

b) Show that the Lyapunov exponents σ_1 and σ_2 for this system are

$$\log \sigma_1 = \alpha \log \frac{1}{\alpha} + \beta \log \frac{1}{\beta},$$

and

$$\log \sigma_2 = \alpha \log a + \beta \log b.$$

c) Define the *Lyapunov dimension* d_L of the system's attractor to be

$$d_L = 1 + \frac{\log \sigma_1}{\log(1/\sigma_2)} \cdot$$

Show that the Lyapunov dimension of the baker's transformation is the integer $d_L = 2$ when $\alpha = \frac{1}{2}$, $a = b = k > \frac{1}{2}$.

14. Consider the one-dimensional quadratic map

$$x_{t+1} = C - x_t^2 \equiv f(x_t),$$

with C a real parameter.

a) Show that for $|C| > \frac{1}{4}$, the orbits all tend to $\pm\infty$.

b) Prove that at $C = -\frac{1}{4}$ a period-doubling regime sets in followed by a chaotic region, with the domain of attraction for the chaotic orbits given by $|x| \leq x_*$, where $x_* = \frac{1}{2} + (\frac{1}{4} + C)^{1/2}$. Show that the period-doubling and chaotic behavior holds for all C in the range $-\frac{1}{4} \leq C < 2$.

c) What happens when $C > 2$?

15. In the quadratic family of Problem 14, assume the parameter C is chosen so that there is exactly one periodic orbit of every period $n > 0$. The *kneading sequence* of the map f is then given by the binary sequence defined by $b_i = 0$ or 1 as $f^i(0)$ is less than or greater than zero.

a) For a particular value of C in the periodic range above, prove that the corresponding kneading sequence for f can be calculated recursively as

$$b_i = \begin{cases} b_{i-2^n}, & \text{if } 2^n + 1 \leq i < 2^{n+1}, \\ 1 - b_{2^n}, & \text{if } i = 2^{n+1}. \end{cases}$$

The section of the sequence from $i = 2^n + 1$ to 2^{n+1} is generated by repeating the first 2^n terms and then changing the last.

b) Prove that the kneading sequence of f^2 is obtained by taking the subsequence of $\{b_i\}$ consisting of terms having even indices, which leads to a sequence that is obtained from $\{b_i\}$ by changing every term. Show that this fact implies that $f(x)$ and $-f^2(-x)$ have the same kneading sequences.

c) Show that if two quadratic maps f and h have the same kneading sequence, there exists a homeomorphism g such that $f \circ g = g \circ h$; i.e., f and h are topologically conjugate. Prove that the map g is the universal Feigenbaum function discussed in the text.

16. Instead of the usual quadratic map with a single hump, consider the cubic map with *two* humps

$$x_{t+1} = ax_t^3 + (1 - a)x_t,$$

where a is a real parameter.

a) For what values of a do the two humps lie within the interval [−1, 1]? Where are the attractors for this system if $a < 0$?

b) Show that the dynamical behavior of this cubic system is given by

$$\begin{aligned}
0 < a < 2 &\quad \text{(a stable point at the origin)},\\
2 < a < 3 &\quad \text{(a stable cycle of period 2)},\\
3 < a < 1+\sqrt{5} &\quad \text{(two distinct cycles, each with period 2)},\\
1+\sqrt{5} < a < 4 &\quad \text{(distinct periodic attractors, as well as aperiodic orbits)}.
\end{aligned}$$

17. Consider the two-dimensional map

$$\begin{aligned}
\theta_{n+1} &= 2\theta_n \mod 2\pi,\\
z_{n+1} &= \lambda z_n + \cos\theta_n,
\end{aligned}$$

where $1 < \lambda < 2$, $0 \leq \theta < 2\pi$.

a) Show that the Jacobian matrix of the map has characteristic values λ and 2, proving that there can be no attractors having z finite.

b) Since almost all initial conditions lead to $z_\infty = \pm\infty$, there must be a function $f(\theta)$ such that initial points (θ_0, z_0) having $z_0 > f(\theta_0)$ lead to $z_\infty = +\infty$, while initial points (θ_0, z_0) with $z_0 < f(\theta_0)$ lead to $z_\infty = -\infty$. Show that such a function $f(\theta)$ is given by

$$f(\theta) = -\sum_{k=1}^{\infty} \lambda^{-(k+1)} \cos(2^k\theta).$$

c) Prove that $f(\theta)$ is nondifferentiable, and has infinite length with a fractal dimension $d = 2 - \log\lambda/\log 2$.

18. a) Let $I = [a, b]$ be a closed interval of the real line and let $f\colon I \to I$ be continuous. Assume that $I \subset f(I)$. Prove that f has a fixed point in I. (Hint: Consider the function $g(x) = x - f(x)$.)

b) Let there be a point $\alpha \in I$ such that: $\alpha < f(\alpha) < f^3(\alpha), f^4(\alpha) = \alpha$, and that f is increasing on $[\alpha, f(\alpha)]$. Prove that f has an orbit of period-3; hence, f has an uncountable number of aperiodic points. (Hint: Consider the image of the interval $[\alpha, f(\alpha)]$ under f^3 and use the result of part (a).)

19. In simple models of two competing technologies, if a represents the annual adoption rate of one of the technologies and f_t is the fractional market share enjoyed by that technology in period t, then the dynamics of adoption of this technology are described by

$$f_{t+1} = \begin{cases} (a+1)f_t - af_t^2, & \text{if } -1 < a \leq 1,\\ (a+1)f_t - [(a+1)/2]^2 f_t^2, & \text{if } 1 < a \leq 3.\end{cases}$$

a) Show that for $a > 2.57$ the above system displays chaotic behavior.

b) Since a represents an adoption **rate**, how can you interpret values of a greater than 1?

c) Do you think that the possibility of chaotic motion for these dynamics in any way invalidates the use of such a model for describing the dynamics of technological substitution?

20. Let $p: \mathbf{C} \to \mathbf{C}$ be a polynomial. Show that if the critical points of p iterate to infinity, then the Julia set is totally disconnected. But if the orbits of all the critical points are bounded, then the Julia set is connected.

21. Consider the system

$$\begin{aligned} \dot{x}_1 &= a x_2 x_3, \\ \dot{x}_2 &= b x_1 x_3, \\ \dot{x}_3 &= c x_1 x_2, \end{aligned}$$

where a, b and c are real, nonzero constants.

a) Show the the functions

$$\begin{aligned} \phi_1(x_1, x_2, x_3) &= \frac{x_1^2}{a} - \frac{x_2^2}{b}, \\ \phi_2(x_1, x_2, x_3) &= \frac{x_2^2}{b} - \frac{x_3^2}{c}, \\ \phi_3(x_1, x_2, x_3) &= \frac{x_2^2}{b} - \frac{x_1^2}{a}, \end{aligned}$$

are each first integrals of the system, i.e., $\phi_i(x_1, x_2, x_3) = c_i$, where c_i is a constant, $i = 1, 2, 3$.

b) From part (a), derive the following closed-form expressions for x_1 and x_2:

$$\begin{aligned} x_1 &= \pm \left(\frac{a}{c} x_3^2 + \alpha \right)^{1/2}, \\ x_2 &= \pm \left(\frac{b}{c} x_3^2 + \beta \right)^{1/2}, \end{aligned}$$

where $\alpha = -ac_3$, $\beta = bc_2$.

c) Substituting these expressions for x_1 and x_2 into the differential equation for $\dot{x}_3$, derive the integral relation

$$t - t_0 = \pm \frac{1}{c} \int \left[\left(\frac{a}{c} x_3^2 + \alpha \right) \left(\frac{b}{c} x_3^2 + \beta \right) \right]^{-1/2} dx_3.$$

In other words, x_1, x_2 and x_3 are elliptic functions of t.

22. (a) The classical Cantor set is obtained by removing the middle third of the interval $[0, 1]$, followed by removal of the middle third of each remaining subinterval and so on. Show that the measure (metric length) of the Cantor set is zero, while its fractal dimension is $D = \log 2/\log 3$. That is, its classical dimension differs from its fractal dimension.

b) The construction of the Koch snowflake is shown in the diagram below, where the sides of an equilateral triangle are continually divided into thirds.

Construction of the Koch Snowflake

Prove that the fractal dimension of this snowflake is $D = \log 4/\log 3$. Moreover, show that the Koch snowflake has an infinite perimeter, yet encloses a finite area.

23. (a) Prove the converse of Sarkovskii's Theorem. That is, show that there exist maps having periodic points of period p and no "higher" periodic points in the Sarkovskii ordering.

b) Show that Sarkovskii's Theorem fails for maps on the circle by displaying such a map having a point of period 3, but no other periodic points.

24. Consider the scalar dynamical system $x_{t+1} = f(x_t)$, and assume there is a repelling fixed point at $x = 0$. If f is an analytic function, then the orbit can be parametrized by $x_t = F(a^t c)$, where $F(z)$ is an analytic function and where $a = f'(0)$. The constant c is determined by the starting point x_0.

The function $F(z)$ is the solution of the *Poincaré functional equation*

$$F(az) = f(f(z)),$$

with $F(0) = 0$, $F'(0) = 1$. (The latter condition is needed in order to make the solution unique.)

a) Show that for $f(x) = ax$, we have $F(z) = z$ and $x_t = a^t c$.

b) For the logistic map, we have $f(x) = ax(1-x)$. Show that in the completely chaotic case when $a = 4$, $F(z) = \sin^2 \sqrt{z}$ and the orbit is given by $x_t = \sin^2(2^t c)$.

c) Prove that for the logistic map in the region $2 \leq a \leq 4$, the Poincaré function satisfies the equation $F(az) = aF(z)(1 - F(z))$. Since $F(z)$ is analytic at the origin, we can expand it into the power series:

$$F(z) = z - c_2 z^2 + c_3 z^3 - c_4 z^4 + \dots .$$

Use the above relation for F to show that

$$c_2 = \frac{1}{a-1}, \qquad c_3 = \frac{2}{(a-1)(a^2-1)}, \qquad c_4 = \frac{a+5}{(a-1)(a^2-1)(a^3-1)}.$$

(Hint: Substitute the power series into the functional relation for F and equate coefficients of like powers of z. Why do we take the coefficients to be $c_0 = 0$, $c_1 = 1$?)

25. Consider the following system, which represents the dynamics of a "toy" atmosphere:

$$\begin{aligned} \dot{x}_1 &= -d_1 x_1 + x_2 x_3, \\ \dot{x}_2 &= \quad g_* x_2 - 2x_1 x_3, \\ \dot{x}_3 &= -d_3 x_3 + x_1 x_2, \end{aligned}$$

where the constants d_1 and d_3 represent dissipation coefficients, while the constant g_* represents energy generation. (Here the x_i can be thought of as the amplitudes of a triplet of resonantly-interacting Rossby waves in the atmosphere.)

a) Show that when $d_1 = d_3 = g_* = 0$, this system has the two constants of motion: $x_1^2 + x_2^2 + x_3^2 = E$ and $x_1^2 - x_3^2 = F$, where E and F are constants depending only on the initial conditions.

b) Prove that regardless of the values of the dissipation and energy generation coefficients, if the system's initial condition is on one of the axes, then the system stays on that axis forever. That is,

$$\begin{aligned} (x_1, x_2, x_3) &= (ae^{-d_1 t}, 0, 0), \\ &= (0, ae^{g_* t}, 0), \\ &= (0, 0, ae^{-d_3 t}), \end{aligned}$$

for some constant a.

26. Assume we observe a time series of numbers $\{a_t\}$. We say that this sequence has a smooth, deterministic explanation if there are smooth functions h and F, together with an initial condition x_0, such that:

i. The solution of the equation $\dot{x} = F(x)$, $x(0) = x_0$, is bounded;

ii. $a_t = h(x_t)$.

Suppose that the sequence $\{a_t\}$ is a set of independent and identically distributed random variables. Prove that the probability that such a sequence admits a smooth, deterministic explanation is zero. (Remark: This Problem shows that the likelihood of "explaining" a random sequence with a deterministic model is negligible.)

27. Consider the scalar map

$$x_{t+1} = x_t + b \mod 1,$$

where b is an irrational number.

a) Show that the solution of this equation is given by

$$x_t = bt + x_0 \mod 1, \qquad t = 1, 2, \ldots .$$

b) Suppose there is a small error Δx_0 in the initial state x_0. Show that the error at time t is then $\Delta x_t = \Delta x_0$. Conclude that there is no exponential growth of error upon iteration; hence, no chaos.

c) Explain why there is chaos when $b = 0$ but not when b is irrational.

Notes and References

§1. For the idea of the Circle-10 system, I am indebted to the excellent treatment of chaos for the layman given in

Stewart, I., *Does God Play Dice?,* Basil Blackwell, Oxford, 1989.

For those who care to sample the delights and fantasies of Borges's imaginary worlds, the following collection of short stories is a good place to start:

Borges, Jorge Luis, *Labyrinths,* New Directions, New York, 1964.

§2. Just about every introductory article and book on chaos discusses the logistic map and its strange behavior. A couple of especially good treatments are

Devaney, R., *An Introduction to Chaotic Dynamical Systems,* 2nd Edition, Addison-Wesley, Reading, MA, 1989,

Rasband, S. Neil, *Chaotic Dynamics of Nonlinear Systems,* Wiley, New York, 1990.

A good introductory account of the period-doubling phenomenon for quadratic maps within the context of insect population studies is given in

May, R., and G. Oster, "Bifurcations and Dynamic Complexity in Simple Ecological Models," *Amer. Naturalist,* 110 (1976), 573–599.

It's worthwhile noting that quadratic maps can always display unpredictable behavior, and it's exactly this class of maps that forms the basis for the well-known Lotka-Volterra-type predator–prey relations that lie at the heart of a large number of ecological analyses. So here is part of the reason why the systems ecology community has been one of the primary consumers of the theoretical results in chaos theory. For a detailed account of the predator–prey relations, see

Peschel, M., and W. Mende, *The Predator–Prey Model,* Springer, Vienna, 1986.

Other works outlining the appearance of nonclassical attractors in population dynamics are

Dendrinos, D., "Quasiperiodicity and Chaos in Spatial Population Dynamics," *Socio-Spatial Dynamics,* 2 (1991), 31-59,

Weidlich, W., "Physics and Social Science—The Approach of Synergetics," *Physics Reports,* 204 No. 1 (1991), 1-163.

§3. A good elementary discussion of the KAM Theorem is found in Appendix 8 of the book

Arnold, V. I., *Mathematical Methods of Classical Mechanics,* Springer, New York, 1978.

For a more detailed account, see the works

Arnold, V. I., "Small Denominators—II: Proof of a Theorem of A. N. Kolmogorov on the Preservation of Conditionally Periodic Motions under a Small Perturbation of the Hamiltonian," *Russian Math Surveys,* 18 (1963), 9–36,

Helleman, R., "Self-Generated Chaotic Behavior in Nonlinear Mechanics," in *Fundamental Problems in Statistical Mechanics,* E. Cohen, ed., North-Holland, Amsterdam, Vol. 5, 1980, pp. 165–233.

A very interesting treatment of the entire problem of "small divisors" and its historical origins in celestial mechanics in connection with the stability of the solar system, is found in

Sternberg, S., *Celestial Mechanics,* Vols. 1 & 2, Benjamin, New York, 1969.

§4. A more detailed discussion of the "tent-map" and its relationship to quadratic functions is found in

Lichtenberg, A., and M. Lieberman, *Regular and Stochastic Motion,* Springer, New York, 1983,

as well as in the Devaney and Rasband volumes cited under §2.

The duality between a quadratic optimization problem and stochastic filtering is treated extensively in the linear system theory literature. Two typical references are

Casti, J., *Linear Dynamical Systems,* Academic Press, Orlando, 1987,

Anderson, B. D. O., and J. Moore, *Linear Optimal Control,* Prentice-Hall, Englewood Cliffs, NJ, 1971.

The problem of hidden variables as a way out of the probabilistic interpretation of quantum phenomena has had a rather rocky history, coming in and out of favor with the work of Einstein, Bohm, Bell, and others. Popular accounts of this problem, as well as many others associated with the epistemological content of the quantum measurement problem, may be found in Chapter Seven of

Casti, J., *Pardigms Lost: Images of Man in the Mirror of Science,* Morrow, New York, 1989 (paperback edition: Avon, New York, 1990),

as well as in the volumes

Herbert, N., *Quantum Reality,* Doubleday, New York, 1985,

d'Espagnat, B., *In Search of Reality,* Springer, New York, 1983,

Davies, P. C. W., and J. R. Brown, eds., *The Ghost in the Atom,* Cambridge University Press, Cambridge, 1986.

§5. The best elementary accounts of chaotic phenomena are the introductory sections of the following volumes, each of which consists primarily of reprints of the benchmark papers in the field:

Hao, Bai-Lin, *Chaos,* World Scientific, Singapore, 1984,

Cvitanović, P., *Universality in Chaos,* Adam Hilger, Ltd., Bristol, UK, 1984.

For additional information on the Feigenbaum number and the question of self-similarity, see

Feigenbaum, M., "Universal Behavior in Nonlinear Systems," *Los Alamos Science,* 1 (1980), 4–27 (Reprinted in the Cvitanović book cited above),

Feigenbaum, M., "The Universal Metric Properties of Nonlinear Transformations," *J. Stat. Physics,* 21 (1979), 669–706.

A fuller account of the famous Period-3 Theorem and its proof is found in the original paper:

Li, T., and J. Yorke, "Period-3 Implies Chaos," *Amer. Math. Monthly,* 82 (1975), 985–992.

Sarkovskii's Theorem, while it applies only to scalar maps, provides a wealth of detail about the orbit structure of such maps. A detailed study of this theorem, as well as several important extensions are discussed in the Devaney book cited above and in the paper

Block, L., "Dynamical Complexity of Maps of the Interval," in *Chaotic Dynamics and Fractals,* M. Barnsley and S. Demko, eds., Academic Press, San Diego, CA, 1986, pp. 113-122.

This latter source, for example, establishes deep connections between the topological entropy of a map and the Sarkovskii stratification. Among other things, this result implies that the entropy of a map f will be nonzero (i.e., the map f will display chaotic behavior) if and only if f has a periodic point whose period is not a power of two.

§6. The notion of a snap-back repellor as the natural analogue in higher dimensions for the scalar period-3 attractor is treated in detail in

Kaplan, J., and F. Marotto, "Chaotic Behavior in Dynamical Systems," in *Nonlinear Systems and Applications,* V. Lakshmikantham, ed., Academic Press, New York, 1977, pp. 199–210.

The Hénon attractor is by far the most well-studied example of a higher-dimensional chaotic process. The original reference is

Hénon, M., "A Two-Dimensional Mapping with a Strange Attractor," *Comm. Math. Physics,* 50 (1976), 69–77.

The actual mathematical proof establishing the strangeness of the Hénon attractor was surprisingly long in coming, and was obtained only with considerable effort and ingenuity. This important mathematical result confirming the computational experiments was announced in

Benedicks, M. and L. Carleson, "The Dynamics of the Hénon Map," *Annals of Math.,* 133 (1991), 73-169.

Other important work on higher-dimensional strange attractors is

Collet, P., J. P. Eckmann and H. Koch, "Period-Doubling Bifurcations for Families of Maps on R^n," *J. Stat. Physics,* 25 (1980), 1–14,

Zisook, A., "Universal Effects of Dissipation in Two-Dimensional Mappings," *Phys. Rev. A,* 24 (1981), 1640–1642.

The example of chaotic patterns in the change of daily moods is adapted from

Hannah, T., "Does Chaos Theory Have Application to Psychology?: The Example of Daily Mood Fluctuations," *Network,* 8 (Fall 1990), 13-14.

The gold price example is from

Frank, M. and T. Stengos, "Chaotic Dynamics in Economic Times Series," *J. Econ. Surveys,* 2 (1988), 103-133.

§7. A much more detailed account of the tests for chaos is given in the work by Lichtenberg and Lieberman cited under §4 above. See also

Nicolis, J., *Chaos and Information Processing,* World Scientific, Singapore, 1991,

Ruelle, D., *Chaotic Evolution and Strange Attractors,* Cambridge University Press, Cambridge, 1989,

Crutchfield, J. J. D. Farmer, N. H. Packard, R. Shaw, R. J. Donnelly, and G. Jones, "Power Spectral Analysis of a Dynamical System," *Phys. Lett.,* 76A (1980), 1–4,

Grassberger, P., and I. Procaccia, "Characterization of Strange Attractors," *Phys. Rev. Lett.,* 50 (1983), 346–349,

Wolf, A., "Quantifying Chaos with Lyapunov Exponents," in *Chaos,* A. Holden, ed., Princeton University Press, Princeton, 1986, pp. 273–290,

Brock, W. A., "Distinguishing Random and Deterministic Systems," *J. Economic Theory,* 40 (1986), 168-195.

§8. The capital accumulation example of how economic chaos *might* arise was first suggested in

Day, R., "The Emergence of Chaos from Classical Economic Growth," *Quart. J. Econ.,* May 1983, 201–213.

If you think the emergence of chaos is due to the somewhat outmoded assumptions built in to the classical theory of economic processes, consider the possibilities given in the following accounts for how strange attractors and chaos could easily emerge in the world of money:

Rosser, J. B, Jr., *From Catastrophe to Chaos: A General Theory of Economic Discontinuities,* Kluwer, Boston, 1991,

Brock, W., "Chaos and Complexity in Economic and Financial Science," Chapter 17 in *Acting Under Uncertainty: Multidisciplinary Conceptions,* G. von Furstenberg, ed., Kluwer, Dordrecht, 1990,

Brock, W., "Causality, Chaos, Explanation and Prediction in Economics and Finance," in *Beyond Belief: Randomness, Explanation and Prediction in Science,* J. Casti and A. Karlqvist, eds., CRC Press, Boca Raton, FL, 1990, pp. 230–279,

Chen, P., "Empirical and Theoretical Evidence of Economic Chaos," *System Dynamics Rev.,* 4, Nos. 1-2 (1988), 81-108,

Day, R., "Emergence of Chaos from Neoclassical Growth," *Geographical Analysis,* 13 (1981), 315–327,

Benhabib, J., and R. Day, "Rational Choice and Erratic Behavior," *Rev. Econ. Stud.,* 48 (1981), 459–471,

Ford, J., "Ergodicity for Economists," in *New Quantitative Techniques for Economic Analysis,* G. Szegö, ed., Academic Press, New York, 1981, pp. 79–96.

The monetary aggregate example of the text follows that presented in the Chen article cited above.

§9. For a detailed account of the Beer Game, see

Mosekilde, E., E. Larsen and J. Sterman, "Coping with Complexity: Deterministic Chaos in Human Decisionmaking Behavior," in *Beyond Belief: Randomness, Explanation and Prediction in Science,* J. Casti and A. Karlqvist, eds., CRC Press, Boca Raton, FL, 1990, pp. 199–229.

The discussion of chaos in stock prices follows the Brock article in the Casti and Karlqvist volume. Further studies along the same lines are reported in

De Grauwe, P. and K. Vansanten, "Speculative Dynamics and Chaos in the Foreign Exchange Market," in *Finance and the International Economy: 4,* R. O'Brien and S. Hewin, eds., Oxford University Press, Oxford, 1991, pp. 79-103,

Scheinkman, J. and B. LeBaron, "Nonlinear Dynamics and Stock Returns," *J. Business,* 62 (1989), 311-337,

Mackey, M., "Commodity Price Fluctuations: Price-Dependent Delays and Nonlinearities as Explanatory Factors," *J. Econ. Theory,* 48 (1989), 497-509.

The above studies notwithstanding, the actual evidence for chaos in financial time series is far from convincing. For an up-to-date summary of the current state of the financial forecaster's art, see

Le Baron, B., "Empirical Evidence for Nonlinearities and Chaos in Economic Time Series: A Summary of Recent Results," preprint #9117, Social Systems Research Institute, University of Wisconsin, Madison, WI, August 1991.

§10. A more complete account of this example is given in

Oster, G., "The Dynamics of Nonlinear Models with Age Structure," in *Studies in Mathematical Biology, Part II,* S. Levin, ed., Math. Assn. America, Washington, DC, 1978, pp. 411–438.

§11. The Lorenz equations are discussed in almost every introductory books dealing with chaos and dynamical systems, including those already mentioned under §2 above. A detailed treatment focusing solely on the Lorenz attractor and many of its extensions and generalizations is

Sparrow, C., *The Lorenz Equations: Bifurcations, Chaos, and Strange Attractors,* Springer, New York, 1982.

For Lorenz's original work, see

Lorenz, E., "Deterministic Non-Periodic Flows," *J. Atmos. Sci.,* 20 (1963), 130–141.

It's interesting to note that chaotic behavior, in the strict mathematical sense, has not yet been proved for the Lorenz attractor. All that seems to be available at present is an extensive set of computer experiments giving all the indications of chaotic motion. This situation is in contrast to that of discrete-time dynamics (maps) for which hard theorems confirming true chaotic behavior are known (e.g., the Period-3 Theorem or the results for the Hénon map cited under §6).

§12. A textbook discussion of the Poincaré map is provided in the volumes cited under §2 above. See also the graduate text

Guckenheimer, J., and P. Holmes, *Nonlinear Oscillations, Dynamical Systems, and Bifurcations of Vector Fields,* Springer, New York, 1983.

A popular exposition of the concept of a strange attractor, together with some historical remarks on the genesis of the idea, is provided by one of the founders of the field in

Ruelle, D., "Strange Attractors," *Math. Intelligencer,* 2 (1980), 126–137.

A popular account for the layman is

Gleick, J., *Chaos: The Making of a New Science,* Viking, New York, 1987.

The idea of using embedding theorems to reduce the problem of finding a strange attractor in R^n to that of the properties of a time series appears to have been first suggested by Floris Takens in

Takens, F., "Detecting Strange Attractors in Turbulence," in *Lecture Notes in Mathematics,* Vol. 898, Springer, New York, 1981, p. 366.

The idea was developed further in

Packard, N., J. P. Crutchfield, J. D. Farmer, and R. Shaw, "Geometry from a Time Series," *Phys. Rev. Lett.,* 45 (1980), 712,

Swinney, H., "Observations of Order and Chaos in Nonlinear Systems," *Physica D,* 7D (1983), 3–15.

Perhaps the most thorough account of how the embedding method is used both for global and local prediction schemes is the article

Farmer, J. D. and J. Sidorowich, "Exploiting Chaos to Predict the Future and Reduce Noise," in *Evolution, Learning and Cognition,* Y. C. Lee, ed., World Scientific, Singapore, 1988, pp. 277-330.

Other excellent sources for how the the embedding method is used for predicting chaotic time series include

Casdagli, M., "Nonlinear Prediction of Chaotic Time Series," *Physica D,* 35 (1989), 335-356,

Farmer, J. D. and J. Sidorowich, "Predicting Chaotic Time Series," *Physical Rev. Ltrs.,* 59, No. 8 (1987), 845-848.

The forecasting scheme described in the text for the tent map was first reported in

Sugihara, G. and R. May, "Nonlinear Forecasting as a Way of Distinguishing Chaos from Measurement Error in Time Series," *Nature,* 344 (19 April 1990), 734-741.

For a more extensive discussion of "shadowing," including an account of the example given of the standard map, see

Grebogi, C., S. Hammel, J. Yorke, and T. Sauer, "Shadowing of Physical Trajectories in Chaotic Dynamics: Containment and Refinement," *Physical Rev. Ltrs.,* 65 (1990), 1527-1530.

§13. The scenarios to turbulence, as well as a more extended version of Fig. 4.20, are discussed in far greater detail in

Eckmann, J. P., "Roads to Turbulence in Dissipative Dynamical Systems," *Rev. Mod. Physics,* 53 (1981), 643–654.

Other treatments of the turbulence problem from the viewpoint of modern dynamical system theory are the Ruelle volume cited under §7, as well as in the works

Procaccia, I., "Universal Properties of Dynamical Complex Systems: The Organization of Chaos," *Nature,* 333 (1988), 618–623,

Ruelle, D., and F. Takens, "On the Nature of Turbulence," *Comm. Math. Physics,* 20 (1971), 167–192.

Some of the classical papers on the turbulence question are

Landau, L., "On the Problem of Turbulence," *Akad. Nauk Doklady,* 44 (1944), 339,

Hopf, E., "A Mathematical Example Displaying Features of Turbulence," *Comm. Pure & Appl. Math.,* 1 (1948), 303–309.

For a layman's account of some of these same ideas in the context of weather and climate forecasting, see Chapter Two of

Casti, J., *Searching for Certainty: What Scientists Can Know About the Future,* Morrow, New York, 1991.

Thought-provoking accounts of chaotic behavior in weather patterns, indicating the existence of a low-dimensional strange attractor, are presented in

Tsonis, A., and J. Elsner, "The Weather Attractor Over Very Short Timescales," *Nature,* 333 (1988), 545–547,

Pierrehumbert, R. T., "Dimensions of Atmospheric Variability," in *Beyond Belief: Randomness, Explanation and Prediction in Science,* J. Casti and A. Karlqvist, eds., CRC Press, Boca Raton, FL, 1991, pp. 110–142.

However, recent work by the founder of chaos, Edward Lorenz, suggests that these low-dimensional attractors are more likely to be artifacts of the relatively small amount of data used in the above studies. For an account of Lorenz's work, see

Lorenz, E., "Dimension of Weather and Climate Attractors," *Nature,* 353 (19 September 1991), 241-244.

§14. The model for "turbulent" behavior in the spread of disease is taken from

Cooke, K. L., D. Calef, and E. Level, "Stability or Chaos in Discrete Epidemic Models," in *Nonlinear Systems and Applications,* V. Lakshmikantham, ed., Academic Press, New York, 1977, pp. 73-93.

§15. By universal acclaim, the "father of fractals" is IBM researcher Benoit Mandelbrot. His already-classic work bringing the subject to the attention of the worldwide scientific community is

Mandelbrot, B., *The Fractal Geometry of Nature,* Freeman, New York, 1983.

Other excellent introductions to the topic include

Schroeder, M., *Fractals, Chaos, Power Laws,* Freeman, New York, 1991,

Feder, J., *Fractals,* Plenum, New York, 1988,

Barnsley, M., *Fractals Everywhere,* Academic Press, San Diego, CA, 1988,

Mandelbrot, B., "Fractal Geometry, and a Few of the Mathematical Questions It Has Raised," *Proc. Int'l. Cong. Math.,* Warsaw, August 1983, 1661-1675,

Peitgen, H.-O. and P. Richter, *The Beauty of Fractals,* Springer, Berlin, 1986,

The Science of Fractal Images, Peitgen, H.-O. and D. Saupe, eds., Springer, New York, 1988.

The latter two volumes, incidentally, are distinguished for their fantastic color graphics, making it possible to truly appreciate the endless complexity of things like the Mandelbrot set.

The spiral chaos example is taken from

Rössler, O., J. Hudson and M. Klein, "Chaotic Forcing Generates Wrinkled Boundaries," *J. Physical Chem.,* 93 (1989), 2858-2860.

River basin networks are a fruitful area of application for all mathematical techniques aimed at teasing out the underlying structure of systems showing self-similarity. For more details, see the Schroeder, Feder and Mandelbrot books cited earlier, as well as

Stark, C., "An Invasion Percolation Model of Drainage Network Evolution," *Nature,* 352 (1 August 1991), 423-425.

Frequencies in time and space that appear to be random on any scale underlie many natural processes, leading to the so-called $1/f$ noise discussed in the text. For a good introductory account of these matters, see the article

West, B. and M. Schlesinger, "The Noise in Natural Phenomena," *American Scientist,* 78 (January-February 1990), 40-45.

The discussion of fractal music follows that given in the Schroeder book above, as well as the account

Browne, M., "J. S. Bach + Fractals = New Music," *The New York Times,* 16 April 1991.

§16. A good layman's account of the Mandelbrot and Julia sets can be found in many books. One of the best is

Peterson, I., *The Mathematical Tourist,* Freeman, New York, 1988.

For more technical discussions, see just about any of the volumes cited under §15, as well as several technical articles by Mandelbrot in the procedings volume

Chaos, Fractals, and Dynamics, P. Fischer and W. Smith, eds., Dekker, New York, 1985.

A layman's account of Shishikura's proof that the boundary of the Mandelbrot set has dimension 2 can be found in

Bown, W., "Mandelbrot Set is as Complex as it Could Be," *New Scientist,* 28 September 1991, 22.

DQ #1. For a deeper account of the quantum measurement problem, see the volumes cited under §4 above, as well as

Jammer, M., *The Philosophy of Quantum Mechanics,* Wiley, New York, 1974,

Wheeler, J. A., and W. Zurek, eds., *Quantum Theory and Measurement,* Princeton University Press, Princeton, 1983,

Foundations of Quantum Mechanics Since the Bell Inequalities, Selected Reprints, L. Ballentine, ed., Amer. Assn. Physics Teachers, College Park, MD, 1988,

Quantum Theory Without Reduction, M. Cini and J.-M. Lévy-Leblond, eds., Adam Hilger, Bristol, UK, 1990.

DQ #3. For a more information about statistical mechanics and its relation to KAM theory, see the Lictenberg and Lieberman book cited under §4.

DQ #7-8. These matters pertaining to the preservation of information and its role as a survival mechanism for biological organisms are taken up in much greater detail in

Nicolis, J., *Dynamics of Hierarchical Systems,* Springer, Berlin, 1986,

Nicolis, J., *Chaos and Information Processing,* World Scientific, Singapore, 1991.

DQ #10. The question of long waves in the ebb and flow of global economic indicators was first put forth by Kondratiev in the classic paper

Kondratiev, N., "The Long Waves in Economic Life," *Rev. Econ. Statistics,* 17 (1935), 105–115.

A discussion of the existence and relevance of such postulated waves is given from several points of view in

Freeman, C., ed., *Long Waves in the World Economy,* Butterworth, London, 1983.

Other excellent sources for speculations about this kind of cyclic behavior in world economic affairs include

Van Duijn, J., *The Long Wave in Economic Life,* George Allen & Unwin, London, 1983,

Goldstein, J., *Long Cycles,* Yale University Press, New Haven, CT, 1988,

Sterman, J., "A Simple Model of the Economic Long Wave," *IIASA Collaborative Paper CP-85-21,* IIASA, Laxenburg, Austria, April 1985,

Senge, P., "The Economic Long Wave: A Survey of Evidence," MIT Systems Dynamics Group Working Paper D-3262-1, MIT, Cambridge, MA, 1982,

Mass, N., *Economic Cycles: An Analysis of Underlying Causes,* MIT Press, Cambridge, MA, 1975.

An excellent source of material about electrochemical waves and stable and unstable oscillators in the human body, with special emphasis on the heart, is

Winfree, A., *When Time Breaks Down,* Princeton University Press, Princeton, NJ, 1987.

DQ #12. The general concept of equifinality as a property of biological systems was introduced in the last century by Driesch, who pointed out that the characteristic feature of equifinality, the independence of the final state from the initial conditions, is completely at variance with the traditional equilibrium view of classical physics. This argument, while far from being a convincing proof of "vitalism," serves to underscore the inadequacy of traditional physics for dealing with biological processes. For a fuller discussion of these matters, see

von Bertalanffy, L., *Problems of Life,* Harper & Row, New York, 1960.

DQ #14. For an informal discussion of the problem of quantum chaos, see

Ford, J., "Chaos: Solving the Unsolvable, Predicting the Unpredictable," in *Chaos, Dynamics and Fractals,* M. Barnsley and S. Demko, eds., Academic Press, San Diego, CA, 1986, pp. 1–52.

DQ #15. For a detailed consideration of this problem, see the article

Feldberg, R., M. Szymkat, C. Knudsen and E. Mosekilde, "Iterated Map Approach to Die Tossing," *Physical Review A,* 42 (1990), 4493–4502.

DQ #16. These views contrasting literary criticism and chaos theory are expounded in

Hayles, N. K., *Chaos Bound,* Cornell University Press, Ithaca, NY, 1990.

DQ #17. For more information about Zipf's Law, see the Schroeder book cited under §15 or the 1991 Nicolis volume noted under DQ #7-8.

PR #5. For a full treatment of the Center Manifold Theorem, see the Guckenheimer and Holmes book cited under §11, as well as

Carr, J., *Applications of Centre Manifold Theory,* Springer, New York, 1981.

PR #9. Rössler's attractor is developed in detail in

Rössler, O., "An Equation for Continuous Chaos," *Phy. Lett.,* 57A (1976), 397.

PR #12. The introduction of a control element into the dynamics opens up the possibility of many additional types of dynamical behavior being *induced* (perhaps inadvertently) by the controller. This problem shows only the tip of the iceberg. It originally appeared in the paper

Baillieul, J., R. Brockett and R. Washburn, "Chaotic Motion in Nonlinear Feedback Systems," *IEEE Tran. Circuits & Sys.,* CAS–27 (1980), 990–997.

PR #15. Kneading sequences play an important role in the study of the behavior of many types of dynamical processes. For an introductory account, see the Devaney book cited under §2, as well as the article

Guckenheimer, J., "Bifurcations of Dynamical Systems," in *Dynamical Systems,* J. Guckenheimer, J. Moser, and S. Newhouse, eds., Birkhäuser, Boston, 1980.

PR #24. For more details on the Poincaré functional equation, together with additional examples of cases when it is solvable in closed form, see the article

Lauwerier, H., "One-Dimensional Iterative Maps," Chapter 3 in *Chaos,* A. Holden, ed., Manchester University Press, Manchester, UK, 1986, pp. 39–57.

INDEX

Page numbers for *The Fundamentals* appear in roman type. Page numbers for *The Frontier* appear in *italic* type.

Page numbers for *The Fundamentals* appear in roman type. Page numbers for *The Frontier* appear in *italic* type.

Page numbers for *The Fundamentals* appear in roman type. Page numbers for *The Frontier* appear in *italic* type.

Page numbers for *The Fundamentals* appear in roman type. Page numbers for *The Frontier* appear in *italic* type.

Page numbers for *The Fundamentals* appear in roman type. Page numbers for *The Frontier* appear in *italic* type.

Page numbers for *The Fundamentals* appear in roman type. Page numbers for *The Frontier* appear in *italic* type.

Page numbers for *The Fundamentals* appear in roman type. Page numbers for *The Frontier* appear in *italic* type.

Page numbers for *The Fundamentals* appear in roman type. Page numbers for *The Frontier* appear in *italic* type.

Page numbers for *The Fundamentals* appear in roman type. Page numbers for *The Frontier* appear in *italic* type.

Page numbers for *The Fundamentals* appear in roman type. Page numbers for *The Frontier* appear in *italic* type.

Page numbers for *The Fundamentals* appear in roman type. Page numbers for *The Frontier* appear in *italic* type.

Page numbers for *The Fundamentals* appear in roman type. Page numbers for *The Frontier* appear in *italic* type.

Page numbers for *The Fundamentals* appear in roman type. Page numbers for *The Frontier* appear in *italic* type.

Page numbers for *The Fundamentals* appear in roman type. Page numbers for *The Frontier* appear in *italic* type.

Page numbers for *The Fundamentals* appear in roman type. Page numbers for *The Frontier* appear in *italic* type.

Page numbers for *The Fundamentals* appear in roman type. Page numbers for *The Frontier* appear in *italic* type.